THE EVOLUTION OF LIFE HISTORIES

THE EVOLUTION OF LIFE HISTORIES

STEPHEN C. STEARNS
Zoology Institute, University of Basle

OXFORD NEW YORK TOKYO
OXFORD UNIVERSITY PRESS
1992

Oxford University Press, Walton Street, Oxford OX2 6DP
Oxford New York Toronto
Delhi Bombay Calcutta Madras Karachi
Petaling Jaya Singapore Hong Kong Tokyo
Nairobi Dar es Salaam Cape Town
Melbourne Auckland
and associated companies in
Berlin Ibadan

Oxford is a trade mark of Oxford University Press

Published in the United States
by Oxford University Press, New York

A catalogue record for this book is available from the British Library

Library of Congress Cataloging in Publication Data
Stearns, S. C. (Stephen C.), 1946–
The evolution of life histories / Stephen C. Stearns.
Includes bibliographical references and indexes.
1. Evolution (Biology) 2. Life cycles (Biology) I. Title.
QH371.S72 1992 575—dc20 91-34726
ISBN 0-19-857741-9

Typeset by Integral Typesetting, Gorleston, Norfolk NR31 6RG
Printed in Malta by Interprint Ltd

To my teachers, colleagues, students, and especially to my family.

PREFACE

This book introduces the evolution of life histories. It discusses both the macroevolutionary framework and the microevolutionary conditions that determine how life histories evolve.

Consider a zygote that is about to begin its life and imagine that all opportunities are open to it. At what age and size should it start to reproduce? How many times in its life should it attempt reproduction—once, more than once, continuously, seasonally? When it does reproduce, how much energy and time should it allocate to reproduction as opposed to growth and maintenance? Given a certain allocation, how should it divide those resources up among the offspring? Should they be few in number but high in quality and large in size, or should they be small and numerous but less likely to survive? Should it concentrate its reproduction early in life and have a short life as a consequence, or should it make less reproductive effort in any given attempt and live longer? Such questions have become important parts of evolutionary ecology in the last two decades and are often taught in courses on ecology and evolution or, as advanced topics, on their own. This book is for advanced undergraduates, for graduate students, and for anyone else who might be interested. I have assumed that the reader has already encountered calculus, statistics, computer programming, ecology, evolution, and genetics.

This book covers classical life history theory, not its extensions to modular organisms and complex life cycles, and it pays more attention to animals than to plants. While regretting those omissions I do not regret that the book has been kept to a reasonable length.

No book reviewing an area of active research should give the impression that everything has been done. I have tried to discuss the open questions while keeping the treatment straightforward. I have not avoided any topic that is interesting or relevant because it might not be 'suitable' for an introduction. The questions and problems are meant either to develop a skill, to stimulate discussion or an essay, or to lead to original research.

The literature in this area has exploded. I could not read it all. If I missed something important, please call it to my attention.

S.C.S.
Basle, June 1990

ACKNOWLEDGEMENTS

Susan Abrams, Anette Baur, Bruno Baur, David Berrigan, Hal Caswell, Hugh Dingle, Dieter Ebert, Martin Gebhardt, Julee Greenough, Barbara Hellriegel, Sabine Henrich, Henry Horn, Jeremy John, Gerdien de Jong, Ted Kawecki, Jacob Koella, Jan Kozlowski, Jeff Matthews, Judith May, Arie van Noordwijk, Pawel Olejniczak, Mark Pagel, Geoff Parker, Linda Partridge, Daniel Promislow, Andy Purvis, Andrew Read, David Reznick, Derek Roff, Michael Rose, Bev Stearns, Rosie Trevalyan, Shripad Tuljapurkar, Dave Wake, the students in Population Biology at the University of Basle, 1988–90, and several anonymous referees made comments or answered letters of inquiry that led to improvements and clarifications. Graham Bell and Ric Charnov read most of the book in manuscript, Dafila Scott illustrated the Prologue, and Bev Stearns and Jacqui Shykoff helped with permissions and figures. I am especially grateful for their time and help. The faults are my own.

I could never have written this book had not the secretaries in the Zoology Institute and at the *Journal of Evolutionary Biology* relieved me of many routine duties. My heartfelt thanks to Rita Gunasekera, Evelyn Argast, Eliane Petitjean, Christine Parry, Monika Kiowski, and Jolanda Bonato for their support and understanding. Christine Müller and Regula Schmid-Hempel helped to look up references and get articles.

I thank the following publishers for permission to reproduce material on which they hold the copyright: Academic Press (Table 5.4), American Association for the Advancement of Science (Table D1.1), American Zoologist (Fig. 6.10), Birkhäuser Verlag AG (Fig. 6.20), Blackwell Scientific Publications Inc. (Figs. 3.10, 3.11, 4.5, 4.6), BLV Verlagsgesellschaft (Fig. 4.2), E. J. Brill (Fig. D1.1), Cambridge University Press (Figs. 3.7, 5.1, 8.16, 8.17), Cold Spring Harbor Laboratory (Fig. 3.8), Ecological Society of America (Figs. 4.10, 7.7, 7.8), Elsevier *Trends* Journals (Figs. 2.3, 3.12, Table 7.1), *Evolution* (Figs. 4.13, 5.9, 6.9, 6.11, 6.12, 6.13, 6.14, 6.15, 6.17, 6.18, 6.19, 7.10, 7.11, Tables D1.1, D1.2, 6.4, 8.6), The Field Museum of Natural History (Fig. 4.4), Fisheries and Oceans Canada (Figs. 4.7, D1.3, 6.6, 6.7, 8.14, 8.15, Table 6.1), *The Florida Entomologist* (Fig. 7.9), The Genetical Society of Great Britain (Table 3.6), Longman Group UK Ltd. (Figs. 3.4, 3.5), Macmillan Magazines Ltd. (*Nature*) (Figs. 4.11, 5.10, Tables D1.3, 7.5), Munksgaard International Publishers (*Oikos*) (Fig. 5.5, Table 6.3), Oxford University Press (Table 5.7), Plenum Publishing Corp. (Fig. 8.4, 8.19), Springer-Verlag (Fig. 7.13), The University of Chicago (Figs. 2.2, 3.6, 4.1, 4.12, 4.14, 5.2, 5.7, 6.2, 6.3, 7.6, 7.12, 8.6, Tables 5.6, 6.5, 7.3), and The Zoological Society of London (Figs. 5.11, 5.12, 5.13).

S.C.S.

CONTENTS

PROLOGUE

THE FOREST

In a forest just south of the Rhine, old and dense with oak and beech, a tree falls in an equinoctial storm. A beech seed germinates in the clearing left by the falling tree. It grows rapidly, escapes the attention of slugs during its first summer, overtops the competing grasses and shrubs, and pushes its crown into the narrowing circle of sky overhead. As its branches near the canopy, it begins to

flower. It is 50 years old. Its growth then decelerates until, at more than 200 years of age, when it is two metres thick at breast height and fifty metres tall, further growth is no longer measurable. Every few years, it flowers heavily and sets abundant seed. Much is eaten by flocks of wintering bramblings. Some is scattered by winter storms. The seeds drop in various directions, more near the tree, a few at some distance. In the 154th year of its life, its pollen enters an ovule on a beech half a kilometre away whose seed survives to flower. In the 268th year of its life, a hornbeam falls about 80 metres away. One seed sprouts in the clearing and starts its climb to the canopy. Of the millions of seeds that the tree produces, only this one will survive to flower. It has produced two offspring that survived to reproduce, one through pollen, one through seed.

As the tree thickens with age, its grey bark wrinkles like the baggy skin of an elephant. It continues to flower and set seed in mast years, but all the seeds and seedlings die, eaten by birds, rotted by fungi, nibbled by slugs, grazed by rabbits and deer. Overshadowed by younger competitors, ageing, it flowers less and fails to repair wounds and resist the fungi that have been evolving within it. For centuries it has been shedding its shaded lower branches, leaving a clear column 10 metres to the lowest limb, but now even its trunk and main branches are invaded by beetle larvae and mould. Woodpeckers visit it frequently. In its 316th year it falls in a storm. An oak seedling invades its space. Eighty metres away, its offspring begins to flower. Half a kilometre distant, the tree produced from its pollen is into its second century of reproduction.

THE REEF

Every afternoon, the male blue-headed wrasse swims to the down-current edge of the patch reef. His harem of females, smaller, dull by comparison, move with him, holding position in the current pushing them out from the Panamanian shore. The male couples with each of the females in succession. A cloud of eggs and sperm drifts into deep water. Some fall on to the edge of the reef and are eaten by corals, hydroids, bryozoans, and polychaetes. The reef is a carpet of open mouths.

Before dawn, the surviving eggs reach their third cleavage, but small predators feast on the developing embryos. In deep water and turning in a huge eddy, the survivors hatch as larvae. Some are pursued and eaten by small fish. The others feed on tiny crustaceans as they are carried back towards the coast. A few, just a centimetre long, make it to shallow water. They reach the reef kilometres apart, separated by the vagaries of the sea. One arrives at a small reef on which a single male blue-headed wrasse lives with a harem of three females. The fingerling grows, feeding on plankton, picking small crustaceans and worms out of the coral, and reaches maturity as a female when it is one year old and ten centimetres long. It mates as a female until, when it is three years old, it is the largest surviving female on the reef. One evening the male is eaten by a fish. Within 24 hours the female has acquired the blue head and swollen testes of a male. Its ovaries shrink. The next day it mates for the first time as a male.

Another larva settles on a reef too large to be defended by a single male. Instead of holding to territories, the wrasses forage freely across the face of the reef. The fingerling grows, feeds, and reaches maturity as a small, dull male when it is one year old and ten centimetres long. It looks like a female. Every evening it moves to the downstream side of the reef, and when a large blue-headed male begins to mate with a female, it darts in and releases its sperm into the spawn. Some fertilize eggs that survive to reproduce. Over the course of its life, the small, dull male manages to sneak into enough spawnings to produce about as many surviving offspring as do the few large blue-headed males on the reef, all of which spent their first several years as females.

On small reefs where territories can be defended and females can be guarded, few young fish mature as small, dull males. On large reefs where large males cannot control territories or harems, many young fish mature as small, dull males and never change sex. Ocean currents are capricious, and

there is no guarantee that a young fish will arrive on a reef of a certain type. None the less, they develop into the adult form that is appropriate for the reef at which they arrive. There is some limited opportunity to change reefs after settling.

THE PLUM

Pushed by a gust of wind, a plum falls to the ground and splits. Into a few drops of juice that enter the cleft, yeast cells topple from the plum's skin. Bacteria float in. The yeast and bacteria multiply. Within an hour, the first faint odours of ethanol waft out, and soon a female fruit fly lands. She walks back and forth along the cleft, pausing frequently to probe with her mouth parts, then lays nine eggs in an irregular row and departs. Many other females visit the plum.

The pearly eggs, driven into the flesh of the fruit by ovipositors, gleam in the light reflected from the ground. The two filaments with which each egg breathes stick like arms into the air. A black ground beetle stumbles into the plum and probes along the cleft, eating every egg it touches. A day later, the remaining eggs begin to hatch. The larvae wander for a few hours over the rotting plum, following concentrations of bacteria and yeast. As they grow, they shed their tough skin and expand into the soft membranes beneath. More females visit the cleft. Large flies and beetles suck the fermenting brew. The population of larvae near the cleft grows dense, but the first-laid larvae move on. On the lower side of the fruit, a fungal mycelium invades, feeding on the juices,

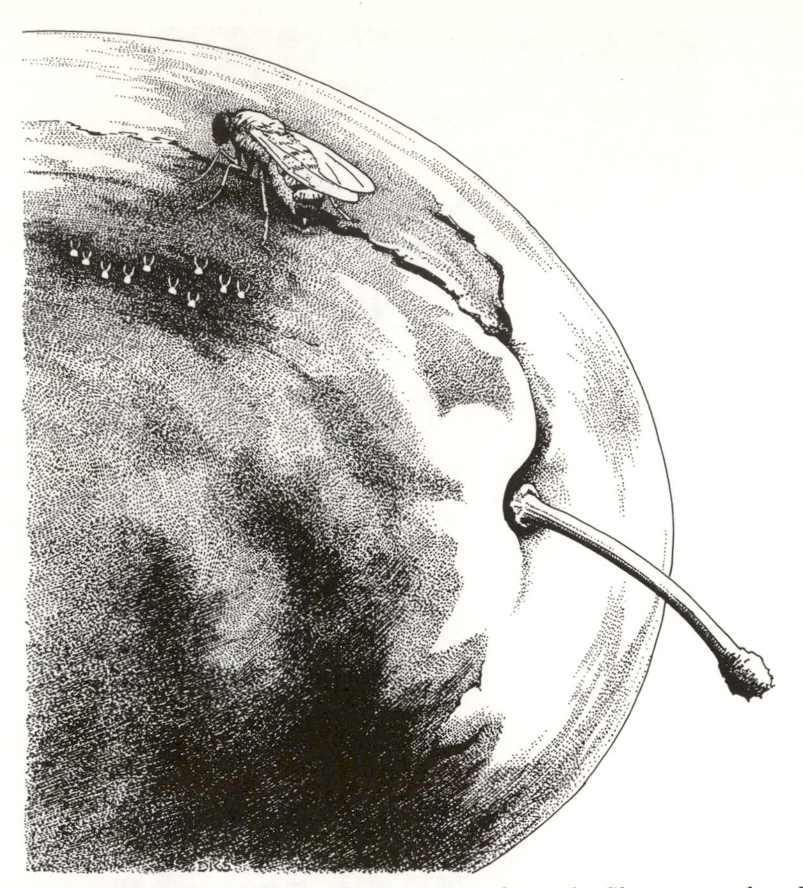

excreting antibiotics. It clears a growing zone in which a fly larva would starve. The lower side of the fruit turns brown and liquefies.

After five days, the first-laid larvae move up out of the flesh and seek well-aerated hiding places among the folds of skin. Forming tough brown cylinders, they pupate. A small wasp alights and drives her needle-like ovipositor repeatedly through the skin of the plum. Every time its tip enters a pupa, she lays one egg. Wasp larvae hatch and devour their hosts.

A few pupae hatch during the night of the twelfth day after the plum fell. When they split the pupal skins and emerge, their bodies are plump and white, their wings are soft and crumpled, and they weight about two milligrams. They move tentatively while their cuticle hardens and their wings expand. Their bodies darken. In the afternoon, they fly out. The next day one of them, a female, finds a fallen apple thick with flies and beset by slugs. She lands on the splotched skin and listens. Soon she hears the characteristic wing beats of a male. She approaches. He sees her and dances to the side, waving his wings. She turns. He mounts and mates, riding her for several minutes after the sperm are transferred.

During the night, she lays thirty eggs on the apple. She moves on, each day laying twenty or thirty eggs into various fruit. Every two or three days, she mates again, and the sperm most recently received fertilize the next eggs laid. Twenty-one days after she left the plum, having laid more than 400 eggs, she slips. A wing gets mired in fruit sap; a beetle eats her.

At the plum, the fungus, having sucked the fruit dry, sends up yellow cups of spores. Some laggard, undernourished fly larvae pupate eighteen days after they were laid. A few escape the wasps and hatch five days later. They weigh about one milligram, and the females among them can lay no more than ten eggs per day. They fly off to find mates, food, and sites to deposit their eggs. Twenty metres away, a fly settles on a crack in the skin of a newly fallen plum.

THE ALBATROSS

Rising on the globe-circling winds of the southern ocean, an albatross leaves its nestling on Kerguelen Island to search for food. Passing the feeding grounds of smaller birds near the island, it scans thousands of square kilometres of water between southern Africa and Australia for squid and fish. Out of sight of land for many days, it shares its feeding grounds with whales and porpoises. After three weeks it returns to a patient chick quite used to waiting for a meal.

The chick grows to the full size of an adult before it fledges. Eight years pass before it first attempts to mate. It does not mate every year, and when it does mate successfully, it lays only one egg. There are no predators on the island, and it is virtually invulnerable in the air. Every second or third year, it mates and raises a chick with its partner. Sometimes it is gone from the nest for up to a month in search of food, travelling up to 16 000 kilometres. During its long life it flies the equivalent of more than 300 times around the world. In its 64th year it dies in a fishing accident. Only two of the 21 chicks it has raised themselves leave offspring that survive to reproduce.

THE PROBLEM

The tree, the fish, the fly, and the albatross all survived and reproduced. Through their diverse life histories run traces of a general pattern created by a common mechanism that expresses the relationships of age and size to mortality and reproductive performance. To analyse the phenotypic variation that produces selection, the expression of genetic variation that enables a response to selection, and the lineage-specific constraints with which selection interacts to produce the observed diversity of life histories is the goal of work on life history evolution.

PART I

THE ELEMENTS OF EVOLUTIONARY EXPLANATION

1

EVOLUTIONARY EXPLANATION

A man who knows how little he knows is well,
A man who knows how much he knows is sick.

Lao Tzu as interpreted by Witter Bynner

CHAPTER OVERVIEW

This book is organized into two parts. Part I presents the tools needed to analyse variation in life history traits; Part II analyses that variation, trait by trait. This chapter introduces Part I. It discusses why life history theory was developed and what role it plays in evolutionary explanation. The major life history traits and the explanatory framework are introduced. That framework could be used to explain any pattern of phenotypic variation. Such explanations are illustrated with an example, human age at maturity.

Life history theory deals directly with natural selection, fitness, adaptation, and constraint. It contributes to evolutionary thought the analysis of the phenotypic causes of variation in fitness and exposes the pervasive tension between adaptation and constraint, brought here into especially sharp contrast by the simultaneous applications of optimality theory, quantitative genetics, trade-offs, and the comparative method to the explanations of the same patterns.

INTRODUCTION

Why study life histories? Life histories lie at the heart of biology; no other field brings you closer to the underlying simplicities that unite and explain the diversity of living things and the complexities of their life cycles. Fascinating in themselves, life histories are also the keys to understanding related fields. Life history theory is needed to understand the action of natural selection, a central element of evolution, the only theory that makes sense of all of biology. It also helps us understand how the other central element, genetic variation, will be expressed. The evolution of life history traits and their plasticities determines the population dynamics of interacting species. Its explanatory power, barely tapped, could reach as far as communities. There is much to be done.

CONTEXT

In *The origin of species*, Darwin (1859) had no convincing mechanism of inheritance, a problem that framed research in evolutionary biology until 1900, when Mendel's laws were rediscovered. Weismann (1885), stimulated by this problem, suggested that germ plasm be distinguished from soma, a forerunner of Johannsen's (1909) distinction between *genotype* and *phenotype*. Today the genotype/phenotype distinction is built deeply into evolutionary explanation.

Two conditions, one genotypic and one phenotypic, are necessary for natural selection to occur. First, heritable variability for the trait in question determines whether there will be a response to selection. Investigation of the sources of heritable variation and the forces that maintain it have constituted the central research programme of population genetics for more than 50 years.

Second, individuals must vary in *fitness*. This is the phenotypic condition. Variation in fitness among individuals *is* natural selection. Analysis of the evolution of fitness components is a new field, life history evolution.

The genotypic and phenotypic conditions for natural selection complement each other. Life history theory analyses what causes differences in fitness among life history variants; population genetics analyses the consequences those fitness differences have for gene frequencies. Life history theory predicts the phenotype at equilibrium; population genetics tells us how fast it should get to that equilibrium—and whether it will get there at all, given genetic constraints.

Population genetics makes the simplifying claim that changes in gene frequencies in populations can be understood by analysing the perturbations of an equilibrium: selection, mutation, drift, gene flow, and non-random mating.

Life history evolution makes the simplifying claim that the phenotype consists of demographic traits—birth, age, and size at maturity, number and size of offspring, growth and reproductive investment, length of life, death—connected by constraining relationships, trade-offs. These traits interact to determine individual fitness. Analysis of the interactions explains phenotypic adaptation. Life history theory explains the broad features of a life cycle—how fast the organism will grow, when it will mature, how long it will live, how many times it will give birth, how many offspring it will have, and so forth.

The connection between genotype and phenotype has traditionally been provided by developmental biology and physiology, fields which have become increasingly molecular. They explain patterns in cells and organs in terms of molecular mechanisms. Such studies do not help much with the problem of how the phenotype is designed for reproduction and survival. That task has been left to the evolutionary biologists. We cannot afford to wait until the molecular analysis of development and physiology has delivered a few mature summary statements relevant to individual variation in fitness, for that will take centuries—if it ever happens at all. We must make our own hypotheses and hope that the molecular connection will come at a later date.

LIFE HISTORY TRAITS

Life history traits figure directly in reproduction and survival. The complexity and interest of life history evolution arises because organisms have evolved so many different ways of combining these traits to affect fitness.

The principal life history traits are:

Size at birth
Growth pattern
Age at maturity
Size at maturity
Number, size, and sex ratio of offspring
Age- and size-specific reproductive investments
Age- and size-specific mortality schedules
Length of life

These traits are bound together by numerous trade-offs, including those between:

Current reproduction and survival
Current reproduction and future reproduction
Number, size, and sex of offspring

This book does not give these traits equal treatment. Charnov (1982) covers sex allocation theory, and the evolutionary ecology of growth and development deserves a monograph itself. Complex life cycles, modular organisms, and frequency-dependent life histories are mentioned but not thoroughly discussed in this book. My aim here is to build a foundation.

Figure 1.1 sketches the life of a European robin (*Erithacus rubecula*) schematically. Robins have two broods per summer of about five eggs each.

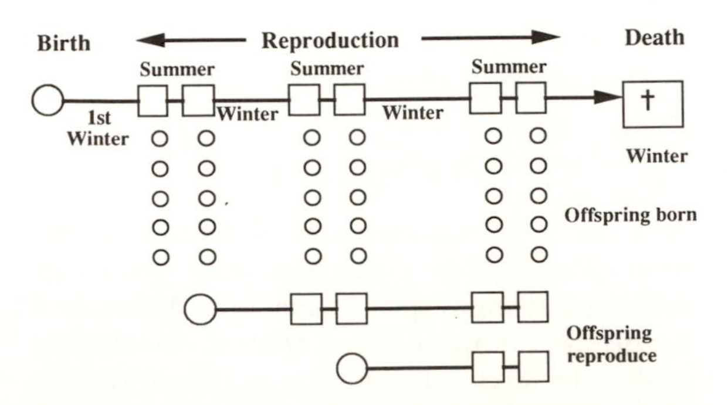

Figure 1.1 An example of a life history. Circles indicate eggs, boxes reproductive events, and the rectangle, death.

Both parents care for the nestlings. This robin is born, grows, survives its first winter, and lays 10 eggs the next summer. At two and at three years, it again lays 10 eggs. Before it reaches its fourth breeding season, it dies. Of the 30 eggs laid, three survive to reproduce, one from each season. They have the same life history as their mother, and some of them reproduce before she dies. The juvenile survival rate is 0.1; all birds that survive to breed then breed for three seasons and die.

This sort of abstraction precedes most demographic analyses. Several aspects are worth noting.

1. Because robins are monogamous and both parents feed nestlings, we counted both males and females: it made no difference. However, often only females are counted. Female fecundity—eggs or offspring—is much easier to count than male fecundity—the number of times a male mates successfully times the fecundity of his mates times his share of the fertilized offspring (if there is multiple mating and sperm competition). Also, because males do not usually limit the reproductive success of females, the number of males present makes little difference in simple models of population dynamics. Both males and females are treated explicitly in sex allocation theory (Charnov 1982) and in Charlesworth's (1980) book on evolution in age-structured populations.

2. Any activity takes time, and activities that reduce the time that can be devoted to reproduction and survival decrease fitness. For example, if the robin's first clutch is clearly going to be a failure, then it pays the robin to abandon it immediately, even though some of the nestlings may still be alive.

3. The internal constraints on the robin are not evident in Fig. 1.1. A robin cannot lay an egg as large as itself, and below a certain size an egg would not contain enough energy to produce a functional hatchling. That the robin lays eggs at all, rather than giving birth viviparously, is itself an ancient lineage-specific trait shared by birds, crocodiles, snakes, turtles, and some dinosaurs.

4. If the robins had tried to raise 15 rather than 10 chicks in their first summer, would they have been able to raise 10 in the second? Would they have been able to survive to reproduce in their third summer? Perhaps not. Such trade-offs are central to the theory but not apparent in Fig. 1.1.

5. Note the abstraction itself. The complexity of the robin's life has been reduced to a few elements. The environment has disappeared from view. No genes are apparent. Such simplification can only be justified if it brings us successful predictions and deeper understanding.

AN EXPLANATORY FRAMEWORK

Science aims at simple explanations of complex things. In life history evolution, the patterns to be explained are the full diversity of life cycles in living things. These range from the familiar cycle of birth, reproduction, and death in birds and mammals, through alternating sexual and asexual generations in cladocerans, rotifers, and some beetles and the modular life histories of many plants and bryozoans, to the complex life cycles of algae, parasites, and corals. In what framework can all life histories be understood as variations of a few general themes? The framework used here contains four elements:

(1) demography
(2) quantitative genetics and reaction norms
(3) trade-offs
(4) lineage-specific effects

These four elements combine to explain life history variation. How they combine is illustrated in the example discussed next. Then their relationships to levels of analysis, and the chapters in which they are presented, are sketched.

An example of an explanation

Simple explanations are rare in life history evolution. If one asks 'Why do humans mature at 16–18 years?' or 'Why do albatrosses lay one egg?', no single approach gives a satisfying answer. Each of the next four chapters examines one approach in some depth. By combining all four—demography, quantitative genetics and reaction norms, trade-offs, and phylogenetic analysis—we can place variation in life history traits into balanced perspective. However, because of the specialization of science and the amount of work involved, no study exists that could be used here to illustrate the power of the combined approach.

Let us imagine what such an explanation would look like so that we can later judge the more specialized explanations offered in Part II. Here is how one could explain variation in age at maturity in humans, given enough information.

We begin with a broad, comparative study of age at maturity in the Class Mammalia and use the methods developed in Chapter 5 to weigh the effects of body mass and lineage. Many mammals that are as large as or larger than humans—bears, lions, and whales, for example—mature earlier. However, within their order, the primates, humans lie close to the average line relating age at maturity to weight. Compared to their closest relatives, the gorillas and chimpanzees, humans do mature relatively late for their body size—about 3–6 years later than an average great ape, at a mean age of 15–16 years for females. (In humans there is a period of adolescent infertility following menarche, which now comes at 10–12 years.) This is the derived condition that needs explanation.

Analysis of phenotypic sources of variation shows that humans have a *reaction norm* for age and size at maturity. Individuals that are fed well and grow rapidly mature younger and larger than individuals that are poorly fed and grow slowly. Some well-fed females achieve menarche at age 10, and some poorly fed females achieve menarche at age 20. Considerable genetic variation underlies the phenotypic variation. Within a single well-fed population, genotypes differ by as much as six years and 10 kilograms at menarche.

Analysis of family pedigrees before, during, and after famines reveals considerable *genotype × environment interaction.* The *heritability* of age at maturity in humans (their genetic variation for this trait) is low when people are starved and large when they are well nourished. The *genetic correlations* between age and size at maturity change from zero under starvation to positive under good nutrition.

Physiological analysis of growth and maturation reveals a network of hormone-mediated feedbacks that determine the allocation of energy between growth and fat reserves and the onset of reproduction. Humans delay maturation until enough fat reserves have been built up to supply milk for several months' nursing; starved females and thin athletes do not menstruate. Genotypes differ in their hormonal concentration at the same nutritional level, in the reactions of their tissues to the same levels of hormones, and in how hormone production differs with food intake. These interactions cause the crossing reaction norms that change genetic covariances between age and size at maturity from positive to zero across a gradient of nutritional conditions. Because the hormones affect many traits, their integrative action results in phenotypic patterns of covariation that extend from growth and maturation to behaviour and lifespan.

Demographic analysis of human reaction norms for age and size at maturity reveals that population mean values are close to those predicted by optimality theory, but the considerable genetic variation maintained in the population means that most individuals are up to six months and one or two kilograms from the optimum when they mature. The reaction norms for age and size at maturity of several human populations differ in a manner consistent with expectation from optimality theory given their historical differences in juvenile mortality rates. Differences among human populations on similar diets are as large as two years and 20 kilograms at maturity.

While no one has actually constructed that explanation, we at least know enough about humans so that the sketch could be given. Remember what it looks like. Similar effects must exist in all lineages, and any explanation relying on a single approach gives only one side of the answer.

The flow of evolutionary causation in this explanation

Evolutionary causation flows in two directions: from variation in the genes through gene expression in development, morphology, and physiology to the phenotype, then back from variation among phenotypes (natural selection) to changes in gene frequencies. This suggests the key elements of organismal structure that should be involved in evolutionary explanation (Fig. 1.2).

Relationship to levels of analysis

Evolutionary causation is complex. Most traits have causes arising at several levels of the biological hierarchy (Fig. 1.2). This book first discusses the phenotypic causes, then the genotypic ones, and then the attempts to connect them (Table 1.1). That discussion is microevolutionary.

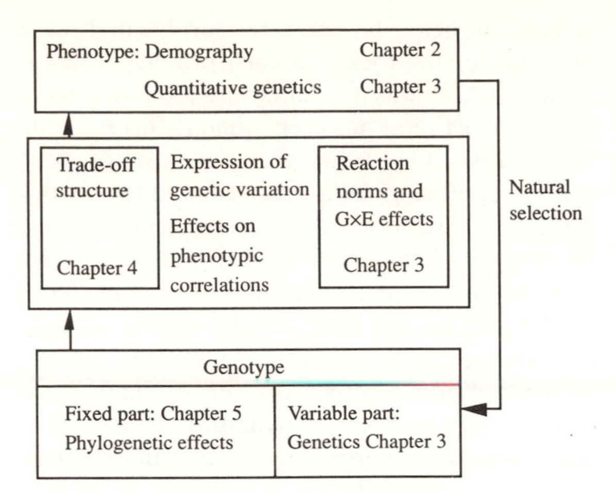

Figure 1.2 A sketch of evolutionary causation.

Macroevolution enters with comparative biology and its methods for weighting the effects of history and phylogeny.

CHAPTER PREVIEWS FOR PART I

Demography (Chapter 2)

Mortality and fecundity vary with age and size. Demography connects age- and size-specific variation in mortality and fecundity to variation in fitness; it compares the strength of natural selection on different traits. However, it does not tell us how the population will respond to selection. To predict the short-term response to selection, we need to know about the genetic variation present and the links between traits.

Quantitative genetics (Chapter 3)

Most life history traits are influenced by many genes of small effect. The overall effect of many genes can be estimated through correlations between relatives and through the response of a population to selection. To make such estimates, one needs a statistical model of many genes acting at once. This model is consistent with Mendelian genetics and with changes in phenotypes in populations under selection. Called quantitative genetics, it is the best tool we have for analysing the genetics of most ecologically important traits.

Quantitative genetics, which originally assumed that environmental factors are held constant, can be extended to natural populations in heterogeneous environments by including the effects of phenotypic plasticity. Phenotypic plasticity has two important roles in evolution. First, by modifying the relationship among traits and of traits with fitness, it changes the selection pressures on traits across environments. Second, by modulating the expression of genetic variation (for single traits) and of genetic covariation (for pairs of traits) it changes the genetic response to selection across environments.

Table 1.1 How levels of analysing organisms relate to the elements of evolutionary explanation discussed in this book

Level of analysis	Traditional discipline	Chapter	Element of explanation
Phenotype	Demography	2	Natural selection (sensitivities of fitness to changes in traits)
	Quantitative genetics	3	Response to selection (genetic variation, correlation)
Intermediate	Development and physiology	3	Phenotypic plasticity (reaction norms)
		4	Energy allocation in trade-offs
		5	Lineage-specific effects resulting from organismal integration
Genotype			
1. Variable loci	Population genetics	3	Response to selection (single locus), pleiotropy
2. Loci fixed in lineages	Systematics, comparative biology	5	Lineage-specific constraints

Physiology and trade-offs (Chapter 4)

Physiology contributes insight into the mechanisms that cause trade-offs. A trade-off exists when a benefit realized through a change in one trait is linked to a cost paid out through a change in another. The benefits and costs are reckoned in the currency of fitness, not in units of energy, nutrients, or time, although a token for fitness, such as energy or growth, is often used for convenience. The answer to a question about life history evolution depends on which trade-offs are important. These details describe the trade-off structure of the organism.

Lineage-specific effects (Chapter 5)

Comparisons among higher taxa raise striking questions. Do organisms that delay maturity live longer? Do most large organisms have relatively few offspring? Why do all octopuses, bamboos, and spiders breed once, then die? To answer such questions, we use methods that weigh the effects of body size and taxonomic group on patterns of variation. They tell us how much of a pattern we should attribute to history and design, and how much we should attribute to microevolutionary processes that operated within the local population in the recent past. We also consider how lineage-specific effects arise and come to differ among lineages.

NATURAL SELECTION AND FITNESS

Life history theory analyses how variation in life history traits leads to variation in fitness among individuals. Fitness and natural selection are concepts invented by biologists to summarize complex interactions. Organisms are born. Some survive, some of those reproduce, and all die. To make sense of the patterns we see in the descendants, it helps to introduce the paired concepts of natural selection and fitness. A definition of fitness implies the part of natural selection on which one is concentrating: the units on which it works and how their differential survival and reproduction result in evolutionary change.

Every analysis of adaptation implies a technical definition of fitness. The definition is chosen because it helps to solve the problem: it is a problem-solving tool. For each problem, there is usually a best way to define fitness. No general definition of fitness has been found (Murray 1990). Whether such a definition exists can provoke spirited discussion. Difficulties enter in two places: first, in the relationship between genotype and phenotype, where the mechanisms of gene action and the role of the epigenetic system in determining life history traits are by no means fully clear (this problem is discussed next), and second, in the decision whether to seek optimal, static solutions or to explain dynamic change. Here density- and frequency-dependence and stochasticity can complicate the calculation of relative fitness. This problem is discussed in Chapter 2 after some background has been developed.

Genotypic and phenotypic fitness

Only organisms reproduce and die. Therefore it might make sense to speak only of the fitness of organisms, but that is inconvenient when there is sexual reproduction. The genotype of a sexually reproducing organism exists only for its lifetime, for its offspring are genetically diverse recombinants. The genes carry the memory of past selection across generations in sexual populations. Thus for some purposes it is better to speak of the fitness of genes. The move from the organism to the genes makes us reconceptualize fitness.

Because the main reason to switch from organisms to genes is sexual reproduction, begin by imagining a population of asexual organisms made up of a set of clones. Since each member of a clone is genetically identical, we speak of the fitness of the clone, not of each individual within it. In your mind's eye, let the individuals in the population be born, survive, reproduce, and die. Do this for many clones. Imagine this happening in a habitat that is heterogeneous in space and variable in time. The population is well mixed and members of all clones encounter all habitats equally often. Because they are phenotypically plastic, the survival and reproduction of the members of a single clone change across habitats, from season to season, and from year to year. Those reactions sum across environments, seasons, and years to give a general measure of how well one clone is doing against the others, and the measures differ among clones.

What is the key feature of this complex process? We could follow the population for a long time to see which clone dominates numerically or endures the longest before vanishing. The same clone might achieve numerical dominance and endure the longest, but a clone that was never dominant might last longer than a number that had been dominant for shorter periods. We are faced with several interesting decisions. Is the numerically dominant clone fitter than the others, no matter how short-lived its dominance, or is the clone that endures the longest before vanishing fitter than the others, no matter how relatively rare it might have been? A good argument can be made for each view, but it is easier to calculate numerical dominance than time to extinction, and that is the approach most used. Unless another definition is mentioned, you can assume that the definition of fitness used is a short-term measure of numerical dominance.

Now introduce sexual reproduction. No longer do the genotypes move through time and space intact, ready for testing across many conditions. With sexual reproduction, recombination breaks up old gene combinations in each generation and creates new ones. Population geneticists think about this by considering the *marginal effect* of a gene substitution on fitness. Imagine that a sexual population exists, like the asexual one just considered, in a similar environment, heterogeneous in space and variable in time. A new mutation occurs, and we want to know whether it will spread, be fixed, and be found at equilibrium. Whether it does so depends on its marginal effects on the survival and reproduction of the organisms that carry it. This depends, in turn, not only on the habitat, the season, and the year, but *also on the other genes with which it is combined*—the genetic background into which it is placed.

This definition of fitness is abstract. We cannot examine all the different genomes present in a population nor measure in each of them the marginal effects of a mutant on reproduction and survival. We have to do a thought experiment in which the mutant is tested against a sufficiently large number of genetic combinations, in a sufficiently large number of habitats, seasons, and years, to give us a reliable measure of its probability of increasing in frequency and being fixed in the population. In population genetics, this process is simplified considerably. The complex biology is reduced to calculating whether or not a mutant increases or decreases in frequency from one generation to the next—a short-term measure of numerical dominance.

When we move from genes to traits, we need a further piece of mental gymnastics. Traits are not inherited; genes are. Traits do not survive and reproduce; organisms do. Organisms and DNA sequences are well defined, but how one defines a trait is not a question that has one clear answer. Traits lie between genes and organisms, determined by a number of the former, being part of the latter. We solve the problem by thinking about the fitness of traits as though traits were genes that could move through populations intact and be tested in many individuals across many environmental conditions. If this were so, then we could speak of the marginal effect of a trait substitution on fitness. This is only fair if we can come up with a method of representing how traits fit together to build organisms and how traits are inherited.

In practice, one makes assumptions about population dynamics, sex ratios, and the mortality and fecundity schedules of the two sexes, and derives a suitable measure of fitness from single locus population genetics by asking what measure predicts the direction of evolutionary change or the equilibrium state of the population. *Then the single locus argument is discarded and the fitness measure is kept, with the claim that it applies to a wide class of problems.* This approach has contributed to the success of sex allocation theory. The physicist Murray Gell-Mann calls such procedures (often used in physics) 'the veal–pheasant approach', after the practice in French cuisine of cooking a piece of pheasant between two pieces of veal. When the dish is done, the veal is discarded and one eats the pheasant. Here the veal is the approach to life histories through demography and quantitative genetics (Chapters 2 and 3), and the pheasant is the optimization approach to phenotypes (Chapters 6, 7, and 8).

Fitness is a composite, relative measure of birth and death rates. The empirical view is that the fitness of a phenotype depends both on the other phenotypes present in the population and on the environmental conditions. Empirical fitness is a property of both an individual and a context. The theoretical measure of fitness most widely used, r, ignores many aspects of empirical fitness. It is called the *Malthusian parameter* in genetics and the *intrinsic rate of natural increase* in demography. Another fitness measure sometimes used is R_0, the per-generation rate of multiplication. They differ in important respects discussed in Chapter 2.

To turn these theoretical measures of absolute growth rate into measures of relative fitness, one divides the fitness of each phenotype by that of the fittest phenotype present. The fittest phenotype then has fitness 1, and all other phenotypes have fitnesses between 0 and 1. Such variation arises from differences among individuals in survival rates, reproductive rates, or both. Two clones or two genotypes could have the same fitness, one having achieved it with low survival and high fecundity, the other with high survival and low fecundity.

ADAPTATION AND CONSTRAINT

These represent the two ends of a continuum of biological explanations. Both are often used naïvely. Traits are a mixture of adaptation and constraint, and methods are available for identifying and weighting their contributions. The tension between adaptation and constraint runs throughout this book.

Definitions of adaptation and constraint can be arranged in sequences of increasing sophistication. The naïve definition is the same for each: all patterns are adaptive, and all patterns are constrained. Biologists exist who believe in one or the other definition; they are mutually exclusive.

More sophisticated definitions of adaptation

The functional definition

According to Williams (1966*a*) and Curio (1973), an adaptation is a *change* in a phenotype that occurs in response to a specific environmental signal and has a clear functional relationship to that signal that results in an improvement in growth, survival, or reproduction. Otherwise, it does not appear. One example is the production of spines and elongate helmets in waterfleas, *Daphnia*, in response to dissolved molecules that indicate the presence of invertebrate predators that prey less effectively on spiny, helmeted *Daphnia.* Helmets and spines are costly, individuals that do not produce them have higher reproductive rates, and therefore when predators are not present, the spines and helmets are not produced. This definition is most appropriately applied to variation within populations.

The phylogenetic definition

Given a phylogenetic tree indicating the relationships of many species and the evolutionary sequence in which traits appeared, Coddington (1988) has suggested that we only apply the word adaptation to homologous traits that are derived, unique to a single species, and functionally associated with a change in habitat and selection pressure that is also unique to that species. This definition is appropriate in the phylogenetic context, but it ignores variation within populations and lineage-specific selection that produces adaptations that can be older than an individual species. It also encounters one problem and creates another (Harvey and Pagel 1991). By requiring that we know both current and ancestral character state and habitat, it asks for details that can almost never be produced; by restricting adaptation to homologues, it rules out convergence as evidence for adaptation.

Adaptation through lineage-specific selection pressures

If entire lineages encounter selection pressures that differ consistently from those of other lineages, one should be able to speak of adaptations older than individual species and detectable by comparisons among lineages. For example, the entire cetacean order lives in aquatic environments that select for a fusiform body that slips efficiently through the water propelled by fins. The entire artiodactyl order moves about on legs in a terrestrial environment that selects for efficient running, jumping, and turning. Active selection of the same habitat by all species within a lineage has similar effects. For example, the planktonic larvae of all species in the Phylum Phoronidae have been selecting shallow beds of marine sand to settle in for hundreds of millions of years. The worm-like adults only encounter selection pressures associated with that habitat.

It is more difficult to generate plausible cases of lineage-specific selection producing adaptation in life history traits. Here the selection pressure that must be common to the whole lineage is not something generally associated with a habitat, but a pattern of age- and size-specific mortality and fecundity. These often vary among species within habitats. However, one can save some generality for lineage-specific selection. The most important part of the environment of a trait is not some aspect of the external habitat but the other traits making up the individual. The traits of a single individual encounter each other reliably in every generation and *must* fit together to produce a well-functioning organism. Such lineage-specific selection is strong and general, for traits move through lineages together. It adapts age at maturity and growth rate to each other and to the fecundity and mortality schedule in a manner that can be interpreted as optimal (Chapter 6). Precise adjustments of several traits to each other are more plausible if the traits have been interacting with each other for a long time—longer than the lifetime of the species, genus, or family.

Two microevolutionary definitions of adaptation

These operational definitions are aimed at the stubborn critic who requires that hypotheses be tested in risky experiments. The first defines an adaptation as the state of a trait predicted to be in that state and no other by an optimality model (Parker and Maynard Smith 1990), but only where that prediction has been tested by using mutations or phenocopies to perturb the phenotype away from the optimal state and thus demonstrate that the fitness of the perturbed phenotypes is lower than the fitness of the optimal one. This has been achieved through clutch size manipulations, e.g. for kestrels by Daan *et al.* (1990).

The second requires one to show that every time a given factor changes in the environment, a trait evolves to the same new state, and that every time the environmental factor changes back to its original state, the trait does also. I know of no good examples where the second definition has been applied in practice.

More sophisticated definitions of constraint

The casual phylogenetic definition

This definition holds that any pattern or state that can be attributed to phylogeny, as opposed to recent microevolution within the currently existing population, is a constraint. For example, all birds in the Order Procellariformes have a clutch of only one egg. The casual definition would take this as evidence that no member of this order *could* evolve a clutch size of two. While the conclusion does not follow from the evidence, it can be used as an alternative to sharpen the naïve adaptationist hypothesis that a clutch size of one is actively maintained by natural selection in every species of the entire order. A more sophisticated definition of constraint can sharpen a less sophisticated definition of adaptation. This also works the other way round.

The biomechanical definition

This definition is relatively uncontroversial. It asserts that organisms are constrained to obey the laws of physics and chemistry. Examples include the need for terrestrial tetrapods to increase the cross-sectional area of their limbs as their body mass increases and the necessity for the surface area of lungs and guts to increase in proportion to body mass rather than some linear measure of size.

Systems definitions of constraint

This is the most interesting definition of constraint, for it suggests a new research programme. The sex ratios of mammals are a good example (cf. Williams 1979). With few exceptions (reviewed in Clutton-Brock and Iason 1986), they are fixed near conception at 1:1. Evolution has evidently not been able to overcome this constraint, for in the many mammal species that have polygynous social systems more flexible sex determination would be advantageous.

The systems definition of constraint comes in several versions. Oster and Alberch (1982, p. 450) see the most basic constraint as the requirement that each stage of development must proceed from where the last one left off. However, their concept of constraint is much more sophisticated than that simple statement would seem to imply. Their central message is that genes act through proteins that change the properties of cells, that cells are the key players in development, and that cells interact in processes constrained by physics in ways that cannot be changed by simple gene substitutions—steered a bit, modified slightly, but not fundamentally changed. Only a small portion of phenotype space is then available for exploration by gene substitutions. This restriction on the exploration of phenotype space is the key element of their definition of constraint. The most convincing support for their approach comes from widespread regularities in the patterns of discontinuous phenotypes that are observed for traits like coat colouring, the development of the vertebrate limb, and the properties of sheets, cylinders, and balls of cells.

Maynard Smith *et al.* (1985, p. 265) give a related definition of developmental constraint that represents a consensus opinion: 'Developmental constraints [are] biases on the production of variant phenotypes or limitations on phenotypic variability caused by the structure, character, composition, or dynamics of the developmental system.' They list several tests of hypotheses based on this definition of constraint.

A third view sees the origin of constraint in the fixation of key traits; it is not at odds with the two definitions just given. Fixation leads to progressive integration and irreversible change (Chapter 5). Once one trait becomes fixed, other functionally related traits are less free to vary. The longer one trait has been fixed, the more traits have been subsequently fixed. Thus constraints are the functional connections between fixed traits—organismal integration—that make it first difficult, then impossible, for evolution to proceed in reverse to the original state. This model suggests two testable hypotheses.

First, if we establish the order in which two traits have evolved (cf. Maddison 1990), the older of the two traits should be more deeply integrated into the organism, more buffered against genetic change. If we induce mutations with radiation or chemical mutagens, the older traits should vary less than the more recent ones in response to mutagenic treatments.

Second, we could compare the same variable trait in two closely related species. In one species, several functionally related traits are fixed. In the other, all those traits vary. If we select the target trait, it should display a larger and more rapid response to selection in the species where all traits still vary than in the species in which some functionally related traits are still fixed.

Comments on adaptation and constraint

Do not assume that biologists have agreed on clear definitions of adaptation and constraint. Those given here do not exhaust the range of opinion. It is a mistake to frame an evolutionary question as a matter of *either* adaptation *or* constraint. The two interact in the causation of every pattern and in the determination of every state of every trait. One should distinguish between patterns that can be interpreted as *adaptive*, including those detected in broad comparisons across taxa, and processes leading to *adaptation* that can be investigated as mechanisms operating within populations. The scope of application of the term adaptive is much broader than that of the word adaptation, which should be reserved for cases satisfying either the phylogenetic or the microevolutionary definition. When a trait has been thoroughly scrutinized with theory and experiment, one often loses interest in the question, Is it an adaptation or is it a constraint? Traits are determined by such intimate interactions of natural selection and internal structure that the manner in which they are determined becomes more interesting and more important than the labels inaccurately describing the causes involved.

RECOMMENDED READING

Buss, L. (1987). *The evolution of individuality*. (Princeton University Press, Princeton). A thought-provoking critique of Weismann's dichotomy and a broad perspective on life-cycle diversity.

Charlesworth, B. (1980). *Evolution in age-structured populations*. (Cambridge University Press, Cambridge). An excellent summary of population genetics with age structure and of life history theory.

Gould, S. J. and Lewontin, R. C. (1979). The spandrels of San Marco and the Panglossian paradigm: a critique of the adaptationist program. *Proc. Roy.*

Soc. Lond. B **205**, 581–98. The most influential critique of the assumption that adaptation is pervasive.

Maynard Smith, J. (1989). *Evolutionary genetics.* (Oxford University Press, Oxford). An excellent introduction to evolutionary theory and good preparation for this book.

Parker, G. and Maynard Smith, J. (1990). Optimality theory in evolutionary biology. *Nature* **348**, 27–33. A recent overview by two masters of their art.

Williams, G. (1966*a*). *Adaptation and natural selection.* (Princeton University Press, Princeton). A deep discussion of adaptation and life histories and a justification for gene-centred explanations. A classic that has had tremendous influence on evolutionary biology.

2

DEMOGRAPHY: AGE AND STAGE STRUCTURE

Connoissant tant le nombre des naissances que des enterremens qui arrivent pendant le course d'une année, trouver le nombre de tous les vivans et leur augmentation annuelle, pour une hypothèse de mortalité donnée.

Leonhard Euler, 1760

CHAPTER OVERVIEW

Demography, the key to life history theory, allows us to calculate the strength of selection on life history traits for many conditions. This chapter presents the tools needed: age, size, and stage-specific schedules of reproduction and mortality; the Euler–Lotka equation relating life history traits to fitness; the matrix representation of that equation that generalizes it to populations classified by stage and size as well as by age; and the sensitivity and elasticity of fitness to changes in life history traits. The chapter is technical, but the insights it offers are worth the effort: why age at maturity is a pivotal trait, how to weigh the contributions to fitness of early and late fecundity, why selection to lengthen life is generally weak, and why life history theory makes no general predictions.

INTRODUCTION

This chapter sketches the demographic theory required to understand life history evolution. Roughgarden (1979), Charlesworth (1980), Emlen (1984), and Caswell (1989a) treat the same topics more thoroughly. The version presented here is the classical case for constant birth and death rates. For recent work on demography in random environments, see Orzack and Tuljapurkar (1989).

Interpreting demography for populations and life history traits

Demography was developed to forecast population growth. In that context, the equations and symbols represent the numbers of whole organisms in the population and their age-specific birth and death schedules. When the same equations and symbols are applied to life history evolution, there is an important shift in interpretation. They no longer refer to whole organisms, but to effects of gene substitutions.

In classical demography, one wants to know how changes in age at maturity affect population growth rate. Delaying age at maturity decreases population growth rate, telling us that motivating people to marry later will buy time to solve population problems. The analysis is of whole organisms.

The same sensitivity analysis is carried out in life history theory. By making small changes in age at maturity in a thought model, and taking account of the effects on growth, fecundity, and mortality, one can predict how age at maturity will evolve. Here one assumes that the small changes in age at maturity are brought about through average effects of gene substitutions. Conceive of a new mutant in a population. It affects age at maturity, and it also affects mortality and fecundity. Will it spread to fixation or not? It is the equations of demography that chart the fate of a gene substitution with age-specific effects.

Here we are dealing with marginal effects of gene substitutions, not with numbers of organisms in the population. The difference in interpretation is critical, for under some circumstances modelling the rate of spread of a gene substitution with a fitness definition that assumes exponential growth can be justified in a population of whole organisms that remains constant in size.

In the first part of his book, Charlesworth (1980) develops the implications of age structure and the presence of two sexes for the rate of spread of gene substitutions under different assumptions about population regulation. There he aims to answer the questions 'What effects does age structure have on changes in gene frequencies?' and 'What if the two sexes have different age structures?'. Those are questions about population genetics; the phenotypic age structures are taken as given, and the consequences for the genes are worked out. Life history theory aims to answer a question posed on a different level: given a diversity of birth and death schedules and organismal designs, what kinds of life histories will evolution produce, and why? To answer that question about phenotypes at equilibrium, one simplifies the genetics to concentrate attention on the phenotype. Charlesworth (1980) takes this approach in his last chapter.

We begin with a simple model of population growth.

Models of population growth

The simplest model of population growth is exponential, without density dependence, frequency dependence, genetic variability, or age structure. If the net reproductive rate is R_0, then we can project population densities from generation to generation with a recursion equation,

$$N_1 = R_0 N_0 \tag{2.1}$$

and if we continue the process for g generations, we get

$$N_g = R_0^g N_0. \tag{2.2}$$

Here the number of organisms after a certain number of generations g, N_g, is the number originally present, N_0, times the net per-generation reproductive rate, R_0, raised to the number of *generations* since time 0. Generations do not overlap: one cohort grows up, matures, reproduces, and dies before its offspring have matured. Examples include annual plants without seed banks (generation time = 1 year) and periodic cicadas (generation time = 13 or 17 years).

Introducing density-dependence to this model leads through the logistic equation to community ecology. Introducing frequency dependence leads to evolutionary game theory. Introducing age structure and overlapping generations leads to classical demography and life history theory. Community and behavioural ecology usually ignore age structure. Life history evolution usually ignores density and frequency dependence. The justification is convenience, not logic or realism.

LIFE TABLES

Consider a population of seasonal breeders with overlapping generations. Keep track of females only: all births are daughters. Usually, one counts females because their fecundity limits the rate of population growth. Follow a *cohort* of newly born daughters (a cohort consists of all organisms in a population born in the same season). As they age and die, the number of survivors decreases. At a certain age, they start to mature and reproduce. Assume they all mature at the same age, in the same season, and that each mature age class has an average birth rate. To keep track of births and deaths, we need the notation given in Box 2.1.

Life tables represent birth and death probabilities[1]. Cohort life tables are measured on individuals born in the same season and followed through time. Horizontal life tables, measured on organisms that are all alive at the same moment, represent a mixture of cohorts and a mixture of

[1] In moving the data gathered in the field to a life table, several problems need attention. They arise from the difference between vertical (cohort) and horizontal life tables, from the choice of age interval, and from sampling problems. Discussions can be found in Caughley (1966), Keyfitz (1968), Lee (1980), Begon *et al.* (1980), Caswell (1989*a*), and Fox (1989).

Box 2.1 Demographic notation for age-structured populations

Symbols

x = age in general
X = age *class* = *age interval x to x* + 1
t = time
T = generation time, average age of mothers giving birth
D_X = number dying during age class X to $x + 1$
S_X = number surviving to beginning of age class X
$N_{t,X}$ = number of individuals in the population of age class X at time t
d_X = probability of dying between age x and age $x + 1$
q_X = d_X
p_X = probability of surviving from age x to age $x + 1$
l_X = probability of surviving from birth to beginning of age class X
m_X = expected number of offspring for a female in age class X
b_X = m_X
c_X = fraction of the population in age class X
v_X = reproductive value of organisms in age class X
α = age at maturity, which here means *age at first reproduction*
ω = age at last reproduction
r = instantaneous rate of natural increase
R_0 = lifetime expectation of female offspring; net reproductive rate
λ = e^r = rate of growth per time unit

Elementary relations:

$l_{X+1} = p_X l_X$
$l_X = p_0 \times p_1 \times p_2 \times \cdots \times p_{X-1}$
$p_X = 1 - d_X$.

age-specific mortality and fecundity rates. Cohort life tables are only equivalent to horizontal ones when birth and mortality rates are constant. For all tables, samples should be taken at the same point in the year—here the end of the reproductive season. This means that newborn offspring that die before the end of the season are simply not counted. A population so modelled is a type of birth–pulse population. Other assumptions require different calculations (Caswell 1989*a*, Ch. 2).

Table 2.1 shows the fate of 1000 newborn females born in one season and followed until all have died. From the number of survivors S_X and of deaths D_X in each age class X, one can calculate the age-specific death rate d_X, the age-specific survival rate p_X, and the age-specific survivorship or probability of surviving from birth to the start of age class X, l_X. The age-specific birth rates, m_X, are included to calculate average lifetime reproductive success, $\sum l_X m_X$.

To forecast the numbers of organisms in each age class from year to year, we need only two relationships:

1. The number of newborn is the sum across age classes of the number of females alive that year times their expected fecundity.
2. The number alive in all older age classes equals the number that were alive in the next younger age class in the previous year, times the probability that they survived one year.

For example (Table 2.2), the number of newborn in year one is the sum across age classes of the probability that a female has survived from the previous breeding season (p_{X-1}) times the age-specific fecundity rate (m_X) times the number of females in the age class in year 0:

$$(0.8 \times 1.2 \times 9) + (0.8 \times 1.4 \times 8) + (0.75 \times 1.03 \times 7) + (0.5 \times 0.96 \times 5) = 25.4075.$$

Table 2.1 Construction of a representative life table for a cohort of 1000 females living 5 years. Numbers of survivors are rounded to the nearest whole number. Number of births is an average for the cohort and therefore can take on fractional values

X (age class)	S_x (survivors)	D_x (dying)	d_x (death rate)	p_x (survival rate)	l_x (survival to age x)	m_x (birth rate)	$l_x m_x$
0	1000	250	0.25	0.75	1.000	0.00	0.000
1	750	150	0.20	0.80	0.750	0.00	0.000
2	600	120	0.20	0.80	0.600	1.20	0.720
3	480	120	0.25	0.75	0.480	1.40	0.672
4	360	180	0.50	0.50	0.360	1.03	0.396
5	180	180	1.00	0.00	0.180	0.96	0.144

$\alpha = 2, \omega = 5$ $R_0 = \sum l_x m_x = 1.932$

Table 2.2 A population projection for a population with the life Table 2.1. Values rounded to the nearest integer

Age class (X)	Year 0	1	2	3	4	50	51	52
0	10	25	23	32	39	1 390 991	1 746 688	2 193 343
1	9	8	19	17	24	830 796	1 043 243	1 310 016
2	8	7	6	15	14	529 290	664 637	834 595
3	7	6	6	5	12	337 204	423 432	531 710
4	5	5	5	4	4	201 401	252 903	317 574
5	3	3	3	2	2	80 194	100 701	126 451
$N =$	42	54	62	75	95	3 369 876	4 231 604	5 313 689
$N_{t+1}/N_t =$		1.29	1.15	1.21	1.27	...	1.26	1.26

We round the answer to 25. The number of one-year-olds in year 1 is the number of newborn in year 0, 10, times the survival rate, 0.75 = 7.5. This is rounded to 8. The calculations in Table 2.2 were made for the first four years and for the 50th, 51st, and 52nd years.

Population growth is rapid. At the start, the multiplication rate per year is variable (from 1.15 to 1.29), but after 50 years the number of organisms in each age class is growing constantly each year by a factor of 1.26. Early on the proportion of organisms in each age class changes from year to year, but after 50 years it is constant (Table 2.3). Tables 2.2 and 2.3 illustrate two properties of constant life tables. *A population with a constant birth and death schedule will attain a stable age distribution and then grow exponentially at a constant rate.* The rate at which it grows can be calculated from the Euler–Lotka equation.

THE EULER–LOTKA EQUATION

This is the basic equation of demography and life history evolution, discovered by Euler (1760) and re-discovered by Lotka (1907). It specifies the relationships of age at maturity (α), of age at last reproduction (ω), of probability of survival to a given age class (l_x), and of number of offspring expected in a given age class (m_x) to the rate of growth (r) of the population.

Recall the two interpretations of demography—for population growth and for life history theory—and imagine an asexual population consisting of many clones. Each clone has different life history

Table 2.3 Successive age distributions for a population with the life table given above

Age class (X)	Year 0	1	2	3	4	→	50	51	52
0	0.24	0.47	0.37	0.42	0.41		0.41	0.41	0.41
1	0.21	0.14	0.31	0.23	0.25		0.25	0.25	0.25
2	0.19	0.13	0.10	0.20	0.14		0.16	0.16	0.16
3	0.17	0.12	0.09	0.06	0.13		0.10	0.10	0.10
4	0.12	0.10	0.08	0.06	0.04		0.06	0.06	0.06
5	0.07	0.05	0.04	0.03	0.02		0.02	0.02	0.02

traits; the growth rates of the clones measure their absolute fitnesses. The Euler–Lotka equation applies to one clone at a time. This is analogous to analysing the marginal effect of a gene substitution with impact on birth and death rates in a sexually outcrossing population. The validity of this analogy has been investigated by Charlesworth (1980).

In life history theory, the solution to the Euler–Lotka equation, r, measures the fitness of a clone or a gene substitution, not the growth rate of the population as a whole, and is most clearly applicable when the population is not regulated by density. The equation is important because it weights the contributions of different age classes to fitness.

Derivation

Assume that the age-specific birth and death rates do not change, and concentrate on the number of offspring born in a time interval t of duration equal to age class X: $N_{t,0}$. These newborn are contributed by females of different age classes. The contribution from females of age class X depends on three things: the number of newborn X time units ago, $N_{t-X,0}$; the probability that they survived to the present time, l_X; and the expected number of offspring that each has at this age, m_X:

$$N_{t,0} = N_{t-X,0} l_X m_X. \tag{2.3}$$

This measures the contribution of females in one age class. We calculate the total number of newborn at time t by summing over all age classes X:

$$N_{t,0} = \sum_{X=\alpha}^{X=\omega} N_{t-X,0} l_X m_X. \tag{2.4}$$

Now try a solution. Assume that the number of newborn grows exponentially:

$$N_{t,0} = K\,e^{rt}. \tag{2.5}$$

This trial solution gives us:

$$K\,e^{rt} = \sum_{X=\alpha}^{X=\omega} K\,e^{r(t-X)} l_X m_X. \tag{2.6}$$

Divide both sides by Ke^{rt} to get the Euler–Lotka equation. Substituting the trial solution eliminates time from the equation.

$$1 = \sum_{X=\alpha}^{X=\omega} e^{-rX} l_X m_X. \tag{2.7}$$

When written as a series, as Euler did in 1758,

$$1 = e^{-r\alpha} l_\alpha m_\alpha + e^{-r(\alpha+1)} l_{\alpha+1} m_{\alpha+1} + \cdots + e^{-r\omega} l_\omega m_\omega \tag{2.8}$$

the age classes appear explicitly as α, $\alpha + 1, \ldots, \omega$.

This equation expresses r, a fitness measure used in life history theory, as an implicit function of life history traits. In the *demographic* interpretation of the Euler–Lotka equation, the following statements are equivalent:

1. The population is in stable age distribution—the proportion of organisms in each age class remains constant in time. The number of organisms in each age class changes at rate r according to the equation:

$$N_{t+1,X} = e^r N_{t,X} \tag{2.9}$$

2. The population grows exponentially with rate r. The total number of individuals in the population is then expressed by the equation

$N_t = N_0 e^{rt}$, in which the growth rate is the same as the solution to the Euler–Lotka equation.
3. The mortality and fecundity schedules remain constant in time.

Demographic theory can be formulated with discrete or continuous mathematics. The technical problems differ, but the approaches converge in the limit. The continuous form of the Euler–Lotka equation is

$$1 = \int_{\alpha}^{\omega} e^{-rx} l_x m_x \, dx \qquad (2.10)$$

The Euler–Lotka equation is always true for a population with fixed birth and death rates in that such a population will have an *intrinsic* rate of natural increase defined by eqn (2.10). It may not actually be growing at that rate if it is not in stable age distribution, but that is the growth rate towards which it is converging. Stable age structure with mortality and fecundity schedules constant in time is less restrictive and unrealistic than it might appear.

1. Although most natural populations are rarely in stable age distribution, moderate deviations from stable age distributions do not often change qualitative predictions. The Euler–Lotka equation captures robust features of demography.
2. The exponential growth referred to is not that of a population but the rate of spread of an allele with marginal effects on a life history. Gene frequencies can change in populations with constant numbers of organisms.
3. Exponential growth describes processes whose rate can be positive, zero, or negative: it refers not to the *quantity* of growth but to its *quality* (smooth, differentiable, monotonic). The model describes the spread of an allele during the phase that determines whether it will be fixed in the population or eliminated. If during that phase deviations from exponential growth would not make any difference to the outcome, the model holds.

CALCULATING r

An equation that cannot be solved does one little good, and the Euler–Lotka equation can only be solved analytically for r under special assumptions. Usually it is solved numerically (see program at the end of the chapter). Lenski and Service (1982) and Service and Lenski (1982) illustrate the calculation of r from experimental data for specific aphid genotypes.

THE STABLE AGE DISTRIBUTION

Most fixed age-specific schedules of births and deaths lead to a *stable age distribution* in which the population grows smoothly and exponentially[2]. A population in stable age distribution can be increasing, not changing in numbers, or declining, but the *proportion* of organisms in each age class is stable. The special stable age distribution where the population is not growing ($r = 0$) is known as a *stationary age distribution.*

Once r is known, the proportion of organisms c_X in each class X of this distribution can be calculated. The number of individuals in a given age class is the number that were born x time units ago, $N_{t-X,0}$, times the probability that they survived to the start of age class X, l_X: $N_{t-X,0} l_X$. The total number of individuals in the population is just this amount summed over all age classes: $\sum N_{t-X,0} l_X$. With the population in stable age distribution, $N_{t-X,0} = e^{r(t-X)} N_{0,0}$. Thus the fraction of organisms in each age class is

$$c_X = \left(e^{r(t-X)} N_{0,0} l_X\right) \Big/ \left(\sum_X e^{r(t-X)} N_{0,0} l_X\right) \qquad (2.11)$$

and we can cancel $e^{rt} N_{0,0}$ to get

$$c_X = \left(e^{-rX} l_X\right) \Big/ \left(\sum_X e^{-rX} l_X\right). \qquad (2.12)$$

For example, for Table 2.1, the stable age distribution is 41 per cent newborn, 25 per cent one-year-olds, 16 per cent two-year-olds, 10 per cent three-year-olds, 6 per cent four-year-olds, and 2 per cent five-year-olds. Compare these values with the age distributions in Table 2.3.

[2] Those in which it does not have *imprimitive life histories* (Caswell 1989*a*) in which the age distribution oscillates without converging to a stable age distribution (e.g. Pacific salmon, periodical cicadas and bamboos, annual plants).

GENERATION TIME

Generation time is defined as the average age of the mothers of newborn offspring in a population in stable age distribution:

$$T = \left(\sum_{x=\alpha}^{x=\omega} x l_X m_X \right) R_0^{-1}. \quad (2.13)$$

For the organisms of Table 2.1, generation time is 2.98 years. It is sometimes convenient to estimate r as

$$r_{est} \approx \frac{\ln(R_0)}{T} \quad (2.14)$$

For Table 2.1, $r_{est} = 0.22$. The estimate is usually within 10 per cent of the right answer. Generation time is often less than 1.5 times age at maturity.

RATES OF INCREASE

Several rates of increase with different meanings have been discussed.

r, the intrinsic rate of natural increase often used in theoretical work, is the per capita instantaneous rate of increase of a population in stable age distribution.

R_0, the number of daughters expected per female per lifetime, is the per-generation rate of multiplication of a population; i.e.

$$N_{T+1} = R_0 N_T. \quad (2.15)$$

λ is the per-*time-unit* rate of multiplication; i.e.

$$N_{t+1} = \lambda N_t. \quad (2.16)$$

$\lambda = e^r$, and $r = \ln(\lambda)$. When $r = 0.0$, $\lambda = 1.0$, $R_0 = 1.0$, and the population neither grows nor shrinks. This is the condition for zero population growth, where each female replaces herself on average with one reproducing daughter.

None of these rates of increase is appropriate for a population in the field, where the best empirical measure of rate of increase is simply N_{t+1}/N_t, or observed rate of multiplication per time unit. If the population is isolated and close to a stable age distribution, this ratio will approximate λ.

Small rates of increase have large effects. A population growing at $r = 0.1$ will double in just under seven time units. If robins weigh 10 g and a population growing at $r = 0.1$ is founded by a single female, then the mass of robins will exceed the mass of the Earth (about 6×10^{27} g) in 594 years.

REPRODUCTIVE VALUE

Reproductive value weights the contributions of individuals of different ages to population growth and compares the sensitivity of fitness to events at different ages (Goodman 1982). If the population in question is stationary ($r = 0$), then the reproductive value of age class A is:

$$v_A = \sum_{X=A}^{X=\omega} \frac{l_X}{l_A} m_X. \quad (2.17)$$

When calculating reproductive value for age class A and beyond, the survival terms needs to be adjusted for the fact that the female has lived to reach age A. The probability of being alive at the start of age class X is l_X; the probability of being alive at age class A is l_A; and the probability of living to age class X, given that one has survived to age class A, is l_X/l_A. Equation (2.17) expresses the idea that the reproductive value of a female of a given age should represent the female's expectation of offspring from that age to the end of her life. Note that $v_0 = R_0$, and when $r = 0$, $R_0 = 1$.

When the population is growing, offspring produced later contribute less to fitness, for offspring produced earlier may already have started to reproduce. Therefore the value of offspring produced later needs to be discounted by a factor that depends on the rate of population growth (r) and on the delay that takes place before they are born ($X - A$). The delay increases in each subsequent age class X, making the discount factor $e^{-r(X-A)}$. Discounting future offspring and correcting for the fact that the female has already survived to age A, we get:

$$v_A = \sum_{X=A}^{X=\omega} e^{-r(X-A)} \frac{l_X}{l_A} m_X \quad (2.18)$$

which can be rearranged as:

$$v_A = \frac{e^{rA}}{l_A} \sum_{X=A}^{X=\omega} e^{-rX} l_X m_X. \quad (2.19)$$

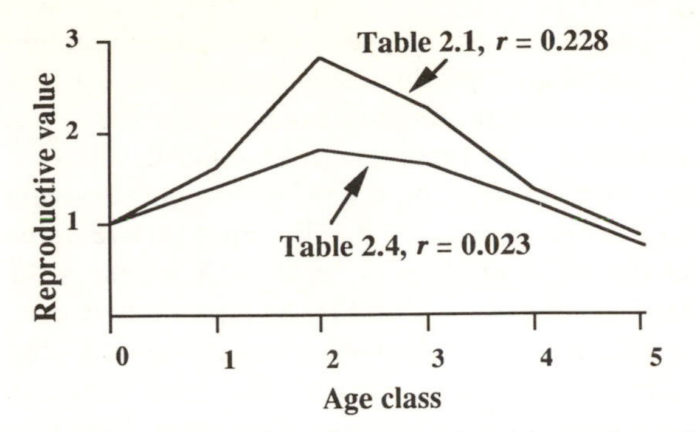

Figure 2.1 Reproductive value curves for the organisms of Tables 2.1 and 2.4.

Here the reproductive value of older females is defined relative to newborn females, and the reproductive value of newborn females always equals 1.0 whether the population is stationary or growing (Fisher 1930). Reproductive value normally increases to the age of peak reproduction, then declines.

Equation (2.19) only holds if we define the newborn as age class 1. If we define the newborn as age class 0, then the Euler–Lotka equation becomes

$$1 = \sum_{X=0}^{X=\omega} e^{-r(X+1)} l_X m_X \qquad (2.7a)$$

and the equation for reproductive value becomes

$$v_A = \frac{e^{r(A-1)}}{l_A} \sum_{X=A}^{X=\omega} e^{-rX} l_X m_X \qquad (2.19a)$$

(Goodman 1982).

For the organisms of Table 2.1, the reproductive value curves have typical shapes (Fig. 2.1) with a maximum at age of maturity.

In a growing population, reproductive value increases up to maturation because the organisms have survived—the longer they survive, the better their chances of reproducing. After maturity is reached, reproductive value usually declines (unless fecundity climbs very rapidly), for the offspring produced in the first reproductive attempt soon reproduce and contribute to the reproductive value of the parents of that first age class.

A flowering plant

The reproductive value of an annual plant, *Phlox drummondii*, has the same basic pattern (Fig. 2.2). *Phlox* exist as dormant seeds for the first 180 days of their lives, start flowering at about 260 days, and begin setting seed at about 300 days. The reproductive value curve for a perennial plant or an iteroparous animal would have saw-teeth if considered on a fine enough time scale.

The reproductive value of an age class weights

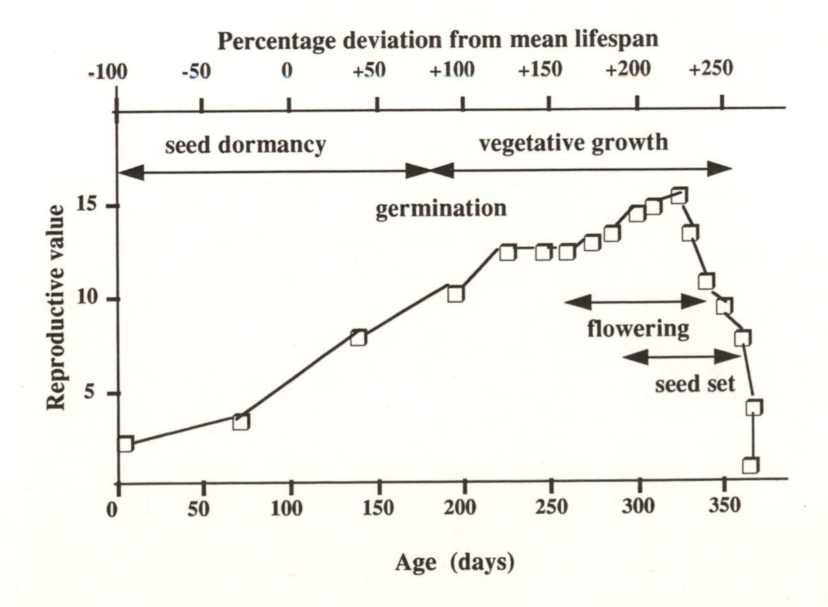

Figure 2.2 The reproductive value curve for an annual plant, phlox, in Texas (after Leverich and Levin 1979).

the contribution of that age class to future generations. Reproductive value also enters the calculation of the selection pressure on an age class (eqns (2.33) and (2.34)).

RESIDUAL REPRODUCTIVE VALUE AND THE COSTS OF REPRODUCTION

Williams (1957, 1966*b*) distinguished between *present reproduction* and *expectation of future reproduction.* He split the reproductive value of a female into two components, her births in this age class A, m_A, and her reproductive value in the next age class, or her *residual reproductive value*, RRV_A. If we isolate the Ath age class from the sum in eqn (2.10), we get

$$v_A = \frac{e^{rA}}{l_A} e^{-rA} l_A m_A + \frac{e^{rA}}{l_A} \sum_{A+1}^{\omega} e^{-rX} l_X m_X. \quad (2.20)$$

The left-hand term in the sum is, after cancellation, just m_A—the expected number of births in the present. The right-hand term is the probability of surviving from A to $A+1$, $p_A = l_{A+1}/l_A$, times a discount term, e^{-r}, necessary because the population may be growing during the delay of one time unit, times the reproductive value of a female aged $A+1$:

$$v_A = m_A + p_A e^{-r} v_{A+1}. \quad (2.21)$$

If we now define residual reproductive value, RRV_A, as $p_A e^{-r} v_{A+1}$, we can write that reproductive value = present births plus future expectations:

$$v_A = m_A + RRV_A. \quad (2.22)$$

With the alternative notation of eqn (2.19*a*), we get

$$v_A = e^{-r} m_A + p_A e^{-r} v_{A+1}. \quad (2.23)$$

To get Williams's equation, we must discount present births by r and define residual reproductive value, RRV_A, as $p_A v_{A+1}$:

$$v_A = e^{-r} m_A + RRV_A. \quad (2.22a)$$

Residual reproductive value is *not* the reproductive value of the next age class.

The distinction between current reproduction and residual reproductive value makes clear that the cost of reproduction takes two main forms. An increase in present reproduction can have a negative influence on the probability of surviving to the next age class, on the reproductive value of the next age class, or on both. The first is the survival cost; the second is the reproductive cost.

One central idea in life history evolution is that current and residual reproductive value should be so balanced in every age class that fitness is maximal. Calculating this balance for all age classes predicts the distribution of reproductive effort across the entire lifetime (Chapter 7).

Box 2.2 The Leslie matrix

Leslie (1945, 1948) developed a convenient matrix representation of life tables that is often used both in life history theory and in age-structured population dynamics. Recall that $N_{t,X}$ is the number of organisms in age class X at time t. Then the distribution of numbers across age classes is a vector,

$$\mathbf{N}_1 = \begin{bmatrix} N_{1,0} \\ N_{1,1} \\ \vdots \\ N_{1,\omega} \end{bmatrix} \quad (2.24)$$

for a population with $\omega + 1$ age classes. How this vector changes with time describes the changes in the numbers of organisms in each age class as the population grows. To get those changes, we multiply the vector by a square matrix of dimension $(\omega + 1 \times \omega + 1)$ called the *Leslie matrix* (eqn 2.25).

$$\mathbf{L} = \begin{bmatrix} 0 & F_1 & F_2 & F_3 & \cdots & F_\omega \\ p_0 & 0 & 0 & 0 & & 0 \\ 0 & p_1 & 0 & 0 & & 0 \\ 0 & 0 & p_2 & 0 & & 0 \\ \vdots & & & & & 0 \\ 0 & 0 & 0 & \cdots & p_{\omega-1} & 0 \end{bmatrix} \quad (2.25)$$

The values in the top row are not simply the age-specific fecundities. When the organisms are counted at the end of the breeding season, a mother must survive from the end of one breeding season to the start of the next if she is to have offspring. Therefore the effective fecundity of females in the ith age class is the probability that they survived from $i - 1$ to i, p_{i-1}, multiplied by the fecundity of i-year-old females, m_i: $F_i = p_{i-1}m_i$ (Caswell 1989*a*). For example, the Leslie matrix corresponding to Table 2.1 is

$$\begin{bmatrix} 0 & 0.96 & 1.12 & 0.77 & 0.48 & 0 \\ 0.75 & 0 & 0 & 0 & 0 & 0 \\ 0 & 0.80 & 0 & 0 & 0 & 0 \\ 0 & 0 & 0.80 & 0 & 0 & 0 \\ 0 & 0 & 0 & 0.75 & 0 & 0 \\ 0 & 0 & 0 & 0 & 0.50 & 0 \end{bmatrix}.$$

The meaning of the parameters in the matrix depends on the timing of reproduction with respect to the boundaries of the age classes. This formulation holds when organisms are counted once per year at the end of the breeding season and the age classes are year classes. One should check that the matrix formulation used represents the process that one wants to model.

To do a population projection, we left-multiply the age-distribution vector by the Leslie matrix, $\mathbf{n}_{t+1} = \mathbf{L} \times \mathbf{n}_t$. How this works can be seen by recalling the rule for matrix multiplication, which is, for the newborn,

$$N_{t+1,0} = F_1 N_{t,1} + F_2 N_{t,2} + \cdots + F_\omega N_{t,\omega}. \qquad (2.26)$$

In brief, the number of newborn in the population at time $t + 1$ is the sum of the expected number of newborn across all age classes. For the rest of the age classes, the expected number in age class X is simply the number in the previous age class in the previous time unit times the probability of survival.

The Leslie matrix is equivalent to a life table and defines a population growth rate. When the population is in stable age distribution, $\mathbf{n}_{t+1} = \lambda \mathbf{n}_t$.[3]

AGE AND STAGE DISTRIBUTIONS

Some organisms are better classified into size or stage classes than into age classes (Lefkovitch 1965). Stages include caterpillars, pupae, and butterflies or seeds, seedlings, juveniles, and adult plants. The type of class chosen should be defined so that the transitions among classes are as clear as possible (van Groenendael *et al.* 1988). The interpretation of stage-classified models is aided by life cycle graphs, exemplified here (Fig. 2.3) for a plant classified by age and a squirrel classified by size. The projection matrix for the plant is a Leslie matrix; that of the squirrel is a size-classified generalization of a Leslie matrix.

The age-classified population (Fig. 2.3(*a*)) has a simple flow diagram. There is only one class of newborn offspring, and age-classified organisms can only get older. All adult age classes contribute to the single pool of newborn offspring. The probabilities of surviving and the expected number of offspring are indicated by arrows labelled with the corresponding elements of the Leslie matrix.

The stage-classified population (Fig. 2.3(*b*)) is more complex. There are two size classes of newborn, newborn offspring can make the transition directly into any of three adult size classes, and all transitions among adult categories are possible. By direct analogy with the Leslie matrix, the survival probabilities and fecundity expectations can be entered into the rows and columns of a stage-classified matrix which can then be used for population projection.

The projection matrices that correspond to the life cycle graphs are

$$(a) \quad \begin{bmatrix} F_{11} & F_{12} & F_{13} & F_{14} & F_{15} & F_{16} \\ p_{21} & 0 & 0 & 0 & 0 & 0 \\ 0 & p_{32} & 0 & 0 & 0 & 0 \\ 0 & 0 & p_{43} & 0 & 0 & 0 \\ 0 & 0 & 0 & p_{54} & 0 & 0 \\ 0 & 0 & 0 & 0 & p_{65} & 0 \end{bmatrix}$$

[3] For those who have had linear algebra: λ is the dominant eigenvalue of the matrix. The stable age distribution is the eigenvector corresponding to the first eigenvalue of $\mathbf{L}$. The vector of reproductive values is the eigenvector corresponding to the first eigenvalue of the transpose $\mathbf{L}'$.

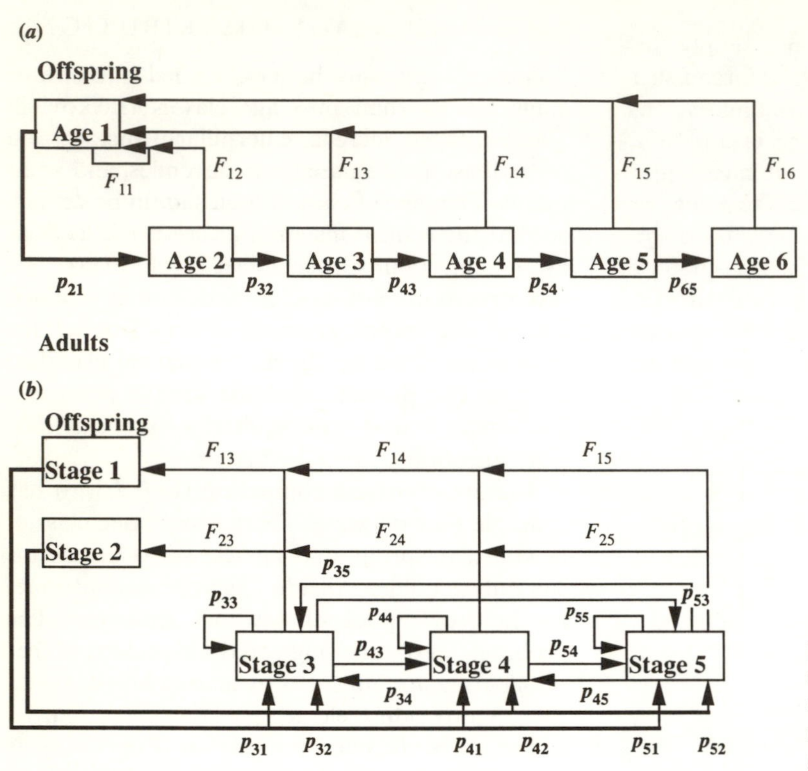

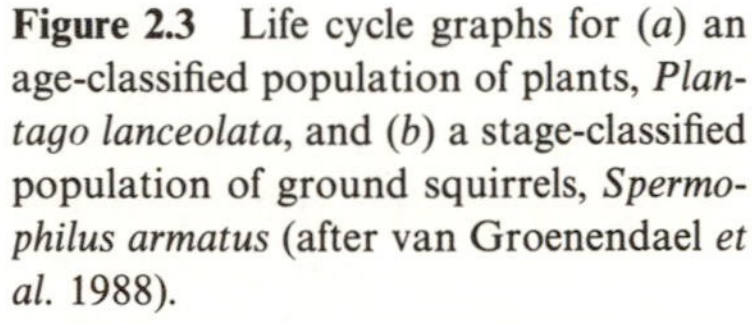

Figure 2.3 Life cycle graphs for (*a*) an age-classified population of plants, *Plantago lanceolata*, and (*b*) a stage-classified population of ground squirrels, *Spermophilus armatus* (after van Groenendael *et al.* 1988).

and

$$(b)\quad \begin{bmatrix} 0 & 0 & F_{13} & F_{14} & F_{15} \\ 0 & 0 & F_{23} & F_{24} & F_{25} \\ p_{31} & p_{32} & p_{33} & p_{34} & p_{35} \\ p_{41} & p_{42} & p_{43} & p_{44} & p_{45} \\ p_{51} & p_{52} & p_{53} & p_{54} & p_{55} \end{bmatrix}.$$

To make a demographic analysis of an age- or stage-classified population:

1. Decide on the type and number of categories to be used and where in the life cycle the census is to be made. The categories must be finer than the periodicity of the environment, especially if one wants to be able to follow seasonal changes in numbers.
2. Sketch the life cycle graph and check each link in it. This should be done before data are gathered to make sure all potential links are included.
3. From the life cycle graph and the data, write down the birth rates and transition probabilities in the form of a projection matrix.

Should one use age classes, stage or size classes, or both? Hughes and Connell (1987) classified colonies of three coral taxa into both age and size classes. Virtually every size class contained both old and young individuals, and there were age effects on colony growth and survival that were independent of size—old colonies grew more slowly and had higher mortality than young colonies of the same size. For organisms with highly variable rates of development, they recommend using both age and size classes.

The projection matrix can be used directly to predict population dynamics, of interest in itself. In life history theory, it is used to calculate the selection pressures operating on the life history traits that are the elements of the matrix (e.g. Ang and De Wreede 1990). First fitness must be defined; then the sensitivity of fitness to changes in life history traits can be calculated.

FITNESS MEASURES

Age structure and stable vital rates but no discounting (R_O)

One encounters several fitness measures in the literature. R_O measures net reproductive rate. Its straightforward combination of age-specific survival and reproductive rates has been favored by field biologists interested in variation in reproductive success (e.g. studies in Clutton-Brock 1988). In using R_O, one makes the implicit assumption that variation in generation length makes no difference to the outcome of selection. The only thing that counts is the number of offspring produced per lifetime, no matter how long it takes to produce them. This is equivalent to assuming that r is near zero. R_O has motivated empirical (e.g. Gross and Charnov 1980) and theoretical research (e.g. Mangel and Clark 1988) and has been used to get interesting results on the evolution of mortality rates (Charnov 1986) and reproductive investment (Kozlowski and Wiegert 1987).

Age structure and stable vital rates with discounting

One cannot compare R_O across populations or among clones, for each usually has a different generation time and R_O is a per generation rate of increase. To avoid these problems, Norton (1928) introduced the analysis of selection in age structured populations, Fisher (1930) developed r to measure the fitness of genotypes, and Cole (1954) and Lewontin (1965) applied r to the evolution of age at maturity and the age distribution of reproductive effort. It has remained the fitness measure most frequently used in life history theory (e.g. Charlesworth 1980, Caswell 1989*a*, *b*). Caswell (1989*a*) shows that the discounting built into r better represents the age distribution of selection pressures on traits than R_O.

However, r is not universally applicable. Strictly speaking, it holds in stable homogeneous environments where density-dependence is not important. It can be applied, with some qualifications, to two sexes with different life tables. Recall that we can regard changes in birth and death rates as marginal effects of gene substitutions.

> For the case of weak selection and random mating with respect to age, the intrinsic rate of increase of a genotype or, more generally, the mean of the male and female intrinsic rates, provides an adequate measure of fitness in a density-independent and constant environment. This parameter can be used to predict, to a good approximation, the rate of change of frequency and ultimate composition of populations with respect to single loci. . . . If selection is strong, however, there is no unitary measure of fitness. ... The intrinsic rate of increase provides a good predictor of the rate of spread of a rare non-recessive gene, but a different measure must be used for a rare recessive gene. Furthermore, in the neighbourhood of a polymorphic equilibrium, the intrinsic rate of increase will not predict the rate of change of gene frequency accurately, and it also gives an inaccurate prediction of the composition of the equilibrium population (Charlesworth 1980, pp. 196–7).

Both R_O and r assume stable birth and death rates and no density-dependent effects on the rate of spread of alleles. One could argue that R_O does represent density-dependent selection when it is used to calculate the joint equilibrium of population size and the phenotype, but this is a rather formal kind of density dependence that is imposed on rather than derived from biological interactions.

For predicting equilibrium phenotypes, R_O functions well in practice. For predicting changes in gene frequencies, r is preferable in many but not all cases.

There are at least two ways to modify the definition of fitness to make it more robust—either account for density dependence, or account for instability in the vital rates, i.e. for heterogeneous and changing environments.

Age structure and discounting in nonequilibrium populations

A good recent summary on fitness in non-equilibrium populations is in the introduction to Orzack and Tuljapurkar (1989). The stochastic growth rate a is a fitness measure appropriate for life history evolution in non-equilibrium populations. If N_t and N_O indicate population sizes at time t and time 0 and $E(\ )$ indicates arithmetic expectation, then the stochastic growth rate is defined in the limit as $t \to \infty$ as

$$a = \frac{1}{t} E(N_t/N_O). \tag{2.27}$$

The predictions of this approach differ from those in which r or R_O is used on several points, including the evolution of delayed maturity (cf. Chapter 6) and the evolution of iteroparity (cf. Chapter 8). However, this more complex measure has not yet had the motivating effect on empirical research programmes that predictions made with the simpler measures r and R_O have had. The reasons are straightforward. It is hard to measure variation in vital rates over stretches of time long enough to be meaningful in the context of a measure like a, and the use of stochastic measures of fitness has not yet produced predictions that are different enough from those of more easily calculated measures to allow the approaches to be discriminated by experiments.

Age structure with density-dependence in some age classes

The standard approach to density dependence in population dynamics assumes that a carrying capacity K defines the population density above which growth rates are negative and below which growth rates are positive. Boyce (1984) reviewed the history of this idea in life history evolution. Applying it to age-specific genetic variation in vital rates, one can determine when K functions as a fitness measure that predicts the direction and outcome of selection. Charlesworth (1980, pp. 200, 203) found

> . . . with weak selection, the genotypic carrying-capacities . . . control the direction of selection in the same way as the intrinsic rates of increase in the density-independent case. In the one-locus case, natural selection tends to maximise the number of individuals in the critical age-group as opposed to the maximisation of the population growth-rate in a density-independent population . . . this result depends on the assumption that the net reproduction rate of each genotype is a strictly decreasing function of the number of individuals in the critical age-group . . . These conclusions about the maximisation of population size depend on the assumptions of frequency-independent selection and a single-locus genetic system.

Demonstrating density-dependent effects on phenotypes does not demonstrate them for alleles, for Prout (1980) has shown that selection on genotypes can be density independent in a population regulated by density dependence. Some would be encouraged by Prout's result to think that they could use r as a fitness definition even in cases where the population dynamics of whole organisms was regulated by density dependence. Others (e.g. Boyce 1984) would conclude that density-dependent effects on life history evolution have not yet been properly analysed. The role of density dependence is certainly an important and unresolved issue in life history evolution on which the experts remain divided—for example, Caswell (1989*b*) strongly defends r as the only reliable fitness measure, whereas Bell (pers. comm.) values both r and K for their simplicity and ready application to single-celled organisms.

Most work on life history evolution has used the Euler–Lotka equation and taken r as the definition of fitness. Life history traits have been *defined* as the things to which r is most sensitive, primarily birth and death rates. If a density-dependent fitness measure is more appropriate, then work should be concentrated on the traits to which it is most sensitive—for example, feeding efficiency, growth rate, and body size—and a book like this would be organized quite differently. This alternative approach has been followed by Roff (1984, 1986) and Kozlowski and Wiegert (1987).

Maynard Smith (1989) noted that both r and K are macroscopic and phenomenological. The microscopic processes are birth and death. Therefore, it makes sense to look at the direct effect of density on the birth rate, the death rate, or both. The equation for exponential population growth is

$$\frac{dN}{dt} = rN \tag{2.28}$$

where r = growth rate = birth rate − mortality rate = $b - m$.

If we introduce density dependence in the birth rate, then we have

$$\frac{dN}{dt} = \{b(N) - m\}N \qquad (2.29)$$

where N = numbers (density), b is the birth rate, and m is the mortality rate. We need a function to represent density dependence in the birth rate that tends to a constant b when N is small and to zero when N tends to infinity. The simplest function is $b(N) = b/(1 - N)$. If we substitute this in the equation above, we get

$$\frac{dN}{dt} = \frac{bN}{1 - N} - mN. \qquad (2.30)$$

If we want to compare two types whose birth rates are equally sensitive to the number of both types present, then we can rewrite the equation as

$$\frac{dN_1}{dt} = \frac{bN_1}{1 - N_1 - N_2} - mN_1. \qquad (2.31)$$

We want to know when population growth will cease as density increases. When $dN_1/dt = 0$, the equation above can be rewritten as $1 + N_1 + N_2 = b_1/m_1$, and when $dN_2/dt = 0$, we get $1 + N_1 + N_2 = b_2/m_2$. The measure determining the winner of competition is b/m, the ratio of the birth rate to the death rate, rather than the difference between the two, $b - m = r$. Differences in logs of the birth and death rates, rather than in the arithmetic values, then determine the outcome of density-dependent selection. Kozlowski (1980) came to the same conclusion.

COMMENT ON FITNESS

Should a fitness measure describe a dynamic process or a static equilibrium? A person interested in the properties of the phenotype at evolutionary equilibrium would choose R_O, for it is easy to measure, easy to calculate, leads to simpler mathematics than other measures, and has been used to make successful predictions about equilibria. Such people are not bothered by the assumption of a stable population ($r = 0$), for this does not strike them as being any more serious than, for example, the assumption of stable birth and death rates (shared by r), or the assumption that allelic substitutions result in marginal effects on life history phenotypes (shared by all other fitness measures). 'I have yet to see anyone use [non-equilibrium measures of fitness] to explain in a believable way some puzzle or pattern of life histories' (Charnov, pers. comm.).

To a person for whom dynamics are more important, R_O does not measure fitness, for it does not take different generation times into account and does not predict changes in gene frequencies (Charlesworth 1980; Lande 1982*a*, *b*). 'The use of R_O as a measure of fitness asserts that timing effects which do not influence mean offspring production are selectively neutral. That is a strong assertion, and I have never seen it defended' (Caswell, pers. comm.).

I have included examples in this book that use several measures of fitness. Some disagreement over fitness measures can be avoided by asking 'What questions do you want to answer and with which assumptions are you satisfied?'.

All fitness definitions are tools invented by scientists to analyse natural selection. They should be judged by how well they do a particular job and replaced if necessary. The most widely used fitness measure is r. Predictions made with r have functioned well in both theory and experiment, but it may not work when environmental stochasticity, frequency dependence, or density dependence are important. There is room for more work on how changing fitness definitions changes predictions about life history evolution.

SENSITIVITY AND ELASTICITY

We now return to the interpretation of life tables and growth rates as marginal effects of gene substitutions. Here projection matrices summarize the age-specific selection pressures encountered by a gene substitution. Recall that fitness is determined by both survival (p_x) and reproduction (m_x). The fitness of a life history variant with a particular life table is given by $\lambda = e^r$. We want to know how sensitive this fitness measure is to changes in each trait in the matrix while all the others are held constant, for that tells us how strong the selection pressures are on the traits. Such *sensitivities* were first calculated by Hamilton (1966) and Emlen (1970). The approach was generalized when Caswell (1978) showed that sensitivities can be calculated directly from the eigenvectors of the projection matrix. This treatment is taken from van Groenendael *et al.* (1988).

Let s_{ij} be the sensitivity of λ to a change in element a_{ij} of the matrix.

$$s_{ij} = \frac{\partial \lambda}{\partial a_{ij}}. \tag{2.32}$$

If w_j is the jth element of the stable age distribution $\mathbf{w}$, and if v_i is the ith element of the reproductive value distribution $\mathbf{v}$, and if $\langle \mathbf{w}, \mathbf{v} \rangle$ signifies the scalar product of the two vectors, then

$$s_{ij} = \frac{v_i w_j}{\langle \mathbf{w}, \mathbf{v} \rangle}. \tag{2.33}$$

By calculating s_{ij} for all the elements of the projection matrix, we estimate the sensitivity of λ to small changes in every trait represented in the matrix.

However, because fecundity and survival are measured in different units and on different scales, it is more informative to compare the *elasticities* of the matrix elements. Elasticities are dimensionless sensitivity coefficients. The elasticity of a matrix element measures the percentage change in λ brought about by a percentage change in mortality or fecundity. Such analysis was pioneered by Cole (1954), Lewontin (1965), and Meats (1971) and can be used for any population for which we can write down a projection matrix, whether it is age, stage, or size classified. The elasticity of a matrix element a_{ij} is given as

$$e_{ij} = \frac{\partial \ln \lambda}{\partial \ln a_{ij}}.$$

SITUATIONAL SENSITIVITY

That sensitivities depend on the rest of the life history is clear from eqns (8.24) and (8.25) (Hamilton 1966; see also Crespi 1990). Here I make the point with numerical examples. Consider two life tables, the first representing a short-lived bird with higher mortality and a larger clutch size, the second representing a long-lived mammal with lower mortality and a smaller litter size (Table 2.4). Think of the bird as a robin or titmouse and of the mammal as a rhinoceros or whale. Both populations are growing slowly.

Table 2.4 Life tables, stable age distributions, and reproductive value distributions for two species with quite different life histories

X	l_X	m_X	c_X	v_X
(a) A small bird: $r = 0.0044$, $\lambda = 1.0044$				
0	1.000	0.000	0.617	1.000
1	0.400	1.570	0.246	2.511
2	0.160	1.825	0.099	2.363
3	0.064	1.350	0.039	1.350
(b) A large mammal: $r = 0.000288$, $\lambda = 1.00029$				
0	1.000	0.000	0.211	1.000
1	0.400	0.000	0.084	2.501
2	0.360	0.000	0.076	2.779
3	0.356	0.000	0.075	2.808
4	0.353	0.000	0.074	2.838
5	0.349	0.600	0.073	2.867
6	0.346	0.000	0.073	2.291
7	0.342	0.875	0.072	2.314
8	0.339	0.000	0.071	1.454
9	0.336	0.950	0.071	1.469
10	0.332	0.000	0.070	0.525
11	0.233	0.750	0.049	0.750

The sensitivities—equivalent to selective pressures (cf. Emlen 1984) and selection gradients (cf. Lande and Arnold 1983)—are given in Table 2.5. Fitness is more sensitive to changes in life history traits of younger organisms than to equivalent changes in older organisms. For the small bird, the decline in sensitivity of fitness to clutch size with age is rapid. For the large mammal, the sensitivity of fitness to changes in litter sizes remains, after the first year, stable to the end of life. The strength of selection acting on a given life history trait depends on the kind of life history in which the trait occurs. Sensitivity is *situational*.

This point can also be made with a technique that allows traits like age at maturity to be analysed. Cole (1954) showed that under certain conditions r is more sensitive to changes in age at maturity than it is to changes in other life history traits. Lewontin (1965) made Cole's point more precise by calculating how large a change in

Table 2.5 Sensitivity of fitness to changes in age-specific fecundity and survival rates for (a) a small bird and (b) a large mammal

Age (X)	Fecundity (m_X)	Survival (p_X)
(a) A small bird		
1	0.407	1.021
2	0.162	0.382
3	0.064	0.087
(b) A large mammal		
1	0.112	0.281
2	0.045	0.125
3	0.040	0.113
4	0.040	0.113
5	0.040	0.113
6	0.039	0.090
7	0.039	0.090
8	0.038	0.056
9	0.038	0.056
10	0.038	0.020
11	0.037	0.028

lifetime fecundity one needed to have an impact on r equal to that of a given change in age at maturity. For fruit flies, changes in age at maturity had greater impact on r than changes in fecundity. Later, MacArthur and Wilson (1967), Meats (1971), Green and Painter (1975), and Snell (1978) got different results with different assumptions. One was not sure to which traits r was most sensitive under which conditions. Because this determines the selection pressures on the life history traits, it is important to get it clear.

Now change the age at maturity and effective fecundity ($l_X m_X$) of the small bird in both slow and rapidly growing populations (Tables 2.6(a)–(f)).

For each of these tables, how much must one change age at maturity (α), all survival terms (l_X), or all fecundity terms (m_X), to produce a 10 per cent increase in r? The answers differ for each life table (Table 2.7). The selection pressure on any life history trait, expressed as the elasticity of fitness to that trait, changes as the life table changes. The key conditions are r itself and age at maturity, a pivotal trait in life history evolution (cf. Chapter 6). When r is near 0, fitness is very sensitive to changes in survival and fecundity rates and not very sensitive to changes in age at maturity (compare the results from tables (c) and (d) with (a) and (b)). When growth rates are high, r is about equally sensitive to changes in all traits for early maturity (table (e)), but when maturity is delayed, then r is much more sensitive to changes in survival or fecundity rates (table (f)). Lewontin (1965) analysed a case like (f).

Caswell and Hastings (1980) reached similar conclusions analytically. They showed that the elasticity of fitness to age at maturity and to fecundity depends on the ratio of rate of increase to average adult survival rate: λ/p_{avg}. In populations that are growing rapidly ($\lambda/p_{avg} > \sim 1.25$), selection on age at maturity is strong and selection on survival and fecundity is relaxed. In populations nearer equilibrium, selection on survival and fecundity rates is strong, and selection on age at maturity is relaxed.[4]

An important element of the selective environment of a life history trait is the state of the other life-history traits, especially age at maturity and the rate at which the population is growing.

Sensitivities describe only the direct selection pressures on single traits; they assume that there are no linkages between elements in the Leslie matrix. But selection pressures are determined both by birth and death schedules and by phenotypic correlations among life history traits. The reactions to selection are determined by the genetic variation available and how it is expressed. These topics are taken up in Chapter 3.

SUMMARY OF INTRODUCTORY DEMOGRAPHY

Elementary demography assumes that age-specific birth and death rates are constant in space and time. Under these conditions the population attains a stable age distribution, in which the ratios of individuals in different age classes remain constant, and then grows smoothly and

[4] One can also do a sensitivity analysis on small changes in trade-offs by examining the Lagrange multiplier associated with each trade-off (Charnov, pers. comm.; Intriligator 1971, Ch. 3).

Table 2.6 Six life tables used in the following elasticity analysis

X	l_x	m_x	$l_x m_x$	r	R_0
(a) r moderate, early maturity, low fecundity					
0.0	1.00	0.000	0.000	0.0495	1.019
1.0	0.25	2.055	0.514		
2.0	0.10	5.055	0.506		
3.0	0.03	2.055	0.062		
(b) r moderate, late maturity, low fecundity					
0.0	1.00	0.000	0.000	0.0495	1.768
11.0	0.25	3.865	0.966		
12.0	0.10	6.865	0.687		
13.0	0.03	3.865	0.116		
(c) r very low, early maturity, low fecundity					
0.0	1.000000	0.000000	0.000000	0.0000495	1.0000779
1.0	0.174865	2.824098	0.493836		
2.0	0.074865	5.824098	0.436021		
3.0	0.024865	2.824098	0.072211		
(d) r very low, late maturity, low fecundity					
0.0	1.0000	0.0000	0.0000	0.0000495	1.003535
11.0	0.1749	2.8244	0.4940		
12.0	0.0749	5.8244	0.4362		
13.0	0.0249	2.8244	0.0733		
(e) r very high, early maturity, low fecundity					
0.0	1.000	0.0	0.000	0.495	2.317
1.0	0.251	3.1	0.778		
2.0	0.201	6.1	1.226		
3.0	0.101	3.1	0.313		
(f) r very high, late maturity, high fecundity					
0.0	1.000	0.0	0.000	0.495	356.725
11.0	0.960	95.0	91.200		
12.0	0.940	190.0	178.600		
13.0	0.915	95.0	86.925		

exponentially at an instantaneous rate r defined as the solution to the Euler–Lotka equation.

Fitness is more sensitive to changes in mortality and fecundity in younger age classes than in older ones. The connection between demography and evolution is direct. Individual variation in survival and reproduction causes variation in fitness and natural selection.

The Euler–Lotka equation can be expressed as a Leslie matrix and used for population projections that can be generalized to size- or stage-classified populations. The sensitivity and elasticity of λ to changes in life history traits tell us about the strength of selection on those traits. The impact on fitness of a given change in a life history trait depends strongly on the values of the other life history traits, particularly age at maturity α and population growth rate r.

RECOMMENDED READING

Caswell, H. (1989). *Matrices in population biology*. (Sinauer, Sunderland, MA). Comprehensive and clear but for the mathematically inclined.

Table 2.7 Sensitivity of r to age at maturity, survival rates, and fecundity

				Change in trait required to increase r 10%		
Table	Maturity	Fecundity	r	α	l_x	m_x
A	Early	Low	Moderate	−15.0%	+0.4%	+1.1%
B	Late	Low	Moderate	− 9.5%	+2.9%	+7.1%
C	Early	Low	Near 0	−16.0%	+0.0004%	+0.001%
D	Late	Low	Near 0	− 9.5%	+0.0028%	+0.007%
E	Early	Low	High	−14.5%	+6.9%	+11.0%
F	Late	High	High	−10.0%	−57%*	+205%

* In this case only, the change required to *decrease* r by 10%, for it is very hard to increase r by 10% for delayed maturity when r is already high.

Emlen, J. M. (1984). *Population biology: The coevolution of population dynamics and behavior.* (Macmillan, New York). An in-depth discussion of the interplay between demography and the evolution of life histories and behaviour.

Keyfitz, N. (1968). *Introduction to the mathematics of population.* (Addison-Wesley, Reading, MA). Everything you wanted to know about the mathematics of demography.

Mertz, D. B. (1970). Note on methods used in life history studies. In *Readings in ecology and ecological genetics* (eds J. H. Connell, D. B. Mertz, and W. W. Murdoch), pp. 4–17. (Harper & Row, New York). A clear introduction to life tables.

PROBLEMS

1. Calculate all the parameters of a vertical life table with five age classes by hand, starting with the raw number of individuals observed alive in each age class at the end of the reproductive seasons. For example, from the zeroth to fourth age class: 12376, 4233, 1749, 340, 268. Age at maturity: 3. Fecundity in third year: 2.1. Fecundity in 4th year: 3.2.
2. What is the formula for the length of time it will take an exponentially growing population to double in size? Use it to get the doubling times of populations with $r = 0.01$, $r = 0.02$, and $r = 0.04$.
3. Show that r is more sensitive to changes in fecundity in age class i than in age class j, where $i < j$, if the population is growing and if the fecundities of different age classes are 'sufficiently' independent of one another. Hint: use the discrete form of the Euler–Lotka equation, $1 = \sum e^{-rx} l_x m_x$. Isolate a single age class from the sum and take the partial derivative of r with respect to m_i.
4. Do the same for survival rate p_i. What does this result, and that of Problem 3, say about the strength of selection on different age classes? Compare your answers with the elasticities given in Tables 2.5 and 2.7.
5. Introduce density dependence into the death rates, rather than the birth rates, in eqns (2.15)–(2.18). What measure now determines the winner of the competition? See Maynard Smith (1989, Chapter 1).
6. Use the literature to construct plausible life tables for a rhinoceros, an oyster, a fruit fly, and an oak tree. Store them in convenient form on a computer.
7. Write a computer program to calculate r, R_O, G, v_x, and c_x given a life table. Run it on the life tables you created and stored in Problem 6. Hint: build on the structure of the programme below that calculates r.
8. Modify the program to calculate the sensitivity and elasticity of λ to the elements in the Leslie matrix. Compare the results. Hint: Change each element by 10 per cent, calculate the new value of λ, then express the change in λ as a percentage of the previous value. This can be done for each element of the Leslie matrix in sequence by building an appropriately indexed loop around the code already given.

9. What are some consequences of different life histories for social and sexual behaviour? Is a long period of survival past sexual maturity a necessary precondition for the evolution of sociality?
10. Suggest a set of experiments that would answer the question, do density-dependent effects on phenotypes translate into density-dependent effects on genotypes? See Prout (1980) and Charlesworth (1980).
11. In what sense are costs of reproduction in terms of survival and future reproduction independent of one another? Argue the case in terms of eqn (2.12).

APPENDIX: A PASCAL PROGRAM TO CALCULATE r

One tries a value for r, and if the sum $\sum e^{-rx} l_x m_x$, which should equal 1, is greater than 1, then r is too small. r is adjusted upwards, and one tries again. If the sum is less than one, r is adjusted downwards. This program assumes that the first line of the life table is the number of age classes, followed by lines consisting of x, l_x, and m_x. It writes out its intermediate results. For (very unlikely) values of r outside the range $-10 < r < 10$, change the values of the variables *uplim* and *lowlim*. For life tables with more than 20 age classes, reset the variable *omega*.

```
program growthrate (input, output);
    const
        omega = 20;
    var
        x, lx, mx : array[1..omega] of real;
        uplim, lowlim, trial, r : real;
        i, last : integer;
begin
    readln(last);          {read number of age classes}
    for i := 1 to last do
        readln(x[i], lx[i], mx[i]);          {read life table}
    uplim := 10.0;          {upper limit for r}
    lowlim := -10.0;          {lower limit for r}
    repeat
        trial := 0.0;
        r := (lowlim + uplim) / 2.0;
        for i := 1 to last do
            trial := trial + exp(-r * x[i]) * lx[i] * mx[i];
        if trial > 1.0 then
            lowlim := r
        else
            uplim := r;
        writeln('trial', trial : 8 : 4, 'r', r : 6 : 3);
    until abs(trial - 1.0) < 0.0001
end
```

3

QUANTITATIVE GENETICS AND REACTION NORMS

The simplest hypothesis ... is that such features as stature are determined by a large number of Mendelian factors, and that the large variance among children of the same parents is due to the segregation of those factors in respect to which the parents are heterozygous

R. A. Fisher, 1918

The plasticity of a character can be (a) specific for that character, (b) specific in relation to particular environmental influences, (c) specific in direction, (d) under genetic control not necessarily related to heterozygosity, and (e) radically altered by selection

A. D. Bradshaw, 1965

CHAPTER OVERVIEW

Whereas demography analyses the phenotypic sources of variation in fitness that produce natural selection, quantitative genetics and reaction norms are used to analyse the genetic variation that conditions the response to selection—how much there is, how it is maintained, and how it is expressed.

This chapter introduces the aspects of quantitative genetics and reaction norms needed to analyse the microevolution of life history traits. Heritability and genetic covariance are defined and methods for measuring them are discussed. The heritabilities of fitness components should be lower than the heritabilities of other traits—and they are—but considerable genetic variation for life history traits is maintained in nature. Five explanations for the maintenance of genetic variation in fitness components are discussed: mutation-selection balance, genotype by environment interactions in heterogeneous environments, negative genetic covariance, gene flow, and flat fitness profiles.

The response to selection depends both on the amount of genetic variation present and on how it is expressed. The expression of genetic variation is mediated by reaction norms. Methods of quantifying plasticity using reaction norms, definitions of the heritability of plasticity, and evidence suggesting that plasticity is under genetic control are discussed.

How phenotypic and genetic variation interact to modulate the expression of genetic covariance is the next theme. The genetic covariance of two traits can be negative, zero, or positive, depending on the environment in which they are measured. These patterns may arise from variation in the reaction norms of different genotypes.

The final section connects quantitative genetics to demography in a model that uses phenotypic and genetic covariance matrices to project evolutionary change from generation to generation.

INTRODUCTION

Why quantitative genetics? Should a book on life history evolution deal with quantitative genetics? It is complex, statistical, and not to everyone's taste. Demography would seem complicated enough. Couldn't we ignore genetics and see how much we can get out of demography by itself? Clearly not. Microevolution has two parts: selection pressures on traits determined by demography and responses determined by genetics. Demography predicts the equilibrium state of the population, but quantitative genetics tells us how fast change should occur and whether the predicted optimum will be reached at all. It helps to understand the evolution of senescence and is essential for the correlation matrix method of analysing the simultaneous evolution of two or more traits. Without quantitative genetics, we cannot understand how life history traits respond to selection in the short term.

Quantitative genetics brings with it several further advantages. It tells us how much the genes and the environment have each contributed to phenotypic variation. It also links life history evolution to one of the great questions of evolutionary biology—how is genetic variation maintained in natural populations? This question has motivated research for half a century. By definition, selection is strongest on fitness itself. The mechanisms that maintain genetic variation for the components of fitness, life history traits, tell us much about how genetic variation is maintained in general.

By using both demography and quantitative genetics, we are led to ask how the two are connected and whether those connections (reaction norms) are themselves an essential part of evolutionary theory. The answer is yes.

Finally, genetics must connect with evology to tell us how evolution works in natural environments. Life history traits have both genetical and ecological significance and thus provide a natural link. Birth and death rates combine with migration rates to determine population density and dynamics and are influenced by competitors, predators, prey, and parasites. Thus birth and death rates both describe life histories and play roles in interactions among species that determine which species will be found together, which will not, and how they interact dynamically. That is their ecological significance.

What is quantitative genetics?

Quantitative genetics deals with traits that vary continuously, such as weights at birth and reproductive investments, in contrast to traits that fall clearly into distinct classes, such as seed coat texture in peas (smooth, wrinkled). Traits segregating into easily recognizable phenotypic classes are called *Mendelian*; continuously varying traits are called *quantitative*, *metric*, or *polygenic*.

> The critical difference between Mendelian and quantitative traits is not the number of segregating loci but the size of the phenotypic differences between genotypes [small] as compared with the individual variation within phenotypic classes [large] (Suzuki *et al.* 1986, pp. 513–14).

Quantitative genetics is the formalization of plant and animal breeding. Its recent application to the evolutionary biology of non-domestic organisms in natural environments (Lande 1979; Barton and Turelli 1989) marks a return to Darwin's (1859) analogy between artificial and natural selection. Much of quantitative genetics deals with attempts to improve the performance of domestic varieties. Such programmes often minimize environmental variation to improve the estimate of genetic variation. The quantitative genetics of organisms living in heterogeneous natural environments is a younger field. Some progress recently has been made on how to incorporate genotype × environment interactions and reaction norms into evolutionary theory (Stearns *et al.* 1991).

Quantitative genetics is also statistical phenomenology. One makes a plausible assumption, convenient for statistical analysis, about the genetics of traits influenced by many genes. To estimate the genetic and environmental contributions to variation, one measures the phenotype, the final product of the assumed causes. One does not see the mechanisms—the physiology and development—that produce the phenotype. These are presumed not to affect the conclusions significantly. Built into the method are assumptions that, if forgotten, contain considerable power to mislead, particularly when one tries to infer causes from statistics (Lewontin 1970).

THE BASIC MODEL

From Mendelian to quantitative variation

In 1900 it was not clear that continuous phenotypic variation could arise in traits influenced by many genes of small effect, each obeying Mendel's laws. Fisher (1918) showed the consistency of Mendelian and quantitative genetics and invented the analysis of variance. Other early contributors to quantitative genetics were Weinberg and Wright (1968–78). Detailed experimental evidence for polygenic inheritance was gathered by East, Morgan and his *Drosophila* group, Nilsson-Ehle, and others (Provine 1986).

Reaction norms

There are two causes of continuous variation in phenotypes: reaction norms and the summed effects of many genes. A reaction norm is the set of phenotypes produced by a single genotype across a range of environmental conditions (Woltereck 1909). If an organism is cloned and the genetically identical offspring are tested across a range of environments, we can measure the reaction norm, for example, of body size as a function of temperature. When a single genotype is cloned and tested in many environments, the reaction norms transform environmental variation into phenotypic variation. A linear reaction norm can change the range and variance of a phenotypic distribution (Fig. 3.1(*a*)). A curved reaction norm can change the shape of the distribution (Fig. 3.1(*b*)). A single distribution of environmental variants is transformed into a phenotypic distribution that is flattened out by a steep reaction norm, compressed by a flat reaction norm, and skewed either to the left or the right by a curved norm.

When several genotypes are present, then environments can differ qualitatively in the distribution of phenotypes. The reaction norms of different genotypes often cross. If the reaction norms are not parallel, that means genotype ×

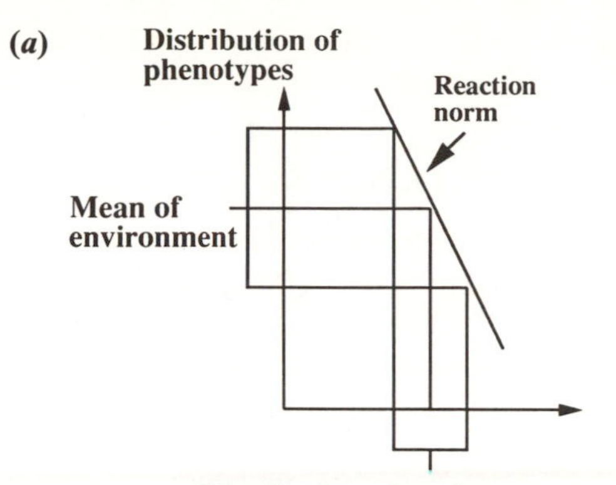

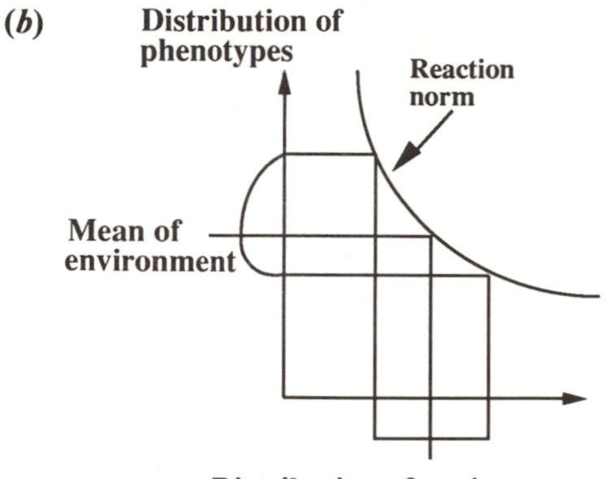

Figure 3.1 Reaction norms transform environmental variation into phenotypic variation. The differences between the two distributions can be considerable. In (*a*), a linear reaction norm broadens the distribution of phenotypes. In (*b*), a curved reaction norm takes a symmetrical distribution of environments and transforms it into a skewed distribution of phenotypes (after Suzuki *et al.* 1986).

environment (G × E) interactions are present (Fig. 3.2). G × E interactions are pervasive (Pani and Lasley 1972; Bell 1991).

Crossing reaction norms—a strong form of G × E interactions—have two important effects on phenotypic distributions (Fig. 3.3).

First, they determine whether or not one can see heritable variation in the phenotype. In the region where the reaction norms cross, the blurring of phenotypic differences among genotypes makes it impossible to assign phenotypes unambiguously to genotypes. Outside that region the genotypic variation becomes clear. The first reason that discontinuous genetic variation can produce continuous phenotypic variation is that

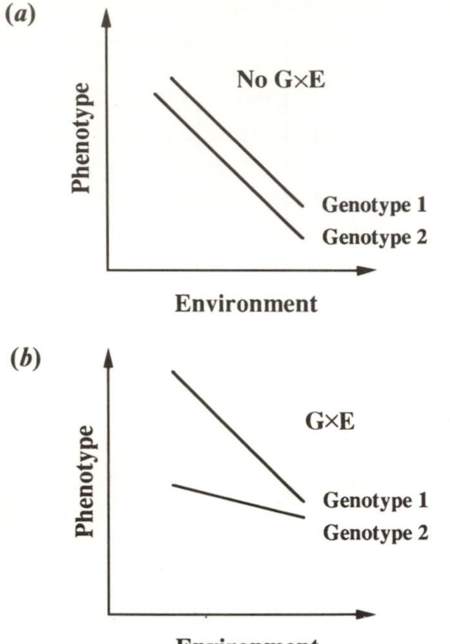

Figure 3.2 (*a*) Only when the reaction norms of the genotypes present run parallel to each other are there no genotype × environment interactions. (*b*) Reaction norms with different slopes are a synonym for genotype × environment interactions. Crossing reaction norms are an especially strong case of such interactions.

reaction norms blur phenotypic differences among genotypes. Further, the expression of genetic variation varies with the environment in which it is measured. Measures of genetic variation in Fig. 3.3 would be positive and large in environments 1 and 3 and insignificant in environment 2.

Second, reaction norms influence the ranking of phenotypes. In environments to the right of the crossing point, the phenotypic ranks of the genotypes are the opposite of those in environments to the left. Under directional selection for higher values one genotype would be favoured to the right of the crossing point and another to the left.

Many genes with relatively small effects

Combinations of many genes of small effect also produce continuous variation in quantitative traits. Suppose a trait is equally affected by alleles segregating at four loci. There are two alleles at each locus and no dominance. One locus adds one unit to the trait (+), the other adds nothing (−). There are three genotypes at each locus, ++, +−, and −−, with effects on the trait of +2, +1, and 0 respectively. Across all four loci there are $3^4 = 81$ diploid genotypes, ranging from ++ ++ ++ ++ to −− −− −− −−. However, the phenotypes fall into only nine classes, 8, 7, 6, . . . , 0, because many genotypes have the same number of + and − alleles. For example, the phenotypic effects of the following genotypes are the same (+4): (a) ++ ++ −− −−, (b) +− +− +− +−, (c) ++ +− +− −−, and so on.

Variation at several loci creates phenotypic classes. The more such loci, the greater the number of phenotypic classes and the closer their spacing. Add to this the effects of reaction norms, and phenotypic variation becomes indistinguishable from continuous variation. With just a few loci the phenotypic distribution is virtually continuous, and with many loci and some plasticity the

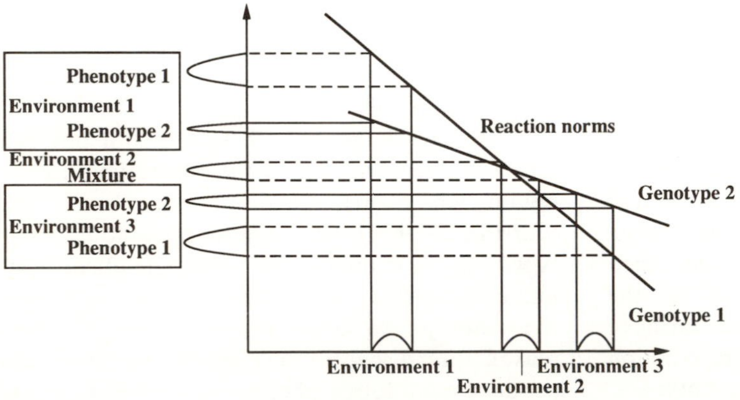

Figure 3.3 Crossing reaction norms convert three similar environmental distributions. In environments 1 and 3 the genotypes can be distinguished in the phenotypes and perceived genetic variation is significantly different from zero. In environment 2, near the crossing point of the norms, the genotypes are indistinguishable in the phenotypic mixture and perceived genetic variation is nil. Between environments 1 and 3 the phenotypic ranking of the genotypes reverses (after Suzuki *et al.* 1986).

approach to a Gaussian curve is quite close. The number of loci affecting the trait need not be especially large if their effects are of the same order as the differences caused by reaction norms. In practice, the number of loci estimated to affect quantitative traits usually falls in the range 5–20. The methods used constrain the number to be smaller than 3–4 times the number of chromosomes, and attempts to estimate it using molecular techniques have not yet been successful (Barton and Turelli 1989).

A model of gene action

Our aim here is to understand the technical definitions of heritability and genetic correlation. To do so we need a series of concepts that move from single genes to properties measurable on whole organisms. We begin by sketching the sequence in reverse, from heritability through additive genetic variance, breeding value, and average effect to genotypic value, so that you can see how the concepts will be built up to define heritability. With that done, the definition of additive genetic covariance and genetic correlation is straightforward.

5. Heritability is the proportion of total phenotypic variance among individuals accounted for by additive genetic variance.
4. Additive genetic variance is the variance among individuals in breeding values.
3. The breeding value of an individual is twice the deviation of its progeny from the population mean when it is mated at random with a sample of the population. It is also equal to the average effects of the alleles that it carries, summed across all segregating loci.
2. The average effect of an allele is the average deviation from the population mean of those individuals that have received the allele from one parent when the allele received from the other parent is a random sample of the population. The average effect of an allele A_2 can also be defined as the change that would occur in the population mean if all A_1 alleles were changed into A_2 alleles. The average effect of an allele is measured in terms of:
1. Genotypic values, which is why we begin there.

Genotypic value

If we cloned individual genotypes and reared them in a constant environment until maternal effects had disappeared, the effect of the environment on their deviations from the population mean would be zero, and the deviation of a clone from the mean of the clones would be its genotypic value. The genotypic values of homozygotes and heterozygotes *at a single focal locus* are modelled by taking the homozygotes as the extremes. The midpoint between the two homozygotes *defines* the reference point for this locus (in Fig. 3.4, 0), and the deviation of the heterozygote from the reference point measures the degree of dominance (in Fig. 3.4, d). If the heterozygote is more extreme than either homozygote, there is overdominance.

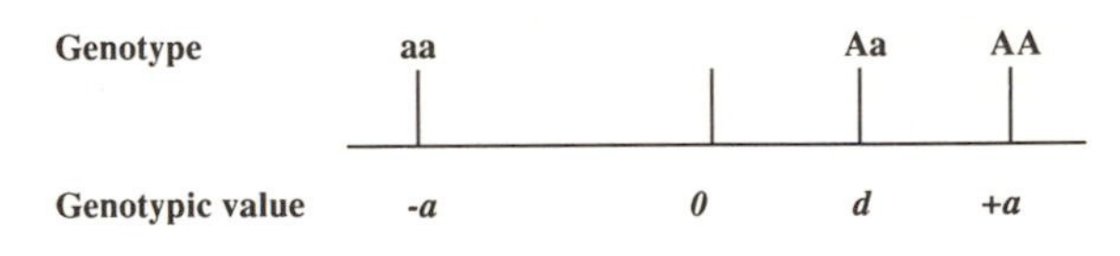

Figure 3.4 Definition of genotypic values for the three genotypes present at a single focal locus when two alleles are present. The phenotypic range contributed by this locus is $2a$ (after Falconer 1989).

If all loci affecting the trait behave like the locus in Fig. 3.4, the total phenotypic range is $2\sum a$. If we started from a population in which the two alleles were present at equal frequency at all loci, and then all alleles increasing the value of the trait were fixed as homozygotes, the mean value would increase by $+\sum a$, and if the alleles decreasing its value were fixed, the mean would decrease by $-\sum a$. The genotypic values of all loci affecting the trait are assumed to have the same structure; a and d are theoretical values used in developing the model. They are rarely estimated in practice because to do so one would have to estimate the number of loci segregating for the trait, then derive a and d as average values, the same for each of the loci. Because estimates of the number of loci affecting a trait are usually unreliable, such an exercise would make little sense.

Average effect

In sexually reproducing populations, offspring get genes, not genotypes, from their parents. Thus we need to relate the concept of 'genotypic effect' to an analogous concept applicable to genes; this is the *average effect of an allele*, defined by Falconer (1989) as the average deviation from the population mean of those individuals that have received the allele from one parent, when the allele received from the other parent is a random sample of the population.

The average effect of a gene substitution, which is the difference between the average effects of the two alleles, depends on the frequency and dominance of the allele. If it is rare and dominant, fixing it will produce a large effect. If it is common and dominant, fixing it will produce a small effect, for most of the individuals in the population already possess it. If it is rare and recessive, fixing it produces an even larger effect that if it were dominant, for the heterozygotes display the phenotype of the dominant homozygotes.

To calculate the overall effect of a substitution in terms of genotypic values, we form an expression for the effect of a dominant allele, then for a recessive allele, and take the difference. If p is the frequency of the dominant allele A_1 and q of the recessive allele A_2, the Hardy–Weinberg proportions are p^2 (dominant homozygotes) + $2pq$ (heterozygotes) + q^2 (recessive homozygotes), and the contribution of this locus to the population mean, $a(p - q) + 2dpq$, can be calculated as follows:

Table 3.1 Genotypes and average effects, one locus, two alleles

Genotype	Frequency	Value	Frequency × Value
A_1A_1	p^2	$+a$	p^2a
A_1A_2	$2pq$	d	$2pqd$
A_2A_2	q^2	$-a$	$-q^2a$
			Sum $= a(p - q) + 2dpq$*

* Recall that because $p + q = 1$,

$p^2 - q^2 = (p + q)(p - q) = p - q$.

The average effect of the dominant allele is the difference between the mean value of genotypes produced when it unites with gametes drawn at random from the population, and the population mean:

$$\alpha_1 = pa + qd \text{ (effect of this allele)}$$
$$- \{a(p - q) + 2dpq\} \text{ (population mean)}$$
$$= q\{a + d(q - p)\}.$$

Similarly, for the recessive allele we get

$$\alpha_2 = -p\{a + d(q - p)\}.$$

The average effect of an allele substitution depends explicitly on allele frequencies and therefore on the population in which the substitutions occurs. Allele frequencies differ among populations in the field and among samples taken from them to estimate heritabilities and genetic covariances. Sampling errors are common in estimates made on natural populations.

Breeding value

Average effects refer to single alleles that usually remain theoretical abstractions for quantitative traits. The alleles combine, however, in an individual, and the average value of an individual, estimated from the mean value of its progeny, is its *breeding value*. The breeding value of an individual is twice the deviation of its progeny from the population mean when it is mated at random with a sample of the population. The deviation is doubled because the parent provides only half the genes in the offspring. The breeding value of an individual is also equal to the average effects of the alleles it carries, summed over loci. Like average effect, it depends on the gene frequencies present in the population, and these frequencies do change. Table 3.2 lists the breeding values for the three genotypes present at a locus for which two alleles are segregating. Note the dependence on gene frequencies p and q.

The variation in the breeding values of individuals in the population is called *additive genetic variation*. *Non-additive* effects include dominance,

Table 3.2 Genotypes and breeding values, one locus, two alleles

A_1A_1	$2\alpha_1$	=	$2q\alpha$
A_1A_2	$\alpha_1 + \alpha_2$	=	$(q - p)\alpha$
A_2A_2	$2\alpha_2$	=	$-2p\alpha$

interactions with other loci (epistasis), and maternal effects in some experimental designs. Additive genetic variation is central to the quantitative genetic view of evolution, appearing in the expressions for both heritability and genetic covariance. The term 'additive' is used because we analyse the inheritance of quantitative traits as if the effects of all genes were independent and simply added up. It is more accurate to speak of variation in breeding values than of additive genetic variation.

Heritability

In quantitative genetics, the term heritability is used in a restricted sense to refer to the degree to which *variation* in the trait is under genetic influence. In everyday language, to say that a trait is heritable means simply that it is under genetic influence. In that sense, the possession of four limbs is a heritable trait in tetrapods, but it is invariant within the lineage. Thus no alleles affecting limb number are segregating, and the *heritability* of the trait is zero.

In the standard model, phenotypic variation is partitioned into genetic and environmental sources plus twice the genotype × environment covariance, and the genetic portion is partitioned into additive, dominance, and interaction effects:

$$\text{Phenotypic variance} = V_p = V_g + V_e + 2\,\mathrm{Cov}_{ge} \tag{3.1}$$

and

$$\text{Genetic variance} = V_g = V_a + V_d + V_i. \tag{3.2}$$

The additive variance, V_a, is the variance in breeding values. The *dominance variance*, V_d, is that portion of the variation that can be attributed to deviations from the breeding value caused by non-additive interactions between alleles at the same locus. The *interaction variance*, V_i, is that portion of the variation that can be attributed to deviations from the breeding values caused by interaction effects among loci. These would occur, for example, when the breeding value of genotype A_1A_2 depended on the frequency of an allele at locus B. In practice, the distinction between dominance and interaction effects is often not made because the experimental conditions that must be arranged to do so are rather special. In many cases, variance is simply partitioned into additive genetic, non-additive genetic, and environmental variance, and the genotype–environment covariance is assumed to be zero.

Broad-sense heritability is the proportion of phenotypic variation under genetic influence: V_g/V_p. It often includes interaction effects, maternal effects, and effects of the common environment shared by sibs. *Narrow-sense heritability*, usually referred to simply as *heritability* and denoted by the symbol h^2, is the smaller proportion of phenotypic variability accounted for by variation in the breeding values of individuals:

$$h^2 = V_a/V_p \tag{3.3}$$

The heritability of a trait is of great interest, for it measures the potential to respond to selection in the short term. It also measures the degree of resemblance among relatives.

The last term in eqn (3.1), $2\,\mathrm{Cov}_{ge}$, is potentially confusing. It resembles a term with quite a different meaning that is important in heterogeneous environments. *Genotype × environment covariance* (Cov_{ge}) arises from different frequency distributions of genotypes over environments. *Genotype × environment interactions* (G × E interactions) arise from the different responses that different genotypes have to the same environments. G × E interactions are described by reaction norms with differing slopes. In standard quantitative genetics, one must either assume that genotypes and environmental deviations are not correlated—that there is no genotype × environment covariance—or else measure the covariance. Both Cov_{ge} and G × E interactions can arise quite independently of one another. The G × E interactions can only be estimated if there is no Cov_{ge}, because the effects caused by Cov_{ge} confound measures of G × E interactions.

All simple estimates of heritabilities based on correlations among relatives assume that there is no genotype by environment covariance and no genotype by environment interaction because they assume that all individuals are measured in the same environment. When environments differ and G × E interactions are significant, then the most important feature of the data is the G × E interactions. In such cases, one can neither partition variation into genetic and environmental components nor speak of additive genetic effects.

One important type of genotype × environment covariance comes under the heading of *maternal effects*. Maternal effects describe all phenotypic effects in the progeny that can be associated with the mother but not attributed to her breeding value. They can arise from cytoplasmic factors in the egg associated with the mother's physiological condition, from the common environment shared by sibs early in life, and from any other mechanism that associates the genotypes of the offspring with the environment of the mother. The crucial effect is the correlation between the environment shared by the offspring and the phenotype of the mother. If the phenotype of the mother is correlated with that environment for any reason—because of habitat selection, because the environment is cold or poor in food and therefore the mother is small and thin as are her offspring—then the correlation of mother and offspring will contain a significant non-genetic element.

Measuring heritability

Two methods are used to measure heritabilities. One relies on resemblance between relatives, usually determined by parent–offspring regression or by the correlations of full- and half-sibs. The other relies on the response to artificial selection. Resemblance of relatives gives more information on maternal and interaction effects, while response to selection measures the item of interest directly but encounters problems with inbreeding and the confounding influence of natural selection. Both take several generations to do properly. There is no royal road to heritability estimates.

Resemblance between relatives

Table 3.3 shows the genetic contribution to phenotypic covariance of relatives on the assumptions that epistatic interactions can be ignored and that only the full-sib correlation contains effects of shared environments. The example demonstrates the measurement of abdominal bristle number in *Drosophila melanogaster* by three different methods. *Mid-parent* means the average of the two parents; b is the slope of the regression line. V_{ec} represents the variation attributable to environmental effects shared by full-sibs. Of these methods, the most reliable are the regression of offspring on father when there is no paternal care and the correlation of paternal half-sibs. The regressions of offspring on mother and of offspring on mid-parent both contain maternal effects. The regression of offspring on father contains paternal effects when fathers care for young. The correlation of full sibs is inflated by dominance effects and effects of common environment. The correlation of maternal half-sibs, not present in the table, is also inflated by maternal effects.

Table 3.3 A comparison of three methods of measuring heritability (from Falconer 1989)

Relatives	Phenotypic covariance	Regression (b) or correlation (t)	Example of h^2
Offspring and one parent	$\frac{1}{2}V_a$	$b = \frac{1}{2}h^2$	
Offspring and mid-parent	$\frac{1}{2}V_a^*$	$b = h^2$	0.51 ± 0.07
Half-sibs	$\frac{1}{4}V_a$	$t = \frac{1}{4}h^2$	0.48 ± 0.11
Full-sibs	$\frac{1}{2}V_a + \frac{1}{4}V_d + V_{ec}$	$t \geqslant \frac{1}{2}h^2$	0.53 ± 0.07

* The phenotypic variance of the midparents is $\frac{1}{2}V_p$.

In the field, one cannot make full-sib/half-sib crosses but often can identify parents and offspring. Van Noordwijk *et al.* (1981) used data on great tits to calculate the heritability of body size from six different parent–offspring regressions (Table 3.4). They only used the mean weights of individuals that had been weighed at least three times outside the breeding and moulting season, and they combined data gathered on one population between 1955 and 1978. They did not

Table 3.4 Heritability estimates for body size in great tits in Holland from six different parent–offspring regressions (from van Noordwijk *et al.* 1981)

	Mother	Father	Mid-parent
Daughter	0.63 ± 0.15 $n = 112$	0.72 ± 0.14 $n = 137$	0.68 ± 0.10 $n = 90$
Son	0.65 ± 0.13 $n = 181$	0.59 ± 0.11 $n = 255$	0.50 ± 0.09 $n = 146$

correct for age, sex, time of day, or season of year. The mid-parent estimates are not independent of the separate regressions on mother and father, nor are the separate estimates based on daughters and sons independent of each other, for siblings shared common environments. Thus there are fewer than six independent estimates of heritability, but all agree that heritability of body size in this population is about 0.6, a high value that may be due to maternal and paternal effects.

One can also mate organisms in a pattern that creates different degrees of relationship among the progeny, then use that fact to estimate heritabilities, dominance and interaction effects, maternal effects, and so forth. Several such breeding designs are available (see Falconer 1989 and Mather and Jinks 1982).

One simple and widely used design is the analysis of full-sib and half-sib families. Several males are each mated to several females. In the simplest design, the males and females are randomly chosen and randomly mated. Several offspring from each female are then measured, and one performs an analysis of variance in which the phenotypic variance of the trait is partitioned into components that can be assigned to differences between the progeny of the different males (the between-father component, σ_f^2), differences among the progeny of the females that were mated to the same male (the between-mother, within-father component, σ_m^2), and to differences among the offspring of a single female (the within-mother component, σ_w^2).

There should be f fathers, m mothers, and k offspring per mother (unequal family sizes have regrettable statistical consequences). The mean square within mothers itself estimates the within-mother variance component, but the between-mother and between-father mean squares both contain contributions from the sources of variation nested below them, as shown in Table 3.5.

From the analysis of variance, one can estimate the variance components for fathers and mothers by subtraction and division. For example,

$$\sigma_m^2 = \frac{1}{k}(MS_m - MS_w). \tag{3.4}$$

Having determined σ_f^2, σ_m^2, and σ_w^2, we use them

Table 3.5 Analysis of variance for a full-sib/half-sib cross

Source	d.f.	Mean square	Composition of mean square
Between fathers	$f - 1$	MS_f	$= \sigma_w^2 + k\sigma_m^2 + mk\sigma_f^2$
Between mothers	$f(m - 1)$	MS_m	$= \sigma_w^2 + k\sigma_m^2$
Within mothers, between offspring	$fm(k - 1)$	MS_w	$= \sigma_w^2$

to calculate h^2. The variation attributable to fathers measures the variance of the means of half-sib families. From Table 3.3 we know that this is $\frac{1}{4}V_a$. Thus $\sigma_f^2 = \frac{1}{4}V_a$, and

$$h^2 = 4\sigma_f^2/\sigma_{\text{tot}}^2. \tag{3.5}$$

The between-mother and within progeny components, σ_m^2 and σ_w^2, contain contributions from dominance and common environment as well. With such data, one can also estimate heritability from the regression of offspring on parents, but because each male has been mated to a series of different females, one should make the estimate as the average regression of offspring on dams within sires. The average of the regression coefficients of offspring on one parent estimates half the heritability (see Table 3.3).

In estimating heritability with a breeding design, one partitions phenotypic variation into components attributable to various sources. One could also build environmental effects into the design or use other methods for generating variation in the degree of relationship of the individuals involved. Each design has different advantages and disadvantages. Because this kind of research involves a lot of work, one should get the best possible statistical advice before proceeding. Actually simulating the experiment on a computer is often worth the effort, for then one can vary, for example, number of fathers, number of mothers per father, number of offspring per mother, or other relevant parameters, and examine the effects on the accuracy of the estimates.

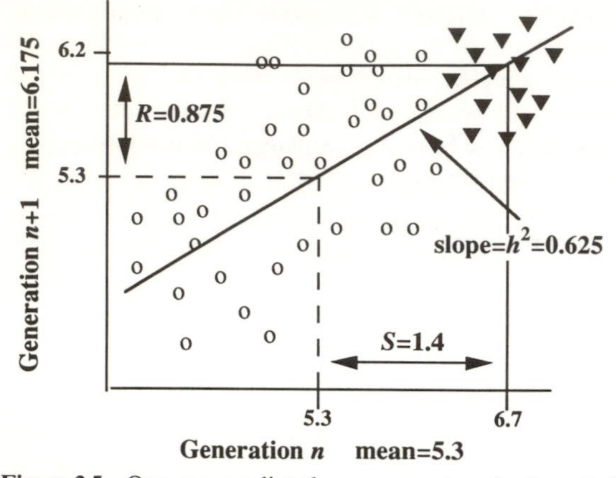

Figure 3.5 One can predict the response to selection, R, if one knows the heritability and the mean value of the parents selected expressed as a deviation from the mean, S. The successful parents and the values of their progeny are here indicated by ▼ (after Falconer 1989).

Selection experiments

Estimates of heritabilities from parent–offspring regressions of full-sib/half-sib designs may be biased. One needs a large sample and cannot always control maternal and other environmental effects to the extent wished. Fortunately, there is a simple relationship between the heritability of a trait and its response to selection. If we can measure the strength of and the response to selection, we can estimate the heritability of the trait. Since we usually want to know the heritability in order to estimate the potential response to selection, a selection experiment gives us the most direct information possible.

For a trait under directional selection, let S represent the deviation from the phenotypic mean of the parents that contribute offspring to the next generation. S is the selection differential. For example, consider selection for number of offspring in dung beetles, which lay a single clutch, care for it, then die (Fig. 3.5). The population mean clutch size is 5.3 eggs, but the beetles reproducing in the next generation came from clutches with a mean of 6.7 eggs. Thus the selection differential is 1.4 eggs.

If R measures the response to selection, the increase in average clutch size in the next generation, then from Fig. 3.5 the ratio of selection response to selection differential, R/S, is the slope of the line, which is also the heritability, h^2. Therefore

$$R = h^2 S. \quad (3.6)$$

To standardize R, we define the selection intensity i as the selection differential divided by the standard deviation of the trait, $i = S/\sigma_p$, so that

$$R = ih^2\sigma_p. \quad (3.7)$$

That completes a brief sketch of the methods used to estimate heritabilities.

HERITABILITIES OF LIFE HISTORY TRAITS

Over the last decade, hundreds of values of heritabilities have been reported for non-agricultural organisms. Roff and Mousseau (1987) compared methods of estimating the heritability of morphological, behavioural, physiological, and life history traits in the fruit fly, *Drosophila*. Estimates made with different methods corresponded only roughly. Mousseau and Roff (1987) analysed 1120 heritability estimates made on wild, outbred animal populations. In their data, the heritabilities of life history traits were lower than those of morphological traits, with behavioural and physiological traits falling in between (see Table 3.6).

Table 3.6 A comparison of heritability estimates for different types of traits measured on wild, outbred animals. From Mousseau and Roff (1987)

	Life history	Physiology	Behaviour	Morphology
n	341	104	105	570
Mean heritability	0.262	0.330	0.302	0.461
S.E.	0.012	0.027	0.023	0.004

Table 3.6 suggests the following conclusions.

1. The heritability estimated for a life-history trait depends, as expected, on the population of origin and on the environmental conditions.
2. The heritabilities of life history traits vary from 0 to 1.0, and values over 30 per cent have frequently been measured in samples taken from natural populations.
3. There is plenty of genetic variation present for life-history traits in natural populations.
4. The heritabilities of life history traits are lower than those of morphological traits.

It is not surprising that heritabilities vary among populations, which have each had a different history of selection and drift. However, the fact that life history traits often display fairly high heritabilities and that heritabilities of life history traits are lower than the heritabilities of other types of traits requires further explanation. Let us first see why one might *not* expect the heritabilities of life history traits to be large.

SELECTION ON QUANTITATIVE TRAITS

Selection reduces genetic variation

Life history traits—components of fitness—are continually under selection. When a quantitative trait is under either directional or stabilizing selection, the genes influencing the trait will be fixed.

Suppose there are 10 loci affecting body weight and that body weight is under directional selection upwards. Selection will fix each locus with the allele that increases body weight, and when all alleles affecting the trait are fixed, the heritability of the trait will have been driven to zero by selection. (However, this can take a long time—a number of generations about half the effective population size minus an amount depending on selection intensity. If the effective population size is large, it will be a very long time before all the loci are fixed, and it is not likely that conditions would remain constant during that time.)

Similar comments hold for stabilizing selection. Consider the case with no dominance ($d = 0$). Heterozygotes are not favoured, for the phenotypic value of a genotype with all 10 loci heterozygous is exactly the same as that of a genotype in which five loci are homozygous for the allele with negative effects and the other five are homozygous for the allele with positive effects. The homozygous genotype has a selective advantage, for the variance in body weights of progeny of heterozygous parents is higher than in the progeny of homozygous parents, and under stabilizing selection very high and very low body weights are selected against. Thus stabilizing selection also reduces genetic variation.

In general, we expect additive genetic variation to decrease under selection but not to vanish. It takes a long time for selection to drive all loci to fixation in a large population, and mutation and migration will provide a steady flow of variation into the population. Furthermore, components of fitness linked by physiology and development may have genetic covariances that prevent selection from reducing the genetic variation in any one component to zero because the consequences for the other components would be maladaptive. Some additive variation should be maintained under selection (cf. Lande 1975; Bulmer 1980).

Gustaffson (1986) estimated the heritabilities of 11 traits in male and 10 traits in female collared flycatchers through male–father and female–mother regressions (Fig. 3.6). The estimates were made on data gathered in the field in a long-term study with large numbers of marked individuals.

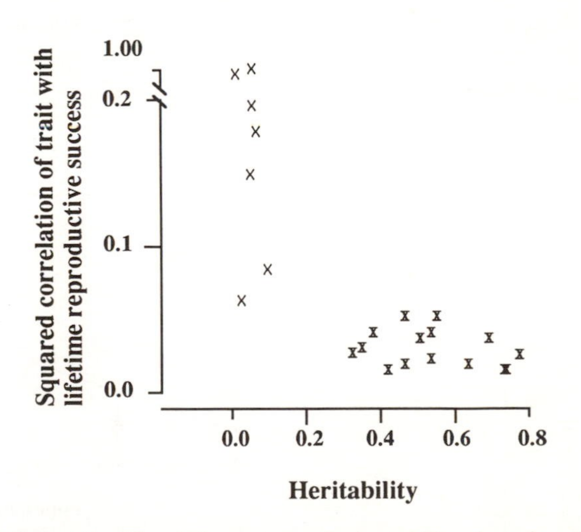

Figure 3.6 The relation between the heritability of a trait and its correlation with lifetime reproductive success is strongly negative in collared flycatchers on Gotland, Sweden (after Gustaffson 1986).

He found a strong negative relation between the heritability of a trait and its correlation with reproductive success.

Selection limits

That selection reduces genetic variation, fixes alleles, and limits the response to further selection has been well confirmed in experiments on *small* populations. When one selects for bristle number in fruit flies or for six-week body weight in mice in populations that have an effective size of 15–30 individuals, one sees a rapid, almost linear response to selection for 10–20 generations, followed by a much slower response that reaches a fluctuating limit. After another 10–30 generations, a second response may or may not result from variation created by new mutations (Hill 1982). Whatever the cause, in small populations a limit to selection is reached after one or two episodes of response that usually occur within the first 30 but in exceptional cases within the first 70 generations (Fig. 3.7(*a*)). The results from larger populations can be dramatically different (Fig. 3.7(*b*)).

Yoo (1980) selected on abdominal bristle number in populations of 50 fruitflies with an intensity of 20 per cent for 86–89 generations. The response continued steadily for at least 75 generations and amounted to more than 36 additive genetic standard deviations in the base population. When selection was relaxed, bristle number declined again and stabilized at two to three times the value of the starting population. Similar protracted periods of steady response to selection with no sign of a limit have been observed in experiments on pupal weight in *Tribolium*, which responded for 75 generations (Enfield 1977), and on oil and protein content in maize, which responded for 76 generations (Dudley 1977).

The theory of selection limits is based on results from experiments done with artificially small populations in which there is relatively little genetic variation. A moderate increase in the size of the population results in a larger and longer-lasting response to selection. In natural populations, where effective population size is often quite large, enough genetic variation is probably

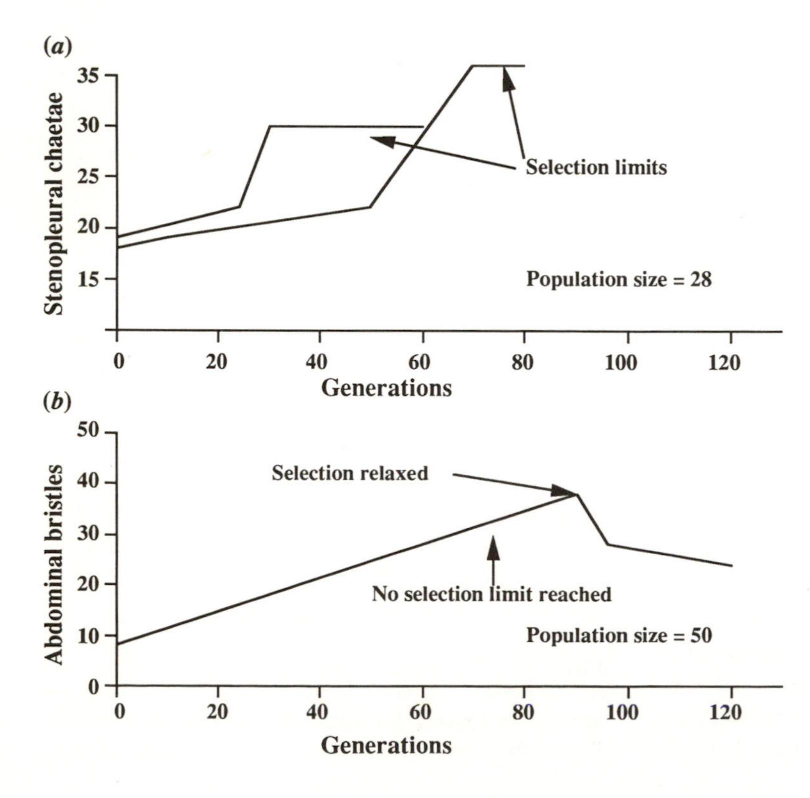

Figure 3.7 A comparison of selection response in small (*a*) and large (*b*) populations of fruit flies. Selection in small populations typically reaches a limit, but in larger populations no limit was reached within 100 generations (after Yoo 1980).

available to allow a steady response to selection over long periods, especially when we recall that mutations for polygenic traits occur frequently in large populations (Hill 1982). The total mutation rate is about 0.01 per polygenic trait per individual per generation (Barton and Turelli 1989).

HOW IS GENETIC VARIATION FOR LIFE HISTORY TRAITS MAINTAINED?

Life history traits are quantitative traits under continual selection, and the genetic variation of such traits should be low compared to traits less directly connected to fitness. Nevertheless, the heritabilities of life history traits are often large. This puzzle has prompted a number of explanations.

Explanation 1. Mutation-selection balance in large populations

Mutations are the ultimate source of all genetic variation, and the larger the population, the more genes will mutate each generation. Lande (1975) suggested that a balance between mutation and selection in large populations would maintain considerable genetic variation, even in homogeneous environments, but he assumed that mutants have a continuous range of effects. The mechanism may not be sufficient (cf. Barton and Turelli 1989 for a review), for conclusions are sensitive to relaxing that assumption. 'If variation is maintained by a mutation–selection balance and the relevant per-locus mutation rates are no higher than 10^{-4}, the variation must be based on rare alleles with appreciable effects ... and the mutation rate per character must be high. ... In general, mutation can account for high heritabilities only if most genes affect most characters ... or if loci affecting quantitative traits are highly mutable. This is possible: mutation is a plausible explanation of quantitative genetic variation if one ignores the consequences of pleiotropy.' (Barton and Turelli 1989)

For traits on which many loci have small effects, per-trait mutation rates can be fairly high because many loci are contributing. Lynch (1985) estimated the rate of accumulation of mutations for life-history characters in *Daphnia* that were reproducing asexually in the laboratory; the rate was high enough to maintain a respectable rate of evolution in a completely asexual population. Most of the spontaneous mutants had pleiotropic effects on the traits he studied. Charlesworth (1987) favours mutation–selection balance as being the most universal mechanism maintaining additive genetic variation in fitness traits.

Explanation 2. Selection in patchy environments

Visualize the environment as a checkerboard in which there is stabilizing selection for one value of the trait on red squares and for another value on black squares. If the populations on the different squares are connected by gene flow, there will be a continuous influx of genes into each population from places where another value of the trait has been selected. Genetic variation will be maintained and gene flow will hinder local adaptation. This view appears to be correct for electrophoretically detectable alleles (see review by Slatkin 1987). Stearns and Sage (1980) suggested that gene flow is strong enough to hold the life history traits of a marginal population of mosquito fish in a state of perpetual maladaptation.

In a model with stabilizing selection for a mean value that changes along an infinite linear gradient, gene flow can maintain large amounts of variation if alleles are common (Barton and Turelli 1989). If alleles are near fixation and rare, spatial heterogeneity will not plausibly maintain genetic variation.

Environments also vary in time. Istock *et al.* (1976) estimated that the heritability of developmental time in the pitcher plant mosquito is about 30 per cent and explained that high value by seasonal variation in the optimal age at hatching. MacKay (1981) kept *Drosophila* populations on food regimes that fluctuated in time and contrasted them with control populations that were always given the same food. She found much higher additive genetic variances for two out of three traits in flies kept on the fluctuating regimes.

Explanation 3. Genotype × environment interaction

This differs from Explanation 2 in that it concentrates on environmental heterogeneity within patches that need not have any immigration. Most quantitative traits investigated display considerable genotype × environment interaction (e.g. Comstock and Moll 1963; Pani and Lasley 1972; Gupta and Lewontin 1982), and G × E interactions are commonly observed in life history traits (Birch *et al.* 1963; Gebhardt and Stearns 1988; Gebhardt 1989; Bell 1991). This means that genotypes react differently to changing environments, even changing ranks across environments.

Heritability measurements depend on the environment. The heritability of life history traits could be low in natural environments but, because of genotype × environment interactions, large in the laboratory. Thus observations of larger-than-expected amounts of genetic variation in life history traits could be a laboratory artefact. While this may partly explain high heritabilities, it cannot be the whole explanation for two reasons. First, heritabilities of life history traits measured in natural populations have yielded some large values. Secondly, when heritabilities of life history traits were measured across a range of conditions in the laboratory, they did change, but even at the lowest values measured remained high for a component of fitness (Gebhardt and Stearns 1988).

G × E interactions could play a role in maintaining genetic variation in the field, for G × E interactions in fitness generate a robust form of balancing selection (cf. Felsenstein 1976). Gillespie and Turelli (1989) have studied a model in which the additive contributions of alleles vary with the environment. They assume that no single genotype will produce the optimal phenotype in all environments, i.e. the reaction norms of fitness cross. Under these conditions, they find that G × E interactions maintain considerable additive genetic variation.

Explanation 4. Negative genetic correlations and pleiotropy

Take the simplest situation: one gene codes for two life history traits, both traits are under directional selection for larger values, but the genotypic value of one allele for the first trait is $+a$ and for the second $-a$. The genotypic values of the other allele are reversed:

	Genotypic value for:	
Allele	Trait 1	Trait 2
A1	$+a$	$-a$
A2	$-a$	$+a$

If one gene increases in frequency to benefit one trait, it decreases the value of the other trait. This frequency-dependent selection is generated by interactions among loci. At equilibrium, both alleles are maintained in a polymorphic compromise.

The other explanation for negative genetic correlations is linkage. Because it can take a long time for recombination to bring all loci into multilocus Hardy–Weinberg equilibrium in a large population, one expects that some closely linked loci behave as though they were alternative alleles at the same locus, with similar effects on genetic correlations.

At equilibrium, all genes whose effects are consistent with selection should be fixed. Only loci with antagonistic effects—or loci closely linked to them—will remain variable, and these will form the basis of the observed heritabilities and genetic correlations. Thus the genetic correlations between traits under selection should be negative (Rose 1983). The evidence for negative genetic correlations between components of fitness in Appendix 1 is mixed. Nevertheless, Barton and Turelli (1989) conclude 'though it is hard to be definite without more direct evidence, our analysis suggests that quantitative genetic variation for most morphological traits is probably maintained by the pleiotropic effects of many diverse polymorphisms'.

Explanation 5. Flat fitness profiles

Selection is directional and strongest on fitness itself. On any component of fitness, selection may be directional or stabilizing, stronger or weaker. Robertson (1955) suggested a method of visualizing such differences: *fitness profiles*. Consider several different quantitative traits (Fig. 3.8).

Traits such as clutch size probably have an intermediate optimum (e.g. Pettifor *et al.* 1988), age at maturity is usually under strong stabilizing selection, and the relation between fitness and age

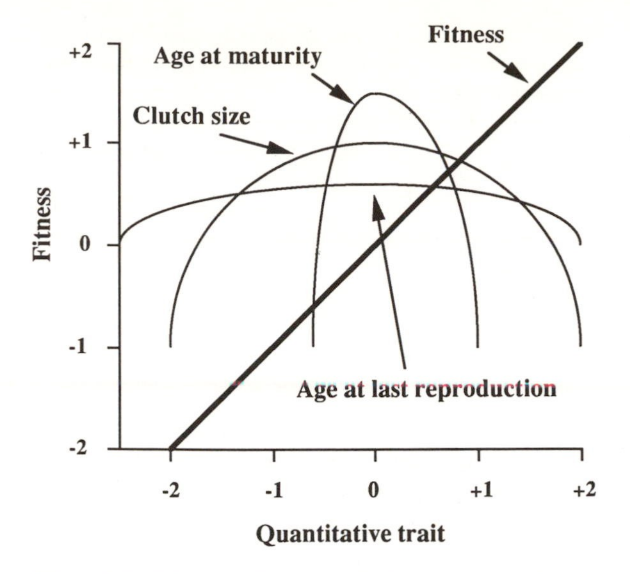

Figure 3.8 Schematic fitness profiles for various life history traits and for fitness itself (idea of fitness profiles from Robertson 1955).

at last reproduction is pretty flat. The fitness profile of one life history trait also changes with the values of the others (Chapter 2). Stearns and Crandall (1981*a*) calculated fitness profiles for age at maturity in salamanders and lizards from the Euler–Lotka equation. They were strongly peaked, with a sharp slope on both sides of the optimum. On the other hand, fitness profiles calculated for seed output in gap-colonizing forest trees were nearly flat (Stearns and Crandall 1981*b*).

When a fitness profile is flat, then genetic variation around the optimum has little impact on fitness and is easily maintained. In contrast, genetic variants causing equivalent deviations in age at maturity, which has a strongly peaked fitness profile, affect fitness in a big way, and selection against them is strong. We expect to see more genetic variation for traits with flat fitness profiles, but because we lack information on the fitness profiles of traits with high heritabilities in natural populations, we do not know how well this explanation accounts for the genetic variation observed. The best information available is on parameters of beak shape in Darwin's finches (Grant 1986). In normal years, the fitness profiles of beak parameters are fairly flat. Only in exceptionally dry years, about once per decade, is selection strong, and the fitness profiles become more strongly peaked.

GENETIC COVARIANCE

Genetic covariances caused by pleiotropy and linkage contribute to the maintenance of genetic variation in natural populations and to trade-offs between life history traits. Here we look at the definition, measurement, and interpretation of genetic correlations before using them in models of the simultaneous evolution of several traits.

Definition

The genetic correlation between two traits is the correlation of their breeding values. We measure the correlation between two traits on the phenotypes, then partition it into a genetic and an environmental component. Many genetic effects contribute to the phenotypic correlation—dominance effects, interaction effects, as well as deviations in breeding values from the population mean. One usually simplifies the problem by concentrating on the correlations of breeding values (additive genetic correlations) and lumping the other genetic sources of covariance together with the environmental sources:

$$\mathrm{Cov}_p = \mathrm{Cov}_a + \mathrm{Cov}_e. \qquad (3.8)$$

Think of Cov_e not as 'environmental covariance' but as the covariance of 'everything else'.

The following symbols are useful (cf. Falconer 1989):

x and y	the two traits
r_p	the phenotypic correlation of x and y
r_a	the genetic correlation of breeding values
r_e	the environmental and non-additive genetic correlation
Cov	the covariance of the two traits, with subscripts p, a, or e
σ^2 and σ	the variance and standard deviation of a single trait, x or y, with appropriate subscripts, e.g. σ_{ay} = the additive genetic standard deviation of trait y
h^2	the heritability, with subscript x or y for the trait
e^2	$= 1 - h^2$

All correlations are ratios of a covariance to the product of two standard deviations. The additive genetic correlation is

$$r_a = \frac{\mathrm{Cov}_a}{\sigma_{ax}\sigma_{ay}}, \tag{3.9}$$

so

$$\mathrm{Cov}_a = r_a\sigma_{ax}\sigma_{ay}. \tag{3.10}$$

A similar treatment of Cov_p and Cov_e allows us to rewrite the equation for phenotypic covariance as follows:

$$r_p\sigma_{px}\sigma_{py} = r_a\sigma_{ax}\sigma_{ay} + r_e\sigma_{ex}\sigma_{ey}. \tag{3.11}$$

Noting that $\sigma_a = h\sigma_p$ and $\sigma_e = e\sigma_p$, then dividing by $\sigma_{px}\sigma_{py}$ gives us an expression for the phenotypic correlation:

$$r_p = h_x h_y r_a + e_x e_y r_e. \tag{3.12}$$

This is how the additive genetic correlation between two traits combines with the correlation caused by 'everything else' to produce the phenotypic correlation.

Patterns of genetic and phenotypic correlations

Equation (3.12) shows that if the heritabilities of x and y are high and the phenotypic correlation is strong, then the phenotypic correlation is dominated by additive genetic effects. The lower the heritabilities of the two traits, the stronger the non-additive and environmental effects on the phenotypic correlation. Because the phenotypic correlation is built from these two components, and because the two components can have different magnitudes and signs, we cannot simply infer the nature of the additive genetic correlation from the phenotypic correlation. Bell and Koufopanou (1986) compared more than 60 estimates of genetic and environmental correlations for various traits in *Daphnia* and found no correlation. In a comparison of 53 estimates of phenotypic and genetic correlations, Roff and Mousseau (1987) found that genetic and phenotypic correlations were positively correlated ($r = 0.35$) overall. Genetic and phenotypic correlations between two morphological traits or a morphological and a life history trait were always positive, but genetic and phenotypic correlations between life history traits were not associated in any clear pattern. For some positive phenotypic correlations, genetic correlations were negative (Table 3.7).

Table 3.7 A comparison of genetic and phenotypic correlations for life history traits (after Roff and Mousseau 1987)

	Genetic correlation	
	Positive	Negative
Phenotypic correlation:		
Positive	16	8
Negative	1	5

Cheverud (1988) confirmed and extended these points. His results suggest why the genetic and phenotypic correlations of life history traits analysed by Mousseau and Roff tend to differ. In 41 pairs of genetic and phenotypic correlations, Cheverud found that the two correlations were nearly equal when effective sample sizes were large (>50) and estimates were reliable. With the small samples that one often has for life history traits, genetic correlations tended to be larger than phenotypic ones and not strongly correlated with them.

When heritabilities of two traits are fairly high, and their phenotypic correlation is fairly strong, a genetic correlation between the two traits is likely to be an important influence on the rate and direction of short-term evolution. Under these circumstances the signs of the genetic and phenotypic correlations are likely to be similar. However, exceptions occur.

Measurement

As with heritabilities, genetic correlations are measured through resemblance of relatives and through selection experiments—in this case, through the correlated response in trait y to selection on trait x.

Resemblance of relatives

To measure genetic correlations through resemblance of relatives, one does an analysis of covariance analogous to the analysis of variance used to measure heritabilities. One starts with the *product* of the values of traits x and y for each individual, where the values of the trait are their deviations from the mean, then partitions the sums of products by source of variation (mothers and fathers in a half-sib design, for example) to estimate the components of covariance. These are then interpreted by means of the statistical model implicit in the breeding design as components of variance, as in Table 3.3. For example, in a half-sib design the between-father component of covariance estimates $\frac{1}{4}\,\mathrm{Cov}_{axy}$ and the between-father component of variance estimates $\frac{1}{4}\sigma^2_{ax}$ and $\frac{1}{4}\sigma^2_{ay}$. The genetic correlation is then:

$$r_a = \frac{\mathrm{Cov}_{axy}}{(\sigma^2_{ax}\sigma^2_{ay})^{1/2}}. \tag{3.13}$$

One can also use parent–offspring regressions, calculating covariance as the product of the value of trait x in the parents and trait y in the offspring. This has to be done in both directions, then averaged. To get the genetic correlation, one divides by the square root of the product of the covariance of parent and offspring for each trait separately:

$$r_a = \frac{\frac{1}{2}(\mathrm{Cov}_{x_p y_o} + \mathrm{Cov}_{x_o y_p})}{(\mathrm{Cov}_{x_p x_o}\,\mathrm{Cov}_{y_p y_o})^{1/2}}. \tag{3.14}$$

Response to selection

To measure genetic correlations through the correlated response to selection, one selects trait x and measures both its response and the response of trait y. One must know the intensity of selection and the heritabilities of the two traits. The correlated response in trait y (CR_y) is then given as the product of the selection intensity i, the heritability of trait x, the heritability of trait y, the genetic correlation between the two traits, and the phenotypic standard deviation of y:

$$CR_y = ih_x h_y r_a \sigma_{py}. \tag{3.15}$$

One measures the other values and solves for r_a. If the selection experiment is done twice, selecting directly once on x and once on y and measuring the correlated response in the other trait, then the measurement of heritabilities can be avoided, and the genetic correlation is given as

$$r_a^2 = \frac{CR_x CR_y}{R_x R_y}. \tag{3.16}$$

MAXIMUM LIKELIHOOD ESTIMATES

The methods given above for estimating heritabilities and genetic correlations are based on least-squares. The estimate minimizes the sum of the squared deviations of the observations. This approach works so long as the parameters are estimated from a balanced design with equal numbers of observations in each of the cells—for example, m males each mated to f females, each of which then gave birth to n offspring, all of which survived and could be measured. Least-squares can deal with slightly unbalanced designs, but for badly unbalanced designs it implies either a fearful amount of algebra or no answer at all. It also cannot yield good estimates from data sets in which the relationships among organisms do not follow a regular pattern.

An alternative, the maximum likelihood approach (Fisher 1922), does not require balanced designs and puts no limits on the genetic relationships. It shares with the least-squares approach the assumption that the traits are distributed normally; if this assumption is violated, one cannot trust the significance tests used for either. The maximum likelihood approach can demand a lot of computation, for the likelihood function can only be maximized analytically in special cases. However, neither shortcoming outweighs its flexibility. It extracts information from data that other methods cannot handle. Shaw (1987) reviews the method, gives examples of its application, and reports on simulation studies of its statistical power.

MEASUREMENTS OF GENETIC COVARIANCES FOR LIFE HISTORY TRAITS

Because genetic correlations can produce responses that would not otherwise be expected, many attempts are being made to measure them (e.g. Via 1984; references in Cheverud 1988) and to measure natural selection on many traits at once (Lynch 1984; Arnold and Wade 1984*a*, *b*; Price and Grant 1984; Grant 1986; Schluter 1988).

Are genetic covariances between life history traits often, seldom, or just sometimes negative?

The answer depends on the organism, the traits measured, the method used, and the environment and stage of development in which the measurements were made. In 18 breeding experiments (see Appendix 1(*a*), (*b*)), the genetic correlations between fecundity and longevity were positive, in four cases they were negative, and in five cases, they were effectively zero. In contrast, in four of eight selection experiments in which fecundity was selected and longevity was the correlated response, there was a negative genetic correlation. In three cases there was no correlation, and in one it was positive, but not very. Estimates of genetic correlations between early and late fecundity from resemblance among relatives differ from those given by selection experiments. Six of eight such experiments reported positive correlations. In 11 of 12 selection experiments, the correlated response was negative.

Clark (1987) discusses several problems with such experiments. For example, if one selects for early fecundity and observes a decrease in late fecundity, one cannot be sure that the correlated response is due to pleiotropic effects. Relaxed selection on late fecundity (cf. Mueller 1986) may simply allow a build-up of mutations that decrease late fecundity. If one measures genetic covariances in the laboratory on the progeny of organisms caught in the field, the genotype by environment interactions involved can render laboratory estimates of the field situation worthless (cf. Giesel 1986).

Both selection experiments and breeding designs are labour-intensive methods of identifying additive genetic covariance. Both approaches are loaded with pitfalls and must be carefully thought through. However, if they are equally well done, then there is one advantage to selection experiments: they measure directly, in the correlated response to artificial selection, an item of central interest—the correlated response to natural selection.

The relative reliability of genetic and phenotypic correlations

Cheverud's (1988) comparison of 41 pairs of phenotypic and genetic correlation matrices led him to these conclusions:

> Phenotypic correlations between characters may be estimated with much greater precision and are also much simpler to obtain than their genetic counterparts. When genetic correlations are well estimated, they tend to be not very different in either magnitude or pattern from their phenotypic counterparts. Thus, when reliable genetic estimates are unavailable, phenotypic correlations and scaled variances may be substituted for their genetic counterparts in evolutionary models of phenotypic evolution. ... These results do not indicate that quantitative-genetic studies are unnecessary, but, rather, that quantitative evolutionary theory can be cautiously applied even when one is so unfortunate as to only have phenotypic data available. (p. 966).

The constancy of the genetic correlation matrix

The assumption of the constancy of the genetic variance–covariance matrix has also been checked. Estimates of genetic correlations are more sensitive to sampling error than estimates of heritabilities. The sample size required to differentiate between a genetic correlation of, say, 0.2 and 0.4 is often out of reach. Because gene frequencies change as a result of selection, any estimate of genetic covariance will change during the course of selection in one population and from population to population (cf. Turelli 1988). Estimates also vary considerably among samples drawn from single populations measured under constant conditions (Hughes and Clark 1988) and among and within taxa (Table 3.8).

There are also theoretical reasons to expect changes in genetic correlations. In de Jong's model (Stearns *et al.* 1991), the sign of genetic covariance between two traits changes in the region of the environmental gradient lying between the environments in which the two traits have their minimum heritabilities. The analysis of such effects is just beginning. While it should be possible to find conditions under which the sign of the genetic covariance does not change across an environmental gradient, under the simplest assumptions—non-parallel linear reaction norms, additive and independent effects of alleles and loci—the change

Table 3.8 Differences in genetic correlations among taxa

Taxonomic group	Traits examined	Differences found	Reference
Deer mice (*Peromyscus*)	15 skull traits	Between 2 species Not between subspecies	Lofsvold (1986)
Wood frogs (*Rana sylvatica*)	Development rate and size at metamorphosis	Between 2 populations	Berven (1987)
Chickens	Early growth rate and weight at maturity	Between 2 commercial races	Soller *et al.* (1984)
Murid rodents	8 pelvic traits	Between mice and rats	Kohn and Atchley (1988)
Milkweed bugs (*Oncopeltus*)	Age at maturity and eggs in first 5 clutches	Between two populations	Groeters and Dingle (1987)
	Wing length and life history traits	Between two populations, migratory and non-migratory	Dingle *et al.* (1988)

from positive to negative genetic covariance must occur.

Developmental effects

Atchley (1984, 1987) and his colleagues (e.g. Atchley *et al.* 1985) have worked on changes in genetic correlations of morphological characters of mice during ontogeny. They find significant changes, for gene expression changes during development. Roach (1986) found that the genetic covariance matrix of life history traits in *Geranium* changed as the plants aged, from generally negative in the early juvenile stage to strongly positive in the adult.

Environmental effects

Changes of *sign* in genetic covariance across environments have been found in fruit flies (Giesel *et al.* 1982; Gebhardt and Stearns 1988) and spadefoot toads (Newman 1988*a*, *b*), and changes of *magnitude* in genetic correlations across environments have been found in fruit flies, herbivorous insects, and milkweed bugs (Via 1984; Service and Rose 1985; Groeters and Dingle 1987; Lynch *et al.* 1989). Scheiner *et al.* (1991), on the other hand, document a case in which genetic correlations did *not* change significantly across environments. Clark and Keith (1988) measured 102 pairs of genetic correlations in two environments of which 95 were positive, one was negative, and only six showed a sign change. Of those six with sign changes, only two seem to differ significantly.

When traits respond differently to environmental change, the expression of genetic covariance is changed by phenotypic plasticity. Such effects can change genetic correlations from positive to negative. The expression of additive genetic variance and covariance changes with environments; the expression of genetic variation can be as heterogeneous as the environments inhabited. For example, in *Drosophila melanogaster* the additive genetic correlation between early-life fecundity and starvation resistance changes from -0.91 ± 0.03 at 25°C on rich medium in continuous light to -0.45 ± 0.18 at 15.5°C on poor medium in continuous dark (Service and Rose 1985). In *Chlamydomonas*, the genetic correlation of r with K changes from 0 to significantly negative when 250 mg/l of $NaHCO_3$ is added to the culture medium (Bell 1990*a*, Table 2). The next section describes how such effects may arise.

Reaction norms modulate the expression of genetic correlations

Recall that heritabilities and genetic covariances vary with the environments in which they are measured. We now seek a method of predicting, for a given lineage, how heritabilities and genetic covariances should change across heterogeneous environments. The expression of additive genetic variance and covariance limits the rate and direction of microevolution, if not its outcome at equilibrium. When there is a change in sign of genetic covariance across environments, a developmental mechanism is modulating the expression of genetic variance and covariance. In some environments, selection for an increase in one trait will cause a correlated *increase* in the other trait. In other environments, selection for an increase in one trait will cause a correlated *decrease* in the other trait. In environments intermediate to those just described, selection on one trait will not be correlated with any response in the other.

What properties of phenotypic plasticity lead to changes in the expression of genetic variance and covariance? The next section discusses some conditions.

A graphical approach using growth curves

Up to this point, we have plotted reaction norms for single traits as functions of the environment. Now plot two traits against each other, one on the abscissa and the other on the ordinate, then let environmental conditions vary across a set of growth curves. In Fig. 3.9, the plot on the left describes the usual way of plotting reaction norms; the plot on the right describes Figs. 3.12 and 3.13.

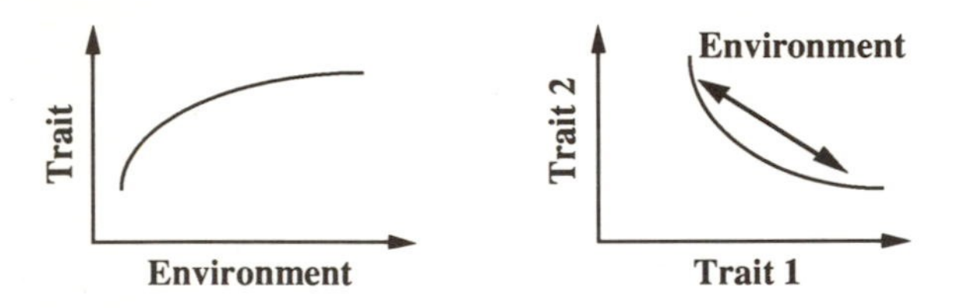

Figure 3.9 A small change in the way reaction norms are plotted helps one to analyse reaction norms for two traits on the same diagram.

Consider two life history traits associated with growth—age and size at maturity—in a population living in an environment heterogeneous for growth conditions. Under some conditions, growth is rapid; under others, slow. Organisms living in such an environment should evolve a reaction norm for the maturation event (see Chapter 6). One common reaction norm for age and size at maturity is maturation early at a large size when growth is rapid and later at a smaller size when growth is slow (Stearns and Koella 1986). Either of two conditions is sufficient to produce the change in sign in genetic covariance.

1. The largest and latest maturing genotype under good growth conditions is the largest but earliest maturing genotype under poor growth

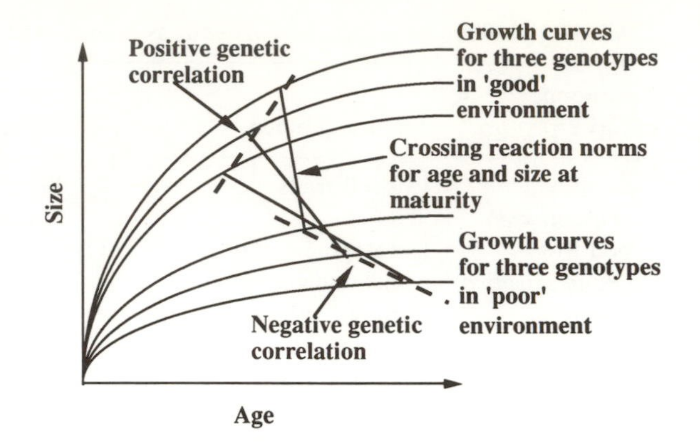

Figure 3.10 The first mechanism for producing a change in sign of genetic covariance: the reaction norms of the genotypes cross. Hypothetical growth curves for three genotypes are drawn for two environments: rapid and slow growth (after Stearns 1989*b*).

conditions. The other genotypes preserve this ordering. The reaction norms intersect, but the growth curves need not (Fig. 3.10).

2. Growth rates are negatively correlated with adult size, i.e. the fastest growing genotypes produce the smallest adults (Fig. 3.11). The extrapolated growth curves of the different genotypes would intersect if followed far enough. The reaction norms need not intersect, but if they do, their intersection strengthens the effect. Figure 3.11 plots growth curves and reaction norms for

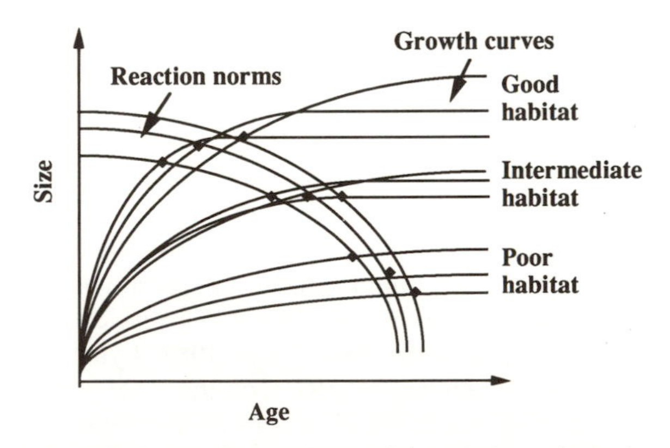

Figure 3.11 A second mechanism for producing a change in sign of genetic covariance: the growth curves cross (growth rate and asymptotic size have negative genetic correlations). Hypothetical growth curves for three genotypes are plotted for each of three environments: fast, intermediate, and slow growth. The heavy dots represent the maturation events of individuals (after Stearns 1989*b*).

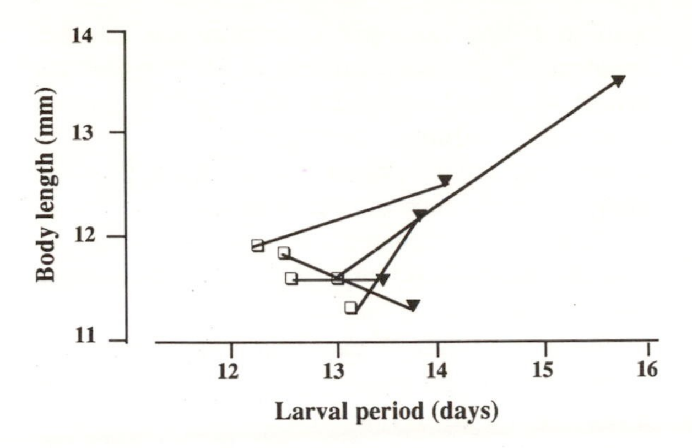

Figure 3.12 Reaction norms for five sibships of spadefoot toads raised either in ponds of short duration (□) or in ponds of long duration (▼) (after Stearns *et al.* 1991).

the maturation event for three genotypes. The genetic correlation between age and size at maturity changes from positive under good growth conditions, at the left, to negative under poor growth conditions, at the right.

Newman's (1988*b*) work on phenotypic plasticity in spadefoot toads provides an opportunity to test these ideas (Fig. 3.12). He measured reaction norms of length at metamorphosis and developmental rate in two environments, ponds of short or long duration, for five sibships. In ponds of short duration, the broad-sense genetic correlation of length at metamorphosis and developmental rate was -0.92 ($r^2 = 0.85$, $p < 0.01$); in ponds of long duration it was $+0.92$ ($r^2 = 0.84$, $p < 0.01$). Broad-sense genetic correlation changed dramatically between two environments.

Thus broad-sense genetic covariance changes from positive to negative when the growth curves of the genotypes cross or when their reaction norms have different slopes and ranges. When growth conditions vary, genetic covariances may vary from positive to negative as well. If it is normal for growth curves or reaction norms to cross, then changes in sign of genetic covariance are frequent. Perhaps even the mechanisms that modulate the expression of genetic variation and covariation have evolved. If so, genetic variation could be released or concealed depending on the environment. A non-adaptationist alternative is that these effects are by-products of developmental systems existing for reasons having nothing to do with genetic variation or microevolutionary adaptation. Further work is required to distinguish between these alternatives (Stearns 1989*c*).

Summary comment on the constancy of genetic correlations

Genetic correlations have complex behaviour. They depend on stage of development (Atchley 1987) and on the environment in which they are measured (Via and Lande 1985). Methods for dealing with these complications are being developed. In general, genetic correlations have both a transient component dependent on gene frequencies and genotype by environment interactions and a deep-rooted component that expresses the constraints of physiology, development, history, and design (see Chapters 4 and 5).

> It is important to separate the fundamental, long-term genetic constraints, which limit species evolution, from the transient genetic correlations that may be found in a particular population. It is unclear whether natural populations ever reach equilibrium conditions, in which the additive genetic variance is depleted. Under many circumstances it is difficult to deduce the past history of selection from the genetic architecture of metrical traits (Mitchell-Olds and Rutledge 1986).

Because genetic correlations can change from population to population, within populations over time as gene frequencies change, during the course of development, and from environment to environment, the responses to selection that are important to life history evolution change in all the same ways. A dynamic theory of evolution based on correlation matrices should include—but has not yet included—the dynamics of the genetic correlation matrix itself. Some find this reason enough to concentrate on predicting the state of the equilibrium phenotype—evolutionary statics. Others are stimulated to plunge further into the complexities of genetic correlations in the hope that simplicity will be found at a deeper level of evolutionary dynamics.

QUANTITATIVE GENETICS IN HETEROGENEOUS ENVIRONMENTS

When the environment is heterogeneous, a single genotype can develop different phenotypes in different environments. How should we quantify genetic effects on plasticity itself? Phenotypic plasticity affects both natural selection on phenotypes and the genetic response to selection (cf. Tables 3.10–3.13). Estimates of heritabilities and genetic covariances that take no account of phenotypic plasticity may confound effects arising in growth and development with effects arising in the genome. An evolutionary theory of phenotypes in heterogeneous environments requires an experimental and theoretical analysis of these interacting influences on phenotypic variation. Some recent attempts are discussed next.

Partitioning variation

The basic model of quantitative genetics breaks total phenotypic variation into three large classes: genetic variance, environmental variance, and genotype-environment covariance.

$$\text{Phenotypic variance} = V_p = V_g + V_e + 2\,\mathrm{Cov}_{ge}. \tag{3.1}$$

One might think that phenotypic plasticity is somehow tied up in the genotype–environment covariance, but in fact that term describes a different effect—the unequal representation of genotypes in different environments. It is not genotype × environment interaction. In many experiments the genotype × environment interactions and the mean response to environmental variation are lumped together in the V_e term.

To quantify plasticity, the components of the V_e term must be separated. This is best done in experiments in which all genotypes are tested across all environments. In plant and animal breeding, one does this to see if genotypes change rank across environments. For example, should farmers plant one broadly tolerant generalist variety or several narrowly tolerant specialist varieties? Experiments answering this question quantify both the mean plastic response and the genetic variation in plastic responses. The design is clearest if clones are used, but one can also work with split-brood designs to estimate the average response to the environment of the part of the genotype shared by siblings. One tests a number of clones, or siblings from a number of families, in different environments, then does a factorial analysis of variance on the results.

When the environments can be represented on a single axis, such as temperature or food quantity, the interpretation of the results begins with a plot of the phenotypes of each genotype across the environmental gradient (Fig. 3.13). Here the scale has been chosen to linearize the reaction norms.

A qualitative impression can be read directly off such a plot: the scope of the mean plastic response, genetic variation for the plastic response in the form of reaction norms with different slopes, and crossing reaction norms that imply changes of rank of genotypes from environment to environment. Keep clear the distinction between

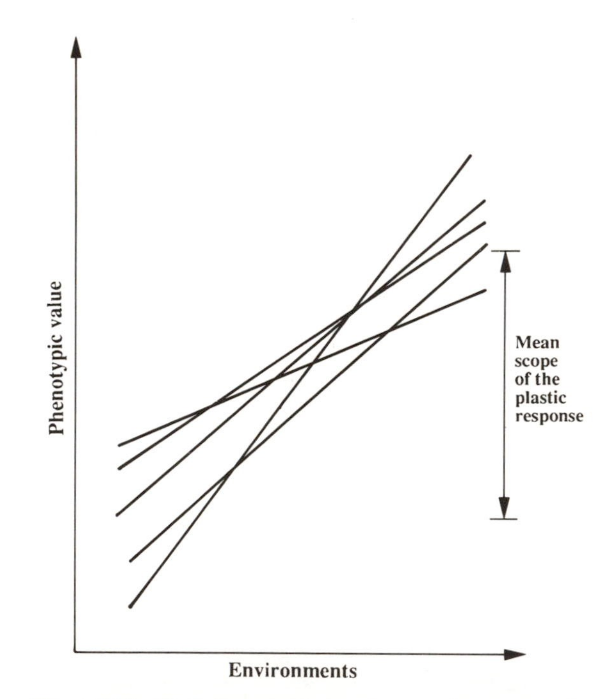

Figure 3.13 Each line represents the reaction norm of a genotype tested across a range of environmental conditions. Three items are of interest. The mean scope of plastic response is represented by the distance between the mean genotypes in the extreme environments. The presence of genotype × environment interactions is indicated by the different slopes of the reaction norms. Changes in rank of genotypes from environment to environment are indicated by reaction norms that cross each other.

the mean scope of plastic response, as measured by the mean reaction norm, and the genetic variation for plasticity, as estimated by variation in slopes of reaction norms. There is a mean plastic response only if the mean (population) reaction norm has a significantly non-zero slope, and there is genetic variation in plasticity only if the lines representing the individual (genotype) reaction norms are nonparallel. This can be tested with the significance of the genotype × environment interaction term in the analysis of variance, and the magnitude of the contribution of this interaction to the variation present can be estimated with analysis of covariance (van Noordwijk 1989).

Bell (1990*a*) presents an experimental design in which the genetic and environmental components of the genotype × environment interaction term can be further decomposed. Statistical methods do not report significant G × E interactions in many cases where the reaction norms of the genotypes do cross. There must be considerable variation in slopes before significance is attained. This property of the statistics suggests that the first step in such analyses must be to plot the reaction norms, as in Fig. 3.13. This will often be as informative as the statistical tests, which must, of course, also be done.

A description like Fig. 3.13, ideal for statistical analysis and biological interpretation, is only possible when genotypes can be cloned, when the response of each genotype to the environmental variation is linear, and when the environment can be represented on a single axis. Difficulties in measuring genetic variation in plasticity arise when these conditions are violated. For example, the response of a trait to different environmental factors—e.g. food and temperature—often differs. A trick allows one to compress combinations of environmental factors on to a single axis. One defines the *environmental value* as the mean phenotype of all genotypes in that environment, then plots the reaction norms of the individual genotypes on the environmental values (Falconer 1989, p. 136. Also see Mather and Jinks 1982, p. 118.) Statistical and biological interpretations are most straightforward when the reaction norms are linear or can be linearized by an appropriate transformation.

Given time and resources, one can do similar experiments for several environmental variables and estimate the scope of plastic response and the importance of genotype × environment interactions for each environmental factor—see, for example, Bell's (1990*a*, *b*) work on *r* and *K* in many strains of *Chlamydomonas* tested in many nutrient environments. Such experiments show that the plastic response of a trait depends on the environmental factor that is varied, and that interaction between environmental factors affects phenotypic means, independent of whether genotype × environment interactions are present.

THE HERITABILITY OF PHENOTYPIC PLASTICITY

A definition of the heritability of plasticity is needed if we want to estimate the potential of the plasticity of a trait for short-term evolutionary change. Such short-term changes do occur. For example, Blackmore and Charnov (1989, their Fig. 2) demonstrated local evolutionary adjustment of a reaction norm for sex allocation. The heritability of plasticity should measure additive genetic *variation* in plasticity, not the mean plastic response, and it should predict the response of plasticity to selection. We now discuss candidate definitions; for linear reaction norms a satisfactory definition is the additive genetic variance in slopes divided by the total variance in slopes.

Scheiner and his colleagues (Scheiner and Goodnight 1984; Scheiner and Lyman 1989, 1991; Scheiner *et al.* 1991) proceed from the definition of phenotypic plasticity given by Bradshaw (1965)—the change in phenotype caused by a change in the environment. Basing their definition on Becker (1984) and Scheinberg (1973), they define the plastic variance as the sum of the variances due to differences in environment and to genotype × environment interaction:

$$\sigma^2_{PL} = \sigma^2_E + \sigma^2_{G \times E}. \quad (3.17)$$

They discuss two ways to look at the heritability of plasticity.

Candidate definition 1

The first is the proportion of total phenotypic variance accounted for by genotype ×

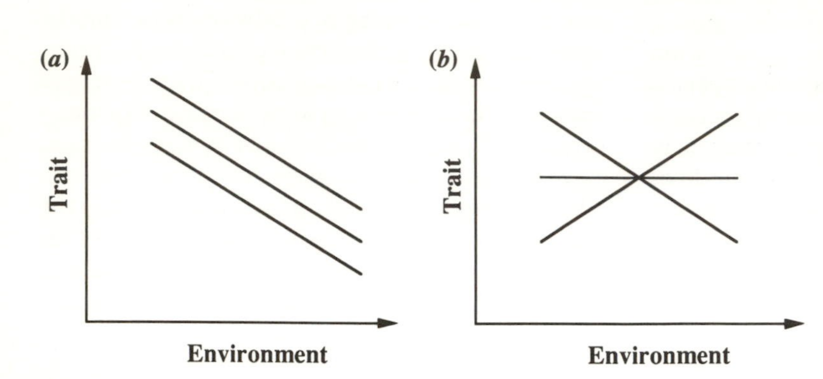

Figure 3.14 A graphical interpretation of why the magnitude of G × E interaction should measure the potential response to selection of plasticity itself as measured by the difference in the mean phenotype between any two environments. On the left the reaction norms are parallel, G × E interaction is zero, and the potential response of the difference in phenotypic mean is zero. On the right, G × E interactions are large, the reaction norms cross, and the potential response ranges from large and negative through zero to large and positive.

environment interactions (Scheiner and Lyman 1989):

$$h^2_{PL} = \sigma^2_{G \times E}/\sigma^2_P. \tag{3.18}$$

To see why this measure should predict the response to selection on plasticity itself, consider Fig. 3.14, in which two situations are contrasted. In the first (Fig. 3.14(*a*)), the reaction norms are parallel and there is no G × E interaction. No matter which reaction norm—or which combination of reaction norms—is selected, the *difference* in the mean phenotypes found in any two environments would remain constant. The scope for response to selection on plasticity is zero, and the G × E interaction is zero. In the second (Fig. 3.14(*b*)), the reaction norms cross, G × E interactions are large, and the scope for a response to selection on slope is large. As depicted, the mean phenotypic difference between the extreme environments is zero. After selection, it could remain zero, be negative and large, or be positive and large, depending on the genotype selected.

The problem with this measure is not the numerator—variance accounted for by G × E interactions—but the denominator, which contains a confusing mixture of terms arising from genetic and phenotypic effects and from genotype–environment covariance (Gebhardt 1989).

Becker (1984) gives methods for calculating $\sigma^2_{G \times E}$ and Pederson (1972) discusses optimal sample sizes for various experimental designs. No least-squares methods are available to estimate standard errors for this measure of heritability, but either maximum-likelihood (Shaw 1987) or bootstrapping (Efron 1981) can be used. The standard errors are often large. Weis and Gorman (1990) discuss the measurement of selection on reaction norms in an experimental system involving a plant and a herbivorous insect.

Such measures of heritability only apply to the *frequency distribution* of environments in which they were measured (Scheiner and Lyman 1989). In an ideal experiment, the different genotypes are cloned and distributed *with equal frequency* across all environments. In nature, different environments are unequally represented in time and space, and the effects of the different environments on the phenotype and on genotype × environment interactions are also unequally represented in the estimates.

A special case of this definition of plasticity is the mean *difference* in performance of offspring or clonal replicates reared in two environments (Eisen and Legates 1966; Scheiner and Lyman 1989), a measure particularly appropriate for split-brood designs. The heritability of plasticity is then the proportion of total phenotypic variance in mean *difference* accounted for by variance among families or among clones. Where it is possible to raise yet another generation, this method can be developed into a parent–offspring regression estimate of heritability in which the mean differences between environments within families of the F2 measure the plasticity of the offspring and the mean differences between environments within families of the F1 measure the plasticity of the parents (Becker 1984; Scheiner and Lyman 1989).

A related index of plasticity is the coefficient of variation of a trait across environments. It would rank traits by plasticity if the traits were all measured on the same organisms to standardize genetic background. One can also estimate the heritability of plasticity through the realized response to artificial selection of the mean difference in phenotypes between environments (Jain 1978; Scheiner and Lyman 1991).

Candidate definition 2—additive genetic variance in slopes

We now consider the framework developed by de Jong (1990*a, b*) and compare it with those developed by Scheiner and Via and Lande (1985). Like Scheiner but unlike Via and Lande, de Jong (1990*a, b*) sees plasticity as a trait in itself. Rather than start with the statistics of quantitative genetics, she begins with a model of one locus with two alleles. She assumes that these alleles affect the slope and intercept of linear reaction norms. To derive the macroscopic behaviour of quantitative genetics from the microscopic properties of a set of such loci, de Jong only considers the case where effects on reaction norms are additive within and among loci. She works out both the two-locus case and the case for polygenes (de Jong 1990*a*) and reaches the conclusion that the heritability of plasticity is the additive genetic variation for slopes of linear reaction norms divided by the total variation in slopes.

Her fundamental result is that optimizing selection on reaction norms can be split into two components: (1) optimizing selection on the value of the reaction norm in the mean environment, and (2) optimizing selection on the slope itself. If environmental variance is low, the first selection component prevails. If it is high, the second prevails. In the first case, one might get a reaction norm with any slope at all, so long as it passes through the optimum value in the mean environment. In the second case, one might get a reaction norm parallel to, but offset from, the one that passes through the optimum value in the mean environment.

Gebhardt (1989*b*) used responses measured between environments to test how well de Jong's model predicts the heritabilities of traits measured within environments. He measured age and dry weight at eclosion in fruit flies raised at three temperatures and three yeast concentrations. The correlations between predicted and observed additive genetic variances ranged from 0.66 to 0.93. He concluded that de Jong's model could be used as a first approximation to predict changes in the slope of the mean reaction norm. Although several of its assumptions were violated, the model performed reasonably well.

Gebhardt's experiments are equally a confirmation of the Via–Lande model discussed next, for the assumption of linear reaction norms is equivalent to the assumption that plasticity is measured by differences in the reaction to two discrete environments (two points define a straight line).

OTHER WAYS TO QUANTIFY PLASTICITY

Cross-environment genetic correlations

The approach to phenotypic plasticity developed by Van and Lande (1985, 1987) yields some insights more easily than other methods, but it does not define the heritability of plasticity because they treat plasticity in quite a different way. Their treatment stems from Falconer's (1952) suggestion that reaction norms be connected to quantitative genetics by taking the expression of a trait *in each environment* as a separate character. Because the realization of the trait in each environment is now a separate trait, no traits can be plastic because no traits are measured in two or more environments.

The genetic correlation between expressions of the trait in discrete environments is called the *cross-environment correlation*. It measures the genetic variance of plasticity, functions like a definition of the heritability of plasticity, and can be directly incorporated into the dynamic evolutionary equations formulated by Lande (1979) and Lande and Arnold (1983).

There is a direct relation (de Jong 1990*b*) between cross-environmental correlations (Via and Lande 1985) and G × E interactions (Scheiner and Goodnight 1984), and there is a direct relation between the magnitude of G × E interactions and the magnitude of additive genetic variation in the slopes of linear reaction norms (de Jong 1989, 1990*b*). The approaches differ not so much in the phenomena analysed or the information included in the analysis as they do in the language—both mathematical and verbal—used to describe plasticity.

Three aspects of Via and Lande's approach strike me as interesting. First, they predict that plasticity will evolve so that the optimal mean phenotype is attained in each environment unless there is a cross-environment genetic correlation of $+1$ or -1, exhaustion of polygenic variation, or a cost to plasticity. These conditions are not likely to be general, and genetic correlations should slow the approach to the optima.

Second, they point out that if the joint optimum can be attained in all environments encountered, and if there is sufficient re-mixing of the genotypes over the whole environment in every generation, then environmental heterogeneity plays no role in the maintenance of additive genetic variation. These global conditions, which strike me as unlikely, were those assumed by Dempster (1955). When there is local population regulation in each environmental type, as assumed by Levene (1953), then environmental heterogeneity implies negative frequency-dependent population regulation that can maintain genetic variation (Bell, pers. comm.).

Third, the distribution of environmental frequencies has decisive influence on the evolution of plasticity. If an environment is rare, the approach to the optimum phenotype in that environment will be slow.

Despite their advantages, the methods developed by Via and Lande have not been adopted by everyone working on plasticity. One reason is the way they define traits. Lynch and Gabriel (1987), Scheiner and Lyman (1989, 1991), Schlichting and Levin (1986, 1990), Stearns and Koella (1986), van Noordwijk and Gebhardt (1987), Gebhardt and Stearns (1988), Sibly and Calow (1989), and de Jong (1989, 1990*a*), among others, prefer to conceive of phenotypic plasticity as a continuous norm of reaction, a trait in itself under the control of different genes from those that determine the mean value of the trait. Further, the Via–Lande approach applies only to characters under stabilizing selection, not directional selection, and to pairs of environments, not to large numbers of discrete environments or to continuously distributed environments (but see Gomulkiewicz and Kirkpatrick, in press, *Evolution*). De Jong's approach contains all the information needed to derive the cross-environment genetic correlations used by Via and Lande (1985), but one cannot work in the other direction, deriving reaction norms and genotype × environment interactions from their cross-environment correlations (de Jong 1990*b*).

Quantifying plasticity as environmental tolerance

Lynch and Gabriel (1987) take another approach to the genotype–phenotype–environment relation. As Levins (1968) emphasized, the fitness of a phenotype varies across an environmental gradient, and for many physical and chemical factors, physiological ecology has shown that fitness is usually highest at an intermediate value and decreases towards the extremes (Huey and Stevenson 1979; Huey and Hertz 1984). Lynch and Gabriel therefore conceive of phenotypic plasticity as the breadth of environmental tolerance. If fitness falls off rapidly from the optimum located, for purposes of convenience, at the centre of the environmental range, the breadth of environmental tolerance is small. If fitness is not sensitive to environments and remains high across a broad range centred on the intermediate optimum, then the breadth of environmental tolerance is large.

Such patterns can be modelled as bell-shaped curves determined by the location of the environmental optimum and the breadth of environmental tolerance. Lynch and Gabriel postulate both genetic and developmental effects on the location and breadth of the curve, then analyse genetic and environmental contributions to the variance in breadth of environmental tolerance. They also give a method for estimating the genetic and developmental parameters affecting the curve.

> Whereas any kind of temporal variance in the environment selects for more-broadly-adapted genotypes, temporal variance within generations . . . plays a more important role than that between generations . . . , which becomes of negligible importance when the spatial component of variance is high. Spatial heterogeneity . . . also often selects for more-broadly-adapted genotypes, but only when it operates in conjunction with temporal variance. More-

over, when [the within-generation component of environmental variation] is much less than [the between-generation component], spatial heterogeneity can actually select for a higher degree of specialisation (Lynch and Gabriel 1987, p. 297).

Comments on quantifying plasticity

The questions addressed, points made, and problems solved by each of the approaches depend on the assumptions made. Each approach is best at answering a different set of questions, and common to them all is an arbitrariness that stems from the natural desire for soluble mathematics.

THE INTEGRATION OF THE PLASTIC RESPONSE

Why should the plastic responses of a suite of traits be integrated, and if they are, what mechanisms might be responsible? All traits dependent on size will show integrated responses if size is plastic. Traits with a functional relationship—the area of roots and leaves in a flowering plant, the absorptive surface in the lung and the volume of muscle in a mammal—will be under strong selection to have a coordinated response to environmental change, and they do have such a response. The constancy of shoot/root ratios is well-known in higher plants. For life history traits, there should have been selection for compensation, so that losses in fitness incurred by plastic changes in one trait are compensated for by increases in fitness due to plastic changes in another trait. This is the essence of integrated plastic responses.

Schlichting and Levin (Schlichting 1989*a, b*; Schlichting and Levin 1986, 1988, 1990) did work on *Phlox* aimed at identifying regularities in the plastic changes of many traits across heterogeneous environments. Schlichting and Levin (1986) and Marshall *et al.* (1985, 1986) have found variation among species in the degree of integration of plastic responses across environments.

Two traits could show integrated plastic responses (1) because they are under the control of the same genes, (2) because they share the same pool of resources (the integrated response would appear as a phenotypic trade-off), and (3) because they are both responding to microenvironmental variation—the source of the integration being not within the organism but between the organism and the environment (Schlichting 1989*b*; Cheverud 1988; Zelditch 1988).

CONNECTING QUANTITATIVE GENETICS TO DEMOGRAPHY

Demography tells us how selection pressures are distributed across life history traits. One of its central messages is summed up in a sensitivity or elasticity matrix (Chapter 2). These matrices tell us how fitness would change if the trait represented by each matrix element were to change. The selection pressure on a trait is the sensitivity of fitness to changes in that trait. It has two components: the trait's direct relation to fitness and the indirect relations created by phenotypic correlations with other traits.

Quantitative genetics tells us how populations will react in the short term to selection. When we consider one trait, we can predict its response to selection if we know the strength of selection and the amount of additive genetic variation. When two or more traits are under selection at the same time—always the case in life history evolution—we must also account for genetic correlations between traits. If these are negative, then the response of one trait is constrained by selection on others and little response to selection will be observed despite the presence of genetic variation.

Lande's (1982*a, b*) framework brings together the demographic causes of selection pressures and the genetic causes of the response to selection. The genetics is represented by a genetic variance–covariance matrix in which each trait is listed once as a column and once as a row. The entries along the main diagonal are the additive genetic variances—the heritabilities. The other entries are the genetic covariances of the two traits found in that column and that row. The selection pressures on these traits are represented by a column vector of sensitivities of fitness to each trait.

The selection gradient

To see how this would work in detail, consider the simple population represented by Table 3.9. The sensitivities of fitness to each of the age-specific

Table 3.9 A life table simplified for purposes of illustration

X	p_X	m_X
0	0.50	0.0
1	0.25	1.5
2	0.00	2.0

fecundities (m_X) and survival probabilities (p_X) are given by eqn (2.33):

$$s_{ij} = \frac{v_i w_j}{\langle \mathbf{w}, \mathbf{v} \rangle} \tag{2.33}$$

where $\mathbf{w}$ is the vector representing the stable age distribution, $\mathbf{v}$ is the vector of reproductive values of age classes, and $\langle\ \rangle$ indicates vector multiplication. However, an adjustment must be made to make the sensitivities calculated in Chapter 2 consistent with those used by Lande. Equation (2.33) gives us the sensitivity of λ to changes in an age-specific fecundity or mortality. In Lande's framework, we need the sensitivities of r to those traits. Since $\lambda = e^r$,

$$\frac{\partial \lambda}{\partial a_{ij}} = \lambda \frac{\partial r}{\partial a_{ij}}. \tag{3.19}$$

Thus for Lande's framework, we divide each entry by λ to get the sensitivities (Table 3.10) that represent the selection pressures operating on each of the traits listed. Lande (1982*a*) refers to it as the *selection gradient* because it can be represented as 'the gradient vector of the intrinsic rate of increase of the population with respect to changes in the mean life history' (p. 610).

Table 3.10 Sensitivity of r to fecundities and survival probabilities given in Table 3.7

Trait	Sensitivity
m_1	0.423
m_2	0.203
p_0	0.878
p_1	0.407

The phenotypic correlation matrix

A further adjustment to the selection gradient is necessary because the traits are phenotypically correlated (here I differ with Lande (1979), who modelled the connections of the traits as genetic but not phenotypic). The strength of selection on a trait is measured both by its direct effect on fitness (Table 3.10), and by the indirect effects on fitness from phenotypic correlations with other traits.

Phenotypic correlations may represent a phylogenetic heritage, where all populations in a clade have similar correlations. For example, time to metamorphosis and relative hind-limb length have a negative phenotypic correlation in at least two frog genera in North America (*Hyla* and *Rana*; Emerson 1986), and interspecific differences in relative hind-limb length may reflect selection on time to metamorphosis rather than on hind-limb length. The phenotypic correlations between two traits also change across environments (cf. Schlichting 1989*b*).

Because of phenotypic correlations, the selection pressure on a trait has direct and indirect components. The indirect effects should include all traits with significant impact on fitness. To show what a difference phenotypic correlations can make, let us consider three phenotypic correlation matrices constructed for purposes of illustration. In the first, $\mathbf{P}_1$, there are no correlations among traits—the major diagonal is 1 and all off-diagonal elements are 0. In the second, $\mathbf{P}_2$, the correlations are predominantly positive. In the third, $\mathbf{P}_3$, they are a mixture of positive and negative. The rows and columns of the matrices are defined by the four traits listed in Table 3.10.

$$\mathbf{P}_2 = \begin{bmatrix} 1.0 & 0.4 & 0.2 & 0.1 \\ 0.4 & 1.0 & 0.4 & 0.2 \\ 0.2 & 0.4 & 1.0 & 0.4 \\ 0.1 & 0.2 & 0.4 & 1.0 \end{bmatrix} \tag{3.20}$$

$$\mathbf{P}_3 = \begin{bmatrix} 1.0 & -0.4 & 0.2 & -0.1 \\ -0.4 & 1.0 & -0.4 & 0.2 \\ 0.2 & -0.4 & 1.0 & -0.4 \\ -0.1 & 0.2 & -0.4 & 1.0 \end{bmatrix}. \tag{3.21}$$

Table 3.11 Sensitivity of r to fecundities and survival probabilities for three different phenotypic correlation matrices

	Sensitivity		
Trait	P_1	P_2	P_3
m_1	0.423	0.721	0.477
m_2	0.203	0.805	−0.236
p_0	0.878	1.207	0.719
p_1	0.407	0.841	0.054

If we now multiply the direct sensitivities of fitness to the traits by the phenotypic correlation matrices, we get the total phenotypic sensitivity to changes in each of the traits, including both direct and indirect effects (Table 3.11). In the first case, the sensitivities are unchanged—all the off-diagonal values are 0. In the second case, the sensitivities are all increased by indirect effects on fitness brought about by positive phenotypic correlations. In the third case, with some negative phenotypic correlations, the sensitivity of fitness to changes in late fecundity (m_2) is negative and near zero to changes in survival from the first to the second year (p_1). Negative phenotypic correlations can change the total selection pressure on a trait in unexpected ways. The difference between the two phenotypic correlation matrices $\mathbf{P}_2$ and $\mathbf{P}_3$ could be something as simple as a plastic reaction to two different environments.

The rate of evolution of a trait

The rate of evolution of a trait z_i is calculated by multiplying the entries in the genetic variance–covariance matrix $\mathbf{G}$ by the selection gradient:

$$\frac{dz_i}{dt} = \sum_{j=1}^{n} G_{ij} \frac{\partial r}{\partial z_j}. \tag{3.22}$$

In this summation, when $j = i$ the entry in the genetic variance–covariance matrix is the heritability of the trait. Thus the response to selection is the selection pressure on the trait times its heritability, plus all the effects that enter through linkages to other traits, measured as genetic correlations, times the selection pressures on those other traits. If some of those genetic correlations are negative, they will subtract from the response to selection.

Note that the selection pressures on the correlated traits have entered this calculation twice—once in calculating the indirect selection pressures that arise through phenotypic correlations and once in calculating the indirect contributions to the response to selection that arise through genetic correlations.

Consider three genetic variance–covariance matrices for the traits listed in Table 3.10. In the first, $\mathbf{G}_1$, the traits are genetically fully independent of one another, with genetic correlations of 0, and have moderate heritabilities. In the second, $\mathbf{G}_2$, the traits are tied together by strong negative genetic correlations. The third, $\mathbf{G}_3$, represents a mixed situation with some positive and some negative correlations.

$$\mathbf{G}_1 = \begin{bmatrix} 0.23 & 0.00 & 0.00 & 0.00 \\ 0.00 & 0.19 & 0.00 & 0.00 \\ 0.00 & 0.00 & 0.37 & 0.00 \\ 0.00 & 0.00 & 0.00 & 0.42 \end{bmatrix} \tag{3.23}$$

$$\mathbf{G}_2 = \begin{bmatrix} 0.23 & -0.45 & -0.05 & 0.00 \\ -0.45 & 0.19 & -0.52 & -0.11 \\ -0.05 & -0.52 & 0.37 & -0.12 \\ 0.00 & -0.11 & -0.12 & 0.42 \end{bmatrix} \tag{3.24}$$

$$\mathbf{G}_3 = \begin{bmatrix} 0.23 & 0.45 & 0.05 & 0.00 \\ 0.45 & 0.19 & -0.52 & -0.11 \\ 0.05 & -0.52 & 0.37 & 0.12 \\ 0.00 & -0.11 & 0.12 & 0.42 \end{bmatrix}. \tag{3.25}$$

Now we can ask how the response to selection will differ in each of the three cases. If the mean values of the three life histories are listed in a column vector $\mathbf{z}$ then

$$\frac{d\mathbf{z}}{dt} = \mathbf{G}\nabla\mathbf{r}. \tag{3.26}$$

In this first case, we use phenotypic correlation matrix $\mathbf{P}_1$, where there are no phenotypic correlations among the traits. To calculate the responses,

Table 3.12 A comparison of the responses to selection of four life history traits on the assumption of three different genetic variance–covariance matrices. The selection gradient used was calculated with phenotypic correlation matrix $\mathbf{P}_1$

		Response to selection given genetic variance–covariance matrix		
Trait	Sensitivity	$\mathbf{G}_1$	$\mathbf{G}_2$	$\mathbf{G}_3$
m_1	0.423	0.097	−0.038	0.233
m_2	0.203	0.039	−0.653	−0.272
p_0	0.878	0.325	0.149	0.229
p_1	0.407	0.171	0.043	0.254

Table 3.13 A comparison of the responses to selection of four life history traits on the assumption of three different genetic variance–covariance matrices. The selection gradient used was calculated with phenotypic correlation matrix $\mathbf{P}_2$

		Response to selection given genetic variance–covariance matrix		
Trait	Sensitivity	$\mathbf{G}_1$	$\mathbf{G}_2$	$\mathbf{G}_3$
m_1	0.721	0.166	−0.257	0.588
m_2	0.805	0.153	−0.892	−0.243
p_0	1.207	0.447	−0.109	0.165
p_1	0.841	0.353	0.120	0.410

we multiply the genetic variance–covariance matrices times the selection gradient and get the results shown in Table 3.12.

1. Where there are no genetic correlations between traits, the response to selection is simply the heritability multiplied by the sensitivity of fitness to that trait. For life history traits, the sensitivity of fitness to the trait can be calculated from the stable age distribution and the distribution of reproductive values.

2. When there are strong negative genetic correlations between traits, the response to selection can be negative even though selection is directional for increasing values. The genetic links among traits can be decisive for their response to selection.

3. The response to weak selection of a trait with strong positive correlations to traits under stronger selection can be stronger than the response of the correlated traits themselves.

Phenotypic correlations also have indirect effects on selection pressures, through them on selection gradients, and therefore on the response to selection. Table 3.13 lists the sensitivity of fitness to these four traits for the phenotypic correlation matrix $\mathbf{P}_2$, where the correlations are all positive, and the responses to selection for each of the three genetic covariance matrices.

Positive phenotypic correlations among life history traits tend to strengthen selection on all traits, but the quantitative interactions are decisive (note the sign change in response of p_0 with genetic correlation matrix $\mathbf{G}_2$).

If the phenotypic correlations are mixed, positive and negative, as in phenotypic correlation matrix $\mathbf{P}_3$, then the response to selection is also mixed.

To help connect the formalism of quantitative genetics with the trade-offs discussed in the next chapter, note that if the genetic variance–covariance matrix $\mathbf{G}$ expresses the trade-offs that are important for the optimization of the phenotype, then the quantitative genetics approach leads to the same evolutionary equilibrium as the optimization approach (Charnov 1989; Charlesworth 1990*a*), and at the evolutionary equilibrium, $\mathbf{G}\nabla\mathbf{r} = 0$ (cf. eqn 3.26).

CLOSING COMMENT

Opinion is divided on the rewards of quantitative genetics. Quantitative genetics is a good tool for weighing genetic and environmental effects and their interaction in determining the phenotype. However, quantitative genetics infers genetic structure from phenotypic variation and attributes to the genes effects that are caused by interactions between physiology, development, the environment, and the genome. It misplaces causation. It can also only predict evolutionary change for about 5–10 generations. There is some truth on both sides. One must know what a tool is good for and when to use it.

CHAPTER SUMMARY

Quantitative genetics models the inheritance of continuously varying traits. The variation is continuous because reaction norms blur the phenotypic differences among genotypes and because quantitative traits are affected by the summed, small effects of many genes. Genetic effects and environmental effects combine to produce phenotypic variation.

Heritability is the portion of phenotypic variance attributed to variation among individuals in breeding value. It can be measured either through resemblance of relatives or through the response to a known selection pressure. The response to selection, the product of the heritability and the selection differential, governs the rate at which a single trait will evolve in the short term.

Life history traits—the components of fitness—are always under selection, which reduces genetic variation (lowers heritabilities). While lower than the heritabilities of morphological traits, the heritabilities of life history traits are larger than expected. Five explanations have been suggested. (1) Mutation–selection balance maintains some genetic variability, but not this much. (2) Heterogeneous environments maintain genetic variation through changes in selection pressure. (3) Heritable variation is expressed in the lab but not in the field, i.e. the heritabilities really are low where it counts. This cannot always be the case. (4) Negative genetic correlations between components of fitness maintain variation. (5) Some traits have very flat fitness profiles and therefore may be almost neutral over part of their range of phenotypic variation.

Genetic covariance between two traits is the portion of phenotypic covariance attributed to covariation in the breeding values of the two traits. The mechanisms underlying it are pleiotropy and linkage disequilibrium. Genetic correlations of life history traits have been negative in most selection experiments but mixed in experiments that estimate resemblance of relatives. Genetic correlations change from sample to sample, from population to population, from environment to environment, over the course of development, and from species to species. When genetic correlations change from positive to negative across environments, they reverse the correlated response to selection.

Plasticity has been quantified as the G × E component of variation and as the slopes of linear reaction norms. It is also the cross-environment correlation of the expressions of the same trait in two environments, and the breadth of tolerance of a genotype to variation in an environmental factor. Each of these approaches is best at answering a different type of question.

Phenotypic plasticity modulates the expression of genetic variance and covariance and can change the sign of the genetic covariance between two traits among the patches of a heterogeneous environment. Such effects play an important role in predicting response to selection. When several plastic traits change in a regular pattern, the overall plastic response is integrated. Compensation among fitness components is an important example of such integration.

Phenotypic correlations affect selection pressures and genetic correlations affect the evolutionary response. With strong negative *phenotypic* correlations between traits, the total selection pressure on a trait can be negative due to the indirect contributions of the correlated traits even though the direct selection pressure is positive. With strong negative *genetic* correlations between traits, the response of all of them to selection can be negative even though selection is directional for increasing values of each of them.

RECOMMENDED READING

Barton, N. H. and Turelli, M. (1989). Evolutionary quantitative genetics. *Ann. Rev. Genet.* **23**, 337–70.

Bioscience. (1989). Vol. 39, No. 7 (July/August).

Bradshaw, A. D. (1965). The evolutionary significance of phenotypic plasticity in plants. *Adv. Genetics* **13**, 115–55.

Charlesworth, B. (1987). In *Sexual selection: testing the alternatives.* Dahlem Conference (eds J. W. Bradbury and M. B. Anderson), pp. 21–40. (John Wiley & Sons, New York).

Falconer, D. S. (1989). *Introduction to quantitative genetics.* 3rd edn. (Longman, London).

Mitchell-Olds, T. and Rutledge, J. J. (1986). Quantitative genetics in natural plant populations: a review of the theory. *Am. Nat.* **127**, 379–402.

Schlichting, C. D. (1986). The evolution of phenotypic plasticity in plants. *Ann. Rev. Ecol. Syst.* **17**, 667–93.
Schmalhausen, I. I. (1949). *Factors of evolution.* (University of Chicago Press, Chicago). (Reprinted 1986).

PROBLEMS

(* = adapted from Falconer (1983) *Problems in quantitative genetics*. Longman Scientific and Technical, Burnt Mill, Harlow, Essex.)

1. Reaction norms transform environmental into phenotypic variation. Work out the transformations implied by linear reaction norms for two genotypes with different slopes. Do the same for two sets of two curvilinear reaction norms with different types of curvature.

2*. For a quantitative trait affected by three loci, each of which has two alleles with complete dominance, what is the frequency distribution of the genotypic classes? Recessive homozygotes reduce the value of the trait by 1 unit. The frequency of recessive alleles at each locus is 0.4.

3*. If the influence of environment and dominance can be neglected, then what is the value of the heritability estimated from:

Regression of offspring on father	= 0.18
Regression of offspring on mother	= 0.22
Regression of offspring on parental mean	= 0.41
Correlation of full-sibs	= 0.27
Correlation of half-sibs	= 0.19
Regression of female offspring on mother's sister	= 0.07
Regression of daughters on mothers within fathers	= 0.11

4*. In Boag and Grant's (1978) study of morphological variation in a population of *Geospiza fortis*, one of Darwin's finches in the Galapagos Islands, he found the following relations for bill depth. Are these data consistent, and how would you interpret them?

Regressions, ±standard errors	
Offspring–mid-parent	0.82 ± 0.15
Offspring–father	0.47 ± 0.17
Offspring–mother	0.48 ± 0.13
Correlations, ±standard errors	
Full-sibs	0.71 ± 0.12
Father–mother	0.33

5*. In 1977, the birds discussed in Problem 4 suffered severe mortality from drought. They are seed-eating birds, and the seeds available during the drought were especially large and hard. The survivors were larger and had larger bills than did average birds before the drought. Mean bill depth before the drought was 9.42 mm ($n = 642$). After the drought it was 9.96 mm ($n = 85$). Given the data in Problem 4 and this degree of selection, what change in bill depth would you predict?

6. Make three plots in each of which you sketch three reaction norms for a single trait. In the first plot, depict the situation with no G × E interactions. In the second, depict G × E interactions that occur when reaction norms do not cross. In the third, do the same for reaction norms that do cross. What is the significance of each for response to selection?

7. What implications do reaction norms and genotype × environment interactions have for the definition of breeding value?

8. If genetic covariances between reproductive effort and adult survival change from positive to negative as the food supply changes from abundant to scarce, how does that affect the measurement of the cost of reproduction?

9. Given that all phenotypic plasticity is a mixture of adaptive genetic modifications imposed on the unavoidable physicochemical properties of living matter, suggest ways in which one might discover what part was genetically modified and what part was unavoidable.

10. Given a positive genetic covariance between litter size and survival rates in mice in the lab, but a negative phenotypic correlation in the field, what would you conclude? What observations or experiments would you make next?

11. What physiological mechanisms produce pleiotropic effects that lead to genetic correlations? How should we interpret the genetic covariance of life history traits whose growth and development are regulated by hormones?

12. Write a program to simulate the quantitative

genetics of a population of 200 titmice (male and females) with linear reaction norms for clutch size. Let 10 loci with two alleles at each locus with no dominance affect the slope of the reaction norms, and another 10 loci with similar effects affect the intercepts. Half the population lives in oak woodlands where the optimal clutch size is 10. The other half lives in mixed pine and beech where the optimal clutch size is 6. Assume that the birds breed once in their first year of life, then die, and that mortality rate is adjusted each year so that precisely the same number of males and females survive to breed each year. Within that constraint, local movement mixes the offspring at random between the two woodlands each year before they mate. Initialize the programme with values for the average effects so that the mean clutch size in both environments is 8 and the frequencies of alleles at all loci is 0.5. Assume a per-trait mutation rate of 0.01.

(a) How long does it take the population to evolve to the phenotypic optimum in both environments?
(b) How does your answer differ for complete dominance?
(c) Estimate the heritability of clutch size from parent–offspring regressions *for those parents and offspring that bred in the same environment*. How do heritabilities differ between environments and over time?
(d) How much genetic variation is maintained in the long run for per-trait mutation rates of 0, 10^{-1}, 10^{-4}?
(e) How do your answers to (a) and (d) differ if only 25 per cent of the birds live in the oak woodland? Only 5 per cent?

4

TRADE-OFFS

It would be instructive to know not only by what physiological mechanisms a just apportionment is made between the nutriment devoted to the gonads and that devoted to the rest of the parental organism, but also what circumstances in the life-history and environment would render profitable the diversion of a greater or lesser share of the available resources towards reproduction

R. A. Fisher, 1930

The optimal animal, born with some amount of energy, proceeds through life gaining and expending energy according to some schedule that maximizes its total reproductive output

T. W. Schoener, 1971

You can't always get what you want, but if you try sometimes you might find you get what you need

M. Jagger, 1969

CHAPTER OVERVIEW

Trade-offs are the linkages between traits that constrain the simultaneous evolution of two or more traits. The chapter opens with examples of well-documented trade-offs, then distinguishes between physiological trade-offs—energy allocations between two or more functions competing for the same resources within a single individual—and evolutionary trade-offs, defined by manipulation experiments on phenotypes and by the correlated response to selection on populations.

Some cautionary remarks are given on the interpretation of evidence. Within one environment, individual variation in energy acquired and in energy allocated can determine whether the phenotypic relation between the two traits is positive or negative. Crossing reaction norms—genotype × environment interactions—can have the same effect when trade-offs are compared between environments.

At least 45 trade-offs are readily defined between life history traits. Of these, the most studied are those between current reproduction and survival, current reproduction and future reproduction, reproduction and growth, reproduction and condition, and number and quality of offspring. Examples of each are given, and the evidence pro and contra is summarized.

INTRODUCTION

Trade-offs have a central role in life history theory. They have been measured as correlations in the field, through experimental manipulations in laboratory and field, as phenotypic correlations in the laboratory, and as genetic correlations. Bell and Koufopanou (1986) recently reviewed the evidence. The measurement of trade-offs has attracted criticism (Tuomi *et al.* 1983; Partridge

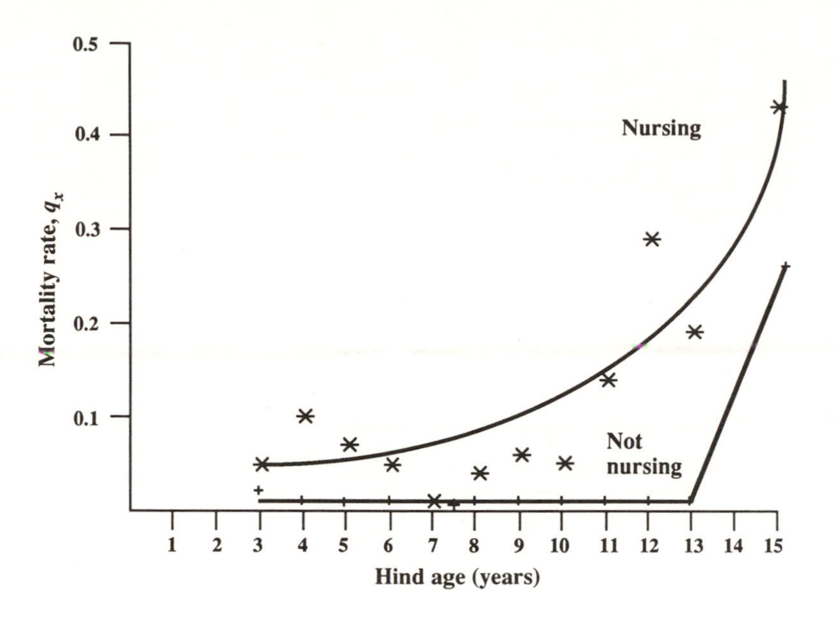

Figure 4.1 The age-specific mortality rates for nursing and non-nursing females are low and not very different for the first ten years of life. After that point, the mortality rates of reproductive hinds climb sooner and more rapidly than do those of non-reproductive hinds (after Clutton-Brock *et al.* 1982).

1987) and controversy (Reznick *et al.* 1986; Bell 1984*b, c*).

Two methods for measuring trade-offs have proven reliable and informative. First, one can select on one trait and measure the correlated responses in other traits. This is the approach taken in artificial selection experiments (Chapter 3 and Appendix 2). Second, one can manipulate the phenotype and study the consequences in the same individuals. This approach is taken in experiments on clutch size and egg size (Chapter 7).

We begin with some examples of suggestive phenotypic correlations.

Red deer (*Cervus elaphus*)

In red deer (Fig. 4.1), adult mortality is higher in females that are nursing offspring (Clutton-Brock *et al.* 1982). Both a physiological and an ecological mechanism are involved. Lactating females have smaller fat reserves and higher over-winter mortality, and hinds with fawns must compete with their offspring for forage after weaning. These effects combine to produce higher mortality among females nursing calves than among those that are not. The effect depends on age and first becomes significant after the females are 10 years old.

Beech trees (*Fagus sylvatica*)

Like many trees, beech have 'mast' years of heavy seed production followed by years in which they set less seed (Fig. 4.2). In mast years, the diameter of growth rings in the wood may be only half that of a normal year. In the case depicted, seed yield was negatively correlated with growth ($r = -0.54$, $p < 0.001$).

Field grasshoppers (*Chorthippus biguttulus*)

Kriegsbaum (1988) compared clutch sizes and egg weights in females caught in the field with females from the same population reared in the laboratory. Clutches were significantly smaller and eggs were larger in the laboratory sample (Fig. 4.3). These grasshoppers die at the end of the season, probably make a maximal reproductive investment, and must divide it between numbers and sizes of offspring. Some cue—perhaps poorer nourishment, here indicated by the lower total mass of eggs produced per female in the lab—may stimulate the production of larger eggs in the laboratory.

Neotropical frogs

Male frogs calling for mates in the neotropics also attract predatory bats. The rate at which

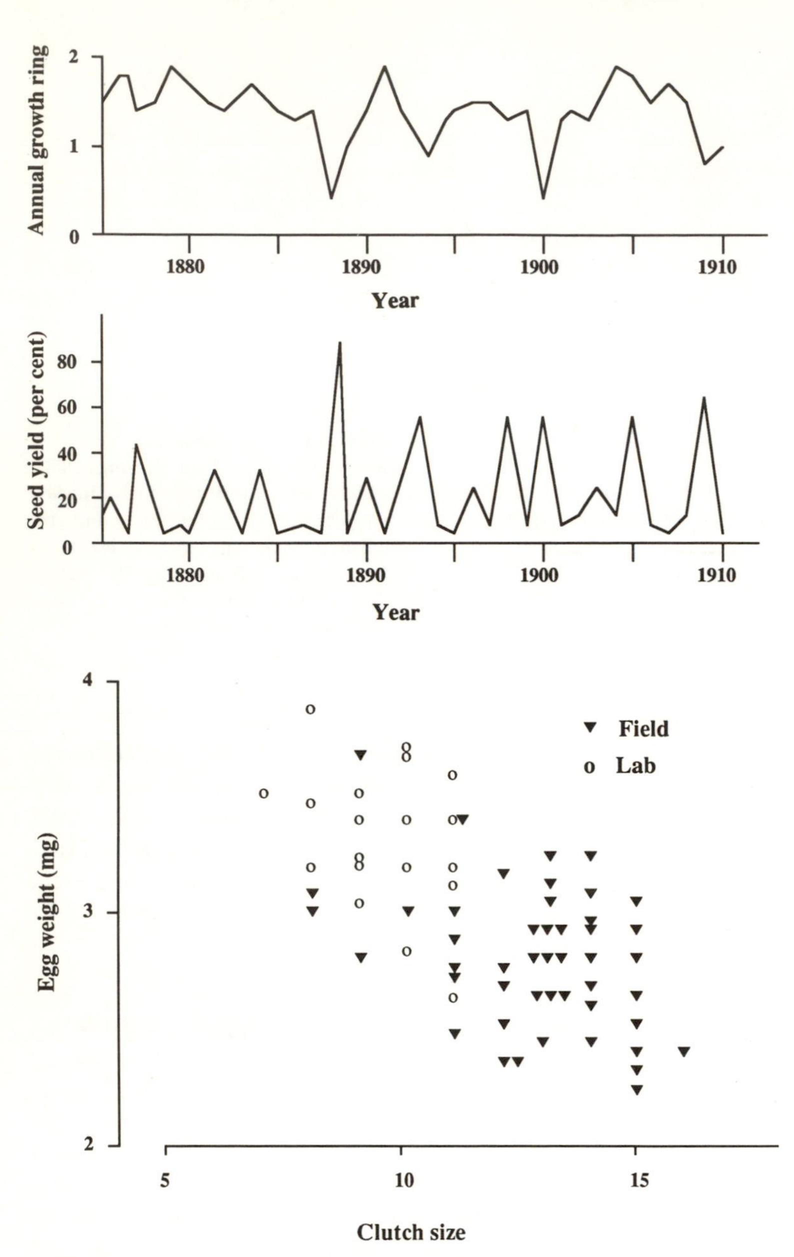

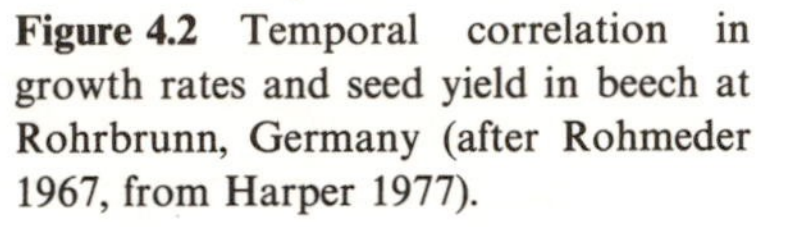
Figure 4.2 Temporal correlation in growth rates and seed yield in beech at Rohrbrunn, Germany (after Rohmeder 1967, from Harper 1977).

Figure 4.3 In the grasshopper *Chorthippus*, individuals from the same population raised under different environmental conditions display a trade-off between egg size and egg number (after Kriegsbaum 1988).

fringe-lipped bats (*Trachops cirrhosus*) capture frogs is higher when the frogs are calling. The bats can discriminate the calls of different species and different sizes of frogs, and they prefer palatable species small enough to handle. The males trade off the attractions of reproduction with the risk of death (Tuttle and Ryan 1981).

Trade-offs are quite diverse.

Physiological, microevolutionary, and macroevolutionary trade-offs

The word 'trade-off' is sometimes loosely used to indicate almost any kind of negative relationship between two traits. To keep clear about what is under discussion, I use words in the following senses.

Physiological trade-offs are caused by allocation

decisions between two or more processes *that compete directly with one another for limited resources within a single individual.* In red deer females, a physiological trade-off causes the mortality cost of reproduction by diverting into milk resources that might have gone into fat reserves for winter. In beech trees, a physiological trade-off causes the growth cost of reproduction by diverting into seeds materials and energy that would have gone into wood and leaves. In field grasshoppers, making larger eggs with limited materials and energy implies fewer eggs.

Microevolutionary trade-offs are broader than physiological trade-offs, which they include. Microevolutionary trade-offs occur in a population under selection when a change in one trait that increases fitness is linked to a change in another trait that decreases fitness. If such trade-offs did not exist, selection would have driven all traits correlated with fitness to limits imposed by history and design. The fact that many life history traits are maintained well within those limits suggests that microevolutionary trade-offs must exist (Stearns 1989*b*).

Microevolutionary trade-offs are defined by the response of populations to selection. Physiological trade-offs are involved in almost all microevolutionary trade-offs, but it is possible for a physiological trade-off to exist without any microevolutionary trade-off in the population. Consider a population of grasshoppers that has a reaction norm for number and size of offspring, like the example above, but with no genetic variation for the reaction norm. When poorly nourished they produce fewer, larger offspring, and when well nourished they produce more, smaller offspring. The physiological trade-off is well defined, but the population would not respond to selection because there is no genetic variation for the reaction norm. A physiological trade-off is one cause of individual plasticity. A microevolutionary trade-off is a population-level response to selection on genetic variation in physiological trade-offs.

This definition requires that one be able to measure a response to selection before calling a trade-off microevolutionary. Whether or not physiological trade-offs are genetically variable today, they may well have been genetically variable in the past but were fixed because they were the optimal allocation rules then. If they are not genetically variable today, they may have been maintained by lineage-specific selection from then until now, or they may have been fixed, integrated into the organism by subsequent evolution, and now can no longer be changed.

Macroevolutionary trade-offs are defined by comparative analysis of variation in traits among independent phylogenetic events (Chapter 5). To take an extreme case, imagine two traits that are not plastic and for which there is no genetic variation—fixed within each species and not varying across environments. Nevertheless, when one compares species within genera, or genera within families, the two traits are negatively correlated. Furthermore, the traits are apparently adaptively associated with habitats. We assume that such patterns could only exist because physiological and microevolutionary trade-offs that existed in the past have left their traces in an entire lineage even though we cannot now measure them within species. (The case described is unrealistic, for most macroevolutionary trade-offs are associated with physiological and microevolutionary trade-offs, but these need not be of the same strength or sign.)

By identifying the comparative pattern within which the intraspecific trade-offs occur, we identify conditions common to whole lineages. This gives greater generality. The macroevolutionary trade-offs also suggest which physiological and microevolutionary trade-offs might need further study, such as the deviations of an intraspecific trade-off from the lineage average.

PHYSIOLOGICAL TRADE-OFFS

Physiological ecology demonstrates the lineage-specific effects that constrain microevolutionary optimization—condition thresholds for breeding (e.g. a critical level of stored fat), growth rates as a function of body size, limits on maximum performance, and the amount of energy that it takes to produce a gram of offspring. All are fairly constant within species but vary among lineages (Drent and Daan 1980). Physiology underlies phenotypic correlations and is the filter through which genetic correlations are expressed. The

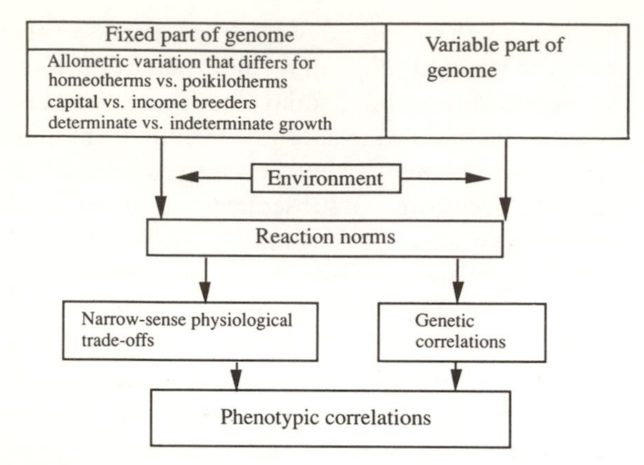

Figure 4.4 The fixed part of the genome carries lineage-specific effects important in physiological ecology. Some of these fixed effects are expressed as physiological trade-offs that combine with genetic correlations coming from the variable part of the genome to produce phenotypic correlations. The type of trade-off expressed depends on allometry, mode of temperature regulation, mode of reproductive investment, and mode of growth.

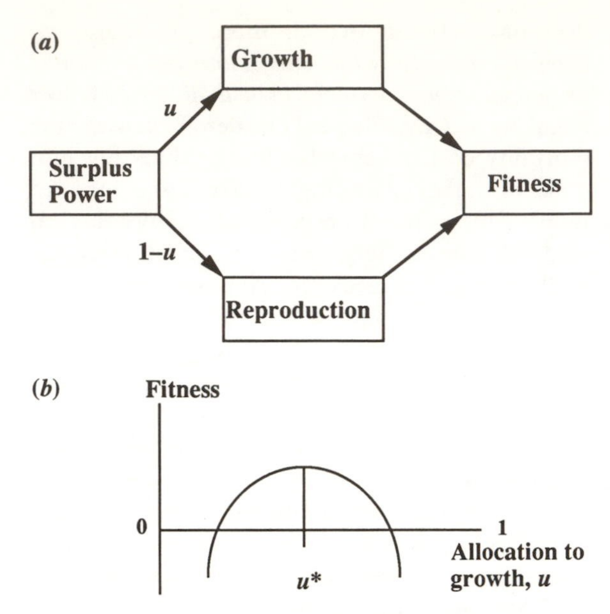

Figure 4.5 (*a*) The simplest model of allocation of surplus energy between growth and reproduction. Each allocation results in a single value of fitness. (*b*) The intermediate allocation that produces the highest fitness is optimal (after Sibly and Calow 1986).

lineage-specific part of physiology constrains evolution (Fig. 4.4).

The genome can be conceptually divided into a part carrying lineage-specific effects characteristic of a species or larger group, and a variable part carrying the differences among individuals measured by quantitative genetics. Physiological trade-offs are often characteristic of large categories, such as poikilotherms vs. homeotherms, capital vs. income breeders, and determinate vs. indeterminate growth.

Physiological tradeoffs constrain adaptation

If an organism can only acquire a limited amount of materials and energy for which two processes compete directly, then an increase in materials and energy allocated to one must result in a decrease in materials and energy allocated to the other. This Principle of Allocation was asserted by Levins (1968) and discussed by Sibly and Calow (1986). The energy left after the costs of standard and active metabolism have been met is *surplus power* (Ware 1980). If a proportion u of the surplus power in food is allocated to growth, then $1 - u$ can be allocated to reproduction. Because growth and reproduction both contribute to fitness (Fig. 4.5), we want to know, *what value of u will maximize fitness?*

The interaction of growth and reproduction in determining fitness can be calculated from the Euler–Lotka equation with appropriate assumptions. Given values of fitness for every pair of values of growth and reproduction, one can plot the fitness values on the growth–reproduction plane and draw contours through points of equal value. These contours are lines of equal fitness. If there were a single combination of growth and reproduction that produced the highest fitness, the contours of fitness values would define a peak on the plane surrounded by slopes of decreasing fitness (Fig. 4.6). Placing the trade-off on to the plane defines a straight line with slope -1, since allocation to growth $= 1 -$ allocation to reproduction. Where the trade-off intersects with the fitness contour of highest value, fitness is optimized.

By measuring the slope of the trade-off and calculating the fitness contours, we can predict an optimal allocation between two traits.

Caveats

At times physiological tradeoffs are not as tight as the Principle of Allocation would suggest:

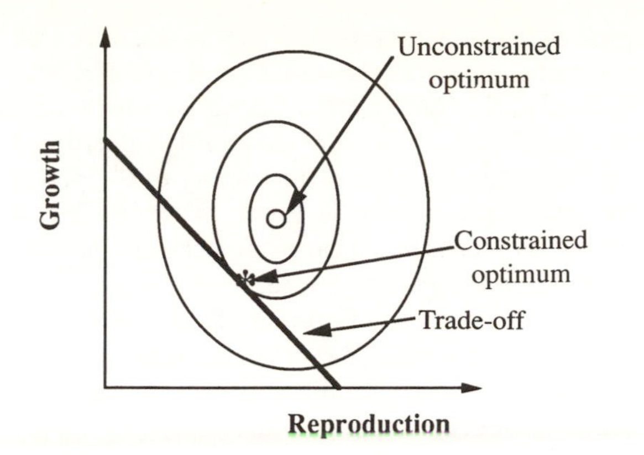

Figure 4.6 On a selective landscape, the trade-off constrains the optimal attainable value. The ovals define fitness contours (after Sibly and Calow 1986).

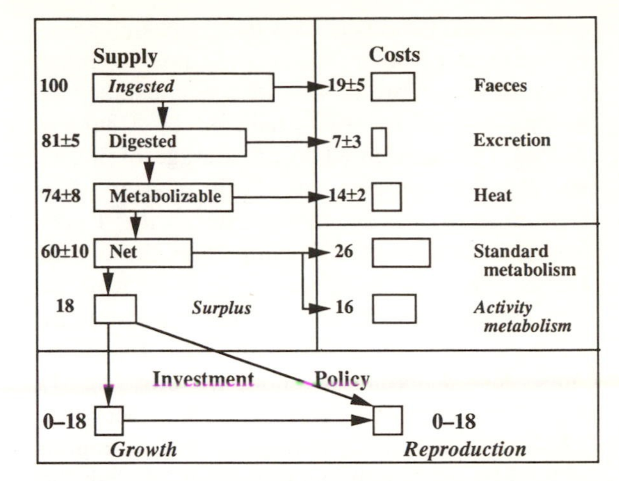

Figure 4.7 The fate of ingested energy. The figures represent average efficiencies and probable ranges for a carnivorous 30 cm fish swimming at optimal foraging speed for 15 hours per day. The processes that are probably fixed within lineages are indicated in normal type. Those that are more likely to vary among individuals within populations and to be locally adapted are indicated in *italics* (after Ware 1982).

1. If a resource other than energy is limiting—time or a mineral nutrient, for example—then energy allocations do not trade off.
2. If energy is limiting, but the functions differ in their sensitivity to variation in the amount of energy allocated, then although the trade-offs are always there, they will be insignificantly small under some circumstances and quite large under others (Tuomi *et al.* 1983).
3. If certain amounts of energy are always devoted to given activities, then allocations are made with what is left over after the high priority needs have been filled. Many trade-offs only appear when food is limited, and a threshold of energy reserves must often be reached before breeding will begin (Drent and Daan 1980).
4. In small, homeothermic species with high metabolic rates and no fat reserves for reproduction, rapid physiological turnover wipes out the traces of reproduction much more rapidly than in large species with fat reserves. Thresholds for breeding condition are more likely in the latter than in the former.

The flow of materials and energy

The physiological model of an organism centres on how materials and energy are acquired, processed, and utilized. It is based on rates—feeding rates, metabolic rates, excretion rates, growth rates, reproductive rates. The energy used per unit of time is the critical measurement. Ware (1982) estimated energy flows for a carnivorous fish, assuming that it swims at its optimal foraging speed (Fig. 4.7). Some parts of physiology are flexible and might adapt the organism to local conditions. Others are not so flexible. Which are flexible, and which are not, depends on the lineage. Digestive efficiency, excretion, and basal metabolism are fairly constant within most groups, but the amount of food eaten, the efficiency with which it is collected, the proportion of net power devoted to activity, the surplus power available for growth and reproduction, and the policy used to allocate that surplus can all vary among individuals.

Feeding constraints and efficiencies

Feeding and activity patterns, often studied in behavioural ecology, determine the amount of energy acquired and connect physiological ecology, behavioural ecology, and life history evolution.

An example of flexible input: foraging in kestrels

Male kestrels are exclusively responsible for feeding the female, and then the nestlings, from about two weeks before the first egg is laid until about two weeks after the eggs hatch. Masman *et al.* (1989) measured the parental effort of male kestrels as total daily flight time and energy expenditure.

Males with broods from 4 to 7 chicks all spent an average of 4.75 hours per day in flight independent of brood size. (Such constancy of reproductive effort was predicted by Charnov (1986)). This amounted to an average of 382 kJ per day for foraging—about a quarter of a medium-sized candy bar. Males with larger broods hunted more efficiently than those with smaller ones and were able to give each chick the same amount of food (63 g per day) while making the same effort as males with smaller broods. Chicks fed *ad libitum* in the lab ate only slightly more than those in natural nests (67 g per day).

Masman *et al.* then manipulated the demand for food by rotating hungry nestlings through the nest (for young nestlings) or by removing food as it was delivered (for older nestlings). The males increased their rate of delivery almost three times, to an average of 646 g per nest per day, and nearly doubled their flight time from 4.5 to 8.4 hours per day. They sustained this effort for up to eleven days with an energy expenditure 3–4 times basal metabolic rate, a level that seems to be an upper limit for many organisms. (Cyclists in the Tour de France sustain that level of energy output.) Even under food stress, half the daylight hours were not used for foraging.

Thus the mean daily energy expenditure of males is adjusted to their individual foraging efficiencies and is well below the maximum they can sustain. Their maximum energy expenditure under food stress is not limited by the hours in the day or by the energy in the environment. Therefore kestrels must incur costs other than energy expenditure—such as parental survival.

Foraging and reproductive success in geese

For more than 10 years, R. H. Drent (pers. comm.) has been studying the relation between foraging efficiency and reproductive success in geese on an island off the north coast of Holland. These geese form pairs before they arrive on their way north to breed in Spitzbergen and Siberia. In Holland, they are caught and banded, then followed individually as they forage in their natural habitat. The efficiency with which a female can graze, and the quality of the forage she eats, both depend on the dominance of her mate. Dominant males protect their females and get them favoured positions in ungrazed areas. Females paired with subordinate males cannot spend as much time foraging because they are frequently interrupted by dominant males, and the quality of the plants on which they graze is lower. The females of dominant males are nearly 10 per cent heavier when they fly north to breed.

This difference in foraging efficiency and fat stores in the spring has two important consequences. Females of dominant males return in the fall with more offspring; the females that were lighter in the spring sometimes return with none. And females paired with subordinate males change mates more frequently from year to year than those paired with dominant males.

Some birds vary individually in foraging efficiency. These differences translate into differences in reproductive success. The standard partitioning of food intake into percentages allocated to different functions must be supplemented by recognizing that individuals vary considerably in the amount eaten and proportion of net power devoted to activity, including acquiring food.

Defining reproductive effort

The notion of reproductive effort is important in life history evolution when it measures the costs of reproduction. Attempts to measure reproductive effort have been variously naïve. The ratio of reproductive biomass to the mass of the parent, or of calories to calories, are static measures that do not represent the proportion of energy flowing through the organism that is devoted to reproduction. As Hirshfield and Tinkle (1975, p. 2230) point out,

(i) Individuals of two species could devote the same quantity of energy to reproduction at equivalent body sizes, but differ greatly in the absolute amount of energy gathered or in the time during which it was gathered. In such cases the true proportional allocation to reproduction (reproductive effort) would be unequal, whereas the ratio of clutch weight or calories to body weight or calories would be identical.

Table 4.1 An incomplete trade-off matrix for life history traits. Parental survival and parental growth: from the current reproductive episode to the next. Parental condition: taken between this reproductive episode and the next, preferably on a standard date critical for survival (e.g. late winter). Size of offspring: to be reported as a series of sizes for species with parental care (birth, fledging, independence). Additional columns could be added to represent the trade-offs associated with maturation. **Intra-individual** trade-offs are indicated in **bold** type. *Intergenerational* trade-offs are indicated in *italics*. From Stearns (1989*b*)

		Trait 2								
Trait 1		PS	FR	PG	PC	NO	SO	OG	OC	OS
Current reproduction	CR	**1**	**2**	**3**	**4**	*5*	*6*	*7*	*8*	*9*
Parental survival	PS	—	**10**	**11**	**12**	*13*	*14*	*15*	*16*	*17*
Future reproduction	FR		—	**18**	**19**	*20*	*21*	*22*	*23*	*24*
Parental growth	PG			—	**25**	*26*	*27*	*28*	*29*	*39*
Parental condition	PC				—	*31*	*32*	*33*	*34*	*35*
Number of offspring	NO					—	*36*	*37*	*38*	*39*
Size of offspring	SO						—	**40**	**41**	**42**
Offspring growth	OG							—	**43**	**44**
Offspring condition	OC								—	**45**
Offspring survival	OS									—

(ii) Even if energy budgets were identical for two species, a comparison of clutch weight/body weight ratios might not provide comparable measures of effort if the species differed in the number of clutches produced in a single season.

When individuals differ in foraging efficiency, they differ in how much it costs them to acquire the same amount of food. They may also differ in their ability to detect predators, in the risks they undertake when they forage, and in their upper sustainable limits of energy output. An energy output that would mean quick death for one might be sustained by another for days. When such factors are important, simply determining the proportion of energy flow devoted to reproduction will not measure reproductive costs. No physiological measure of reproductive effort takes all these factors into account, but measuring energy expenditures with doubly-labelled water on marked individuals followed for long periods in the field brings much insight (cf. Masman *et al.* 1989).

Given the problems with measuring reproductive effort physiologically, must one measure it at all? From the point of view of demographic theory, the answer is no—we need the quantity of reproduction and the cost of reproduction but not reproductive effort (Bell, pers. comm.). The reason to retain the term and continue work on reproductive effort lies in the potential of physiology to explain the constraints on life history variation. Understanding the relation between reproductive cost, measured in changes in birth and death rates, and reproductive effort, measured in physiological allocations (e.g., Kenagy *et al.* 1990), remains worthwhile.

MICROEVOLUTIONARY TRADE-OFFS: INTRA-INDIVIDUAL AND INTERGENERATIONAL

By listing the major life history traits in parents and offspring, once as rows and once as columns, we can build a matrix of trade-offs between traits taken two at a time. The entries in Table 4.1 number the pairwise trade-offs for convenient reference. Some entries are not independent (current reproduction = number of offspring × size of offspring), not all possible trade-offs are listed, and no such general description can cover all organisms. Of the 45 possible trade-offs among these 10 traits, only five have received much attention: numbers 1 (current reproduction vs.

survival), 2 (current vs. future reproduction), 3 (current reproduction vs. parental growth), 4 (current reproduction vs. parental condition), and 36 (number vs. size of offspring).

Others could also produce important effects. Does parental growth trade off with offspring condition (number 29)? That would be plausible in organisms with indeterminate growth and parental care and would be a type of parent–offspring conflict. In seabirds in the family Alcidae, the nest is an environment of good survival and poor growth; the sea is an environment of poorer survival and better growth (Ydenberg 1989). This is an example of trade-off 44.

Because of the historical emphasis on energy flow within individuals rather than among the members of a family, *intra-individual* trade-offs have played a much larger role to date than *intergenerational* trade-offs, which can be quite real. In the poeciliid fish *Heterandria formosa*, where number of broods and size of offspring trade off, larger offspring have a better chance of survival (Henrich 1988). Similar effects exist in wild radishes (Stanton 1984). Here an *intra*-individual trade-off between brood interval and offspring size becomes an *inter*generational trade-off between brood interval and offspring survival.

WHY MIGHT WE OBSERVE THE 'WRONG' TRADE-OFF?

Most studies that failed to report trade-offs where one might expect to find them worked with variation among individuals, genetic or phenotypic (Appendix 2). There are four reasons why one might not find the trade-off expected.

Phylogenetic fixation

If the trade-off has been fixed in the species as an invariant physiological mechanism, it will constrain evolution through phenotypic but not genetic correlations. It will be observed when measured across an environmental range that elicits plastic responses in both traits.

Income vs. capital breeders

In some organisms, such as small birds and mammals, there is no pool of shared resource because high metabolic rates rapidly eliminate the physiological traces of reproduction. A humming-bird must use what it has eaten in the last few hours to produce its eggs. It uses current income for reproductive investment. If great tits are income breeders, it is not surprising that Pettifor *et al.* (1988) found no evidence for a trade-off between clutch size and cost of reproduction in great tits and more surprising that Nur (1984*a*, *b*) did find such evidence in blue tits. Sibly and Calow (1986) refer to income breeders as having 'direct costing'—the cost of reproduction being drawn out of current income.

Capital breeders store energy to use later for reproduction. Current and future reproduction are linked through the shared energy store. Sibly and Calow (1986) refer to capital breeders as having 'absorption costing'—the cost of reproduction is drawn against physiological savings. Most life history theory assumes absorption costing. The direct costing used by income breeders relates more directly to behavioural decisions, such as how long to forage each day, than to costs of reproduction.

Income and capital breeders are the ends of a continuum; some species mix the two modes. The poeciliid fishes have the full spectrum from 'capital breeders', such as guppies (*Poecilia*) and swordtails (*Xiphophorus*), to 'income breeders', such as the least killifish, *Heterandria formosa* (Turner 1937). Comparative studies within this family could suggest reasons for the evolution of capital and income breeding (J. Travis, pers. comm.).

Variation in acquisition and allocation of energy

Third, individuals vary in energy acquired and how it is allocated. Van Noordwijk and de Jong (1986) point out a simple interaction that can lead to positive, zero, or negative correlations between the two traits in a physiological trade-off. Suppose that the energy acquired in food, A, is allocated within each individual only to reproduction, R, or to survival, S:

$$A = R + S. \tag{4.1}$$

This does not make explicit the ever-present variation among individuals. When individuals vary in the amount of energy acquired, A, and in the fraction allocated to reproduction, B, and we

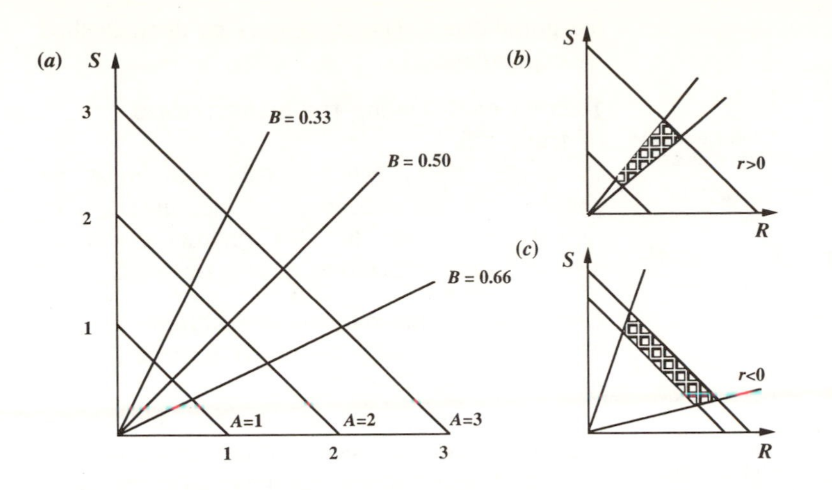

Figure 4.8 Consider a trade-off between reproduction (R) and survival (S). The total amount of energy acquired is set by A, where $A = R + S$. The fraction of energy allocated to reproduction is determined by B, so $R_i = B_i A_i$. In (*a*), the general possibilities are sketched. In (*b*), the variation among individuals in energy acquired (A) is large, but the variation in fraction allocated to reproduction (B) is small. The result is a positive correlation between reproduction and survival. In (*c*), the variation in A is small but the variation in B is large. The result is a negative correlation between reproduction and survival.

indicate each individual by a subscript i, then we have

$$R_i = B_i A_i \tag{4.2}$$

$$S_i = (1 - B_i)A_i. \tag{4.3}$$

Van Noordwijk and de Jong saw that the sign of the correlation between R and S depends on the degree to which each varies among individuals (Fig. 4.8).

They explain this result by an analogy with economics: 'If the budget is fixed, people spending more on housing should spend less on cars. In fact, the amount of expendable income is variable, and in many situations positive correlations are observed between the per-family expenses on housing and on cars. There is little problem in identifying rich and poor families. ... Where biologists have observed positive correlations between life history traits, they have often also identified individuals that perform well or poorly'.

If we observe a positive correlation where theory expects a negative one, this analysis suggests that we ask 'Do the organisms vary more in the amount of energy acquired than in the fraction of energy allocated to each function?'.

Genotype by environment interactions

We might also observe no trade-off or an unexpected one because of genotype × environment interactions. Microevolutionary trade-offs in life history traits result both from genetic correlations and from their expression. The mechanisms involved in the expression of genetic correlations include physiological trade-offs that allocate resources among reproduction, growth, maintenance, and storage. These allocations change across environments; they have reaction norms.

All three levels contribute to how a trade-off works. Natural selection acts on phenotypes; phenotypic correlations determine the trade-offs presented to natural selection. The response to selection depends on genetic variation; genetic correlations determine whether, and in what direction, a response to selection will occur. Genotype × environment interactions change genetic correlations across environments.

Strong, negative, phenotypic correlations between two traits with zero heritability produce no genetic response to selection. When a genetic correlation is positive under some conditions and negative under others, then only in the latter would we perceive a trade-off (Gebhardt and Stearns 1988). When an environmental factor affects one trait more than another, the phenotypic correlation between the two traits changes across environments and alters the selection gradients. For example, the phenotypic correlation between developmental rate and size at metamorphosis in frogs changes from positive to negative within a full-sib family when food level and distribution vary (Travis 1984). The environment can determine whether the trade-off appears at all.

Genotype by environment interactions could change the expression of genetic correlations between reproductive investment vs. survival (Fig. 4.9). Where genetic variation exists for reaction

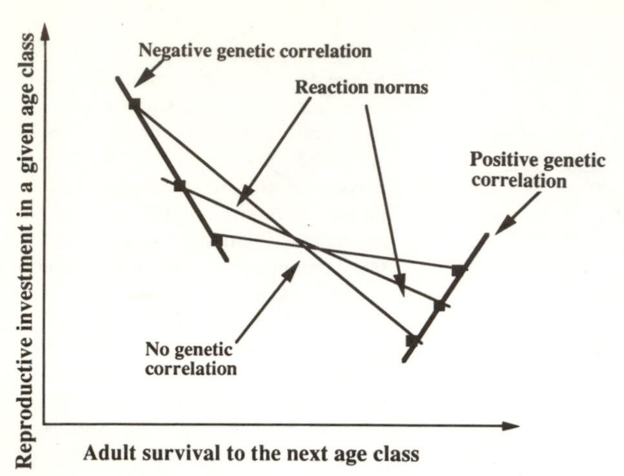

Figure 4.9 If one measures a trade-off between reproductive investment at one age and survival to the next across a range of environmental conditions, the overall phenotypic correlation may be negative, but within that phenotypic variation the expression of genetic correlations may be negative in some environments, zero or non-significant in some, and positive in others (after Stearns 1989*a*).

norms for traits like reproduction and survival, only the crossing norms should be maintained. Any that lay above the others in all environments would have been fixed by selection. The overall phenotypic trade-off is negative—it has a major axis running from the upper left to the lower right. Within that ellipse, however, are hidden important details.

In the left-hand portion of the diagram, with low survival and high investment, the genetic correlation between reproductive investment and survival is negative. In the right-hand portion of the diagram, with high survival and low investment, the genetic correlation is positive. In the middle, there is no genetic correlation between the two traits, heritabilities are low, and there will be no direct or correlated response to selection in such environments.

Figure 4.9 depicts one of the simplest types of genotype × environment interactions. In real data sets, genetic correlations may change back and forth from positive to negative repeatedly as an environmental factor changes. In some species, genetic correlations will be negative in well-fed and positive in poorly fed populations. In species with different physiologies, genetic correlations will be positive in well-fed and negative in poorly fed populations. These can only be distinguished by experiment.

Factors confounding the measurement of trade-offs

Partridge (1991) listed factors that must be controlled to get unambiguous measurements of reproductive costs; her list (below) could be applied to any measurement of a trade-off.

Genotype by environment interaction

The individuals being compared are reared in an environment to which they are not equally suited, or they lived in different environments. This causes problems when correlations are measured on individuals with different genotypes if the genotypic differences are not accounted for and when correlations between two life history traits are estimated in the field or under inadequately controlled conditions in the laboratory (cf. Fig. 4.9).

Phenotypic variation

The individuals differ phenotypically for both traits due to different experiences that have changed total reproductive potential.

Genetic correlations resulting from linkage disequilibrium

The individuals compared have different life history traits for independent genetic reasons because they came from different populations.

Caveat

Bear these comments in mind while reading the following descriptions of measuring trade-offs. These are good examples, but only rarely were confounding factors controlled. No one has measured in a single study the effects on trade-offs of individual variation in energy acquired and allocated and of genetic variation in allocations expressed across environments.

CURRENT REPRODUCTION VS. SURVIVAL

This was the first trade-off to be analysed: '... selection will adjust the amount of immediate reproductive effort in such a way that the cost in physiological stress and personal hazard will be justified by the probability of success' (Williams

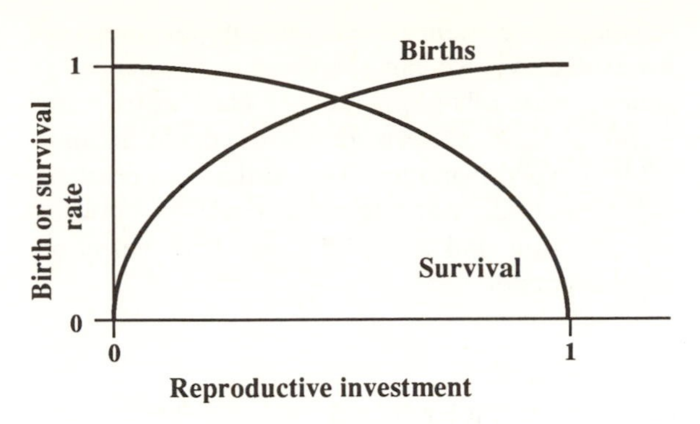

Figure 4.10 In Schaffer's model of reproductive trade-offs, reproductive investment varies from 0 to 1, at which point the organism dies (survival rate = 0) while producing the maximum possible number of offspring (birth rate = 1).

1966*a*, p. 172). Schaffer (1974*a*) plotted reproductive investment at one age against the probability of surviving to the next (Fig. 4.10).

Example 1: Rotifers

Snell and King (1977) measured age-specific fecundity and lifespan in 30 individuals from each of 21 clones of rotifers (*Asplanchna*) at three temperatures. Short- and long-lived rotifers differed in duration of reproductive period but not in age at maturity or length of post-reproductive life. Short-lived rotifers produced more offspring per day, but for fewer days, than did long-lived rotifers. The long-lived clones produced more total offspring per lifetime than the short-lived ones, but the offspring produced later contributed less to *r*. They concluded (p. 889) 'that reproduction is deleterious to future survival'.

Example 2: Male fruit flies

Partridge and Farquhar (1981) compared the longevity of males kept with eight virgin females with that of males kept with eight inseminated females. Males expend more effort on courting and mating virgin than mated females. The results were plotted against size to account for effects of size on survival. Males kept with virgins suffered higher mortality (Fig. 4.11).

Summary of the evidence

Many have tested the effect of mating on longevity. One suppresses reproduction and compares the longevity of virgins or individuals with few matings with the longevity of mated individuals

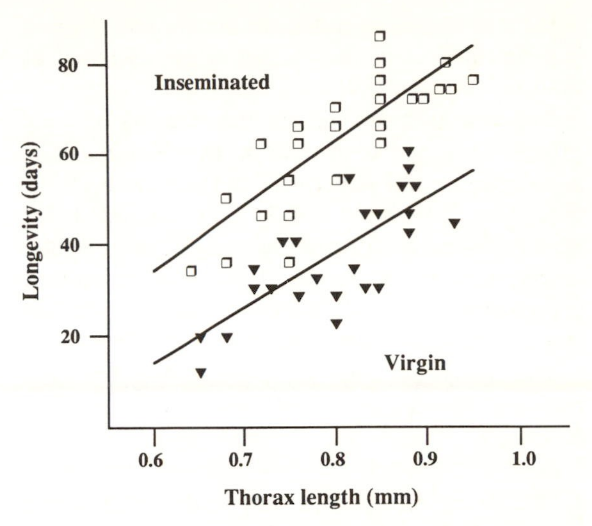

Figure 4.11 Male fruit flies kept with 8 virgin females per day lived about 20 days less than males kept with 8 inseminated females per day (after Partridge and Farquhar 1981).

or individuals with many matings (Appendix 2A). In arthropods and nematodes, virgins and individuals that mate less live longer (23 cases). In only one of four cases in rats and mice did virgins live longer. Results may depend on environmental conditions. Calow (1973) found that virgin *Corixa* lived longer than mated ones on a restricted food supply, but the effect vanished on an ample food supply.

The trade-off between reproduction and survival can also be measured as the survival rates of parents with different fecundities. Results have been mixed (Appendix 2B). In 22 cases where the correlation between fecundity and survival was measured in the laboratory, it was insignificant in 10, negative in seven, and positive in five. Of the 24 cases in unmanipulated field populations, it was insignificant in three, negative in 17, and positive in four. In 16 manipulated field populations, it was negative in six and insignificant in 10.

The most reliable studies manipulate field populations, but not all manipulations are equally relevant. When clutch sizes are manipulated, eggs are added to or removed from nests, increasing or decreasing the amount of work the parents have to do to fledge the clutch. If individuals vary in quality and lay the clutch that is optimal for them, then manipulations are sensible only when made relative to the clutch the female actually laid (e.g. Pettifor *et al.* 1988), not when assigned at random (e.g. Nur 1988).

Reproduction can also increase the risk of being eaten—documented in cladocerans (Mellors 1975; Koufopanou and Bell 1984) and copepods (Winfield and Townsend 1983) and suggested for lizards (Shine 1980), aquatic snakes (Shine 1988), and garter snakes. Brodie (1989) has shown that pregnant female garter snakes are less mobile, but they compensate by adopting more cryptic behaviour

The mixed evidence for a *phenotypic* trade-off between reproduction and survival suggests the following.

1. The trade-off appears only under food stress; the correlation was often negative when food was limiting and was zero or positive otherwise. Phenotypic correlations mislead when some common uncontrolled factor influences several life history traits in the same direction simultaneously (Partridge and Harvey 1985; Reznick 1985).
2. Phenotypic trade-offs tell us something about natural selection but little about the genetic correlations between fecundity and survival that determine the response to selection. A single genotype can be superior in a number of fitness components. When a new sample of fruit flies is brought into the laboratory from a natural population, some genotypes have higher fecundities *and* live longer than others. The phenotypic correlations between reproduction and survival are positive. The genetic correlations could still be negative when measured through breeding or selection experiments.
3. Genotype by environment interactions may change the genetic correlations from positive in some environments to negative in others.
4. Some species pay higher costs of reproduction than others because differences in their lifestyle more often generate constraints on time or energy. For example, bird species that have several broods each year are more likely to encounter conflicts with the requirements of moult, migration, and establishment of winter territories than single-brooded birds whose seasonal activities are well separated (Lindén and Møller 1989).

Since the evidence from phenotypic trade-offs is not decisive, we can return to the evidence for genetic trade-offs (Appendix 3). Here again there is no general pattern that one could assume without experimentation. The results from breeding experiments do not suggest a trade-off between reproduction and survival; those from selection experiments do.

CURRENT VS. FUTURE REPRODUCTION

The direct influence of current on future reproduction is just as important a determinant of fitness as the effect of reproduction on survival.

Example 1: Annual meadow grass

Law *et al.* (1979) found a strong negative relation between the number of inflorescences produced in the first and second seasons of reproduction in annual meadow grass, *Poa annua* (Fig. 4.12). Plants flowering heavily in their first season produced fewer inflorescences in their second season than plants flowering only lightly, if at all.

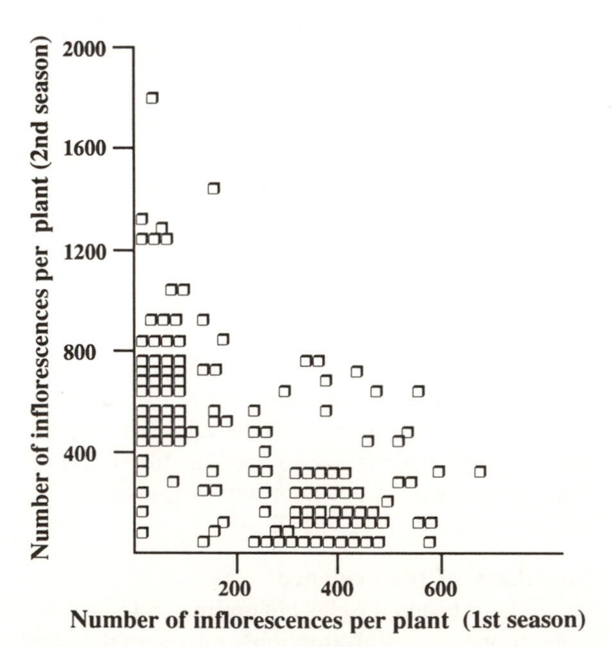

Figure 4.12 Flowering entails reproductive costs in annual meadow grass (after Law *et al.* 1979).

Example 2: Water fleas

Daphnia are capital breeders despite their small size. In *Daphnia galeata mendotae*, Goulden and Hornig (1980) were able to measure energy reserves as oil droplets. Females with larger reserves had higher fecundities during subsequent periods of food shortage, indicating that the resources acquired during one reproductive attempt were later shared with the next reproductive attempt.

Example 3: Collared flycatchers

Gustaffson and Pärt (1990) manipulated the clutch sizes of one-year-old collared flycatchers in a ringed population and followed their reproductive performance for the next three years. Birds whose clutches were enlarged in their first year of life had clutches smaller than the controls for the next three years. There was a link between early effort and late performance.

Summary of the evidence

Of 13 cases in which the correlation between early and late fecundity was measured under favourable conditions in the laboratory, it was insignificant in one, positive in nine and negative in three (Appendix 2(*c*)). Of five cases in unmanipulated field populations, it was negative in four and positive in one. In manipulated field populations, it was negative in six cases and not significant in two. The evidence for a trade-off at the phenotypic level is mixed, but when manipulations have been made, a trade-off has usually been found. Nutritional state and genetic variation affect the results and must be measured.

The trade-off between early and late reproduction is widespread in iteroparous (polycarpic) plants. 'The fruit grower ... knows that after a particularly good crop he may expect one or two years of poor crop and that by reducing the load of fruit in a bumper year he can increase the chance of a good crop in the following year' (Harper 1977, p. 654).

Breeding experiments have rarely found a negative genetic correlation between early and late reproduction. Selection experiments found one 11 out of 12 times (Appendix 1).

REPRODUCTION VS. GROWTH

Investment in reproductive activity can reduce growth and thereby future reproductive success. Larger organisms usually produce more offspring, offspring of better quality, and avoid predation better than smaller organisms.

Example: Blue-headed wrasses

On small reefs, young initial phase males do not spend much time reproducing because the large terminal phase males guard harems of females. On large reefs, small initial phase males can sneak between a spawning pair more frequently. Thus the reproductive investment of initial phase males is lower on small reefs than on large reefs. Warner (1984) measured the relation between reproductive activity and growth in initial phase males (Fig. 4.13)

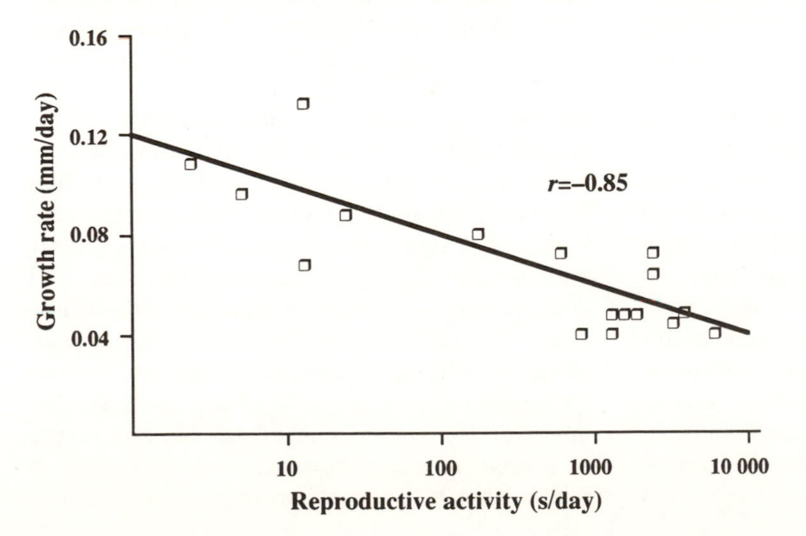

Figure 4.13 Each point is the mean for a reef. Sample size for small reefs (to the left); 3.9 males; for large reefs (to the right): 14.3 males (after Warner 1984).

and found a significant negative correlation. Because Fig. 4.13 measures the trade-off as a phenotypic correlation within a population but across a range of environmental conditions, it is not conclusive.

Summary of the evidence

A negative correlation between reproduction and growth was found in 10 of 11 species studied in the field; the exception has determinate growth. Of two laboratory results, one found no correlation and one was negative (Appendix 2D). The trade-off between growth and reproduction is the best-confirmed broad-sense phenotypic trade-off.

CURRENT REPRODUCTION VS. CONDITION

The impact of reproduction on physiological condition can be dramatic, even in species with indeterminate growth. In a single clone of vetch aphids, Brough and Dixon (1989) sectioned individuals that varied in number of ovarioles and measured the percentage of area occupied by gonads and by fat: the more gonad, the less fat. After spawning, some flatfish have 'jellied flesh' caused by the withdrawal of resources to reproductive tissue (Roff 1982). Red deer nursing offspring have lower kidney fat reserves than females that are not nursing (Clutton-Brock *et al.* 1982). In bighorn sheep, ewes that raised sons, and lactating ewes, had a higher faecal output of lungworm larvae than did ewes that raised daughters or that were not lactating (Festa-Bianchet 1989). Reproduction seems to decrease resistance to parasites; if so, resistance is costly. In Canada geese, females whose broods had been enlarged lost weight, moulted later, and bred later the following year than females whose broods had not been enlarged (Lessells 1986). The trade-off between flowering and carbohydrate storage in sugar cane makes it worth using hormones to inhibit flowering and increase sugar yield.

These examples illustrate physiological effects more subtle than growth with significant consequences for future reproduction. But not everyone has found effects of reproduction on condition. Finke *et al.* (1987) manipulated brood sizes in house wrens and found that females with increased brood sizes did not lose weight, produced just as many fledglings as did normal broods, did not have a longer interbrood interval, and did not produce smaller second clutches.

NUMBER VS. SIZE OF OFFSPRING

Range of seed size in plants

Seed weight varies across 10.5 orders of magnitude among flowering plants. Beeches, with seeds weighing a tenth of a gram, are larger than the double coconut with its 15 kg seed. Orchids and parasitic plants, whose seeds weigh about a millionth of a gram, are just as large as weeds and crop plants whose seeds weigh 0.1–1.0 g. The ratio of offspring size to size of parent in plants varies from about 1 000 000 000:1 for a 100 ton beech down to nearly 1:1 for asexual reproduction in some succulents. In flowering plants, offspring size seems more tightly bound to lineage and way of life than to size of parent or module.

Range of offspring size in animals

Among macroscopic animals, oysters and large fish with small eggs achieve the highest fecundities per unit weight, with eggs of the order of micrograms to milligrams and body weights of the order of 100 grams to 100 kilograms. Here the ratio of offspring to parent size is up to 100 000 000:1. The largest animal offspring are blue whales, at 10–12 tons as large as an adult African elephant, and the largest offspring relative to its mother's size is probably the kiwi's—a third as heavy as its mother.

Example: Offspring size in primates

In the primates (Fig. 4.14), viviparity and the pelvic girdle combine to create a tight correlation between mean offspring weight and mean parental weight. The small species have litter sizes greater than one, but all haplorhine species (monkeys and apes) much larger than 1 kg and all strepsirhine species (lemurs, tarsiers, galagos) much larger than 100 g typically have single offspring. Human offspring are larger and gorilla offspring are smaller than expected for their parents' size. The importance of lineage is clear: monkeys and apes share a relationship between offspring and parental size

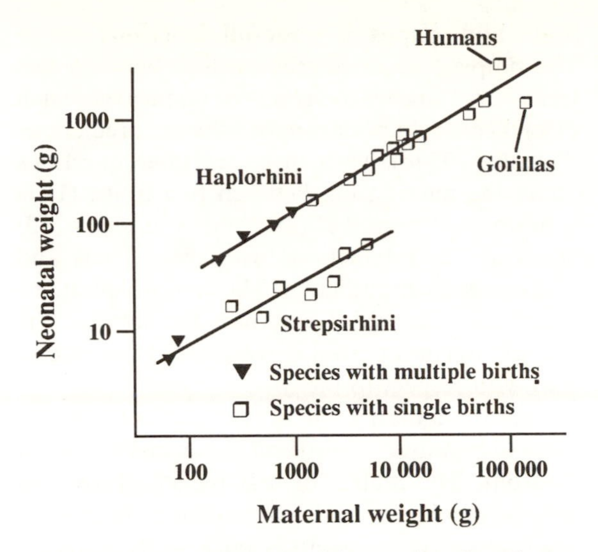

Figure 4.14 Mean neonatal weight as a function of mean maternal weight in 25 haplorhine primates (monkeys and apes) and 9 strepsirhine primates (lemurs and bushbabies) (after Leutenegger 1979).

that is quite different from that shared by lemurs, galagos, tarsiers, and their kin.

One might think that to produce litters larger than one in which a trade-off between number and size could occur, one would first have to select for small body size. However, humans have natural multiple births at low frequency, and multiple births can be induced with drug therapy. Thus body size does not constrain litter size in primates as strongly as Fig. 4.14 suggests. In human multiple births, the trade-off between offspring size and number is strong: twins are smaller than singlets, triplets are smaller than twins, and quadruplets and quintuplets are normally so small and premature that they require special support systems. Within each of the points in Fig. 4.14 lies hidden a potential trade-off between offspring size and number that could vary and be adjusted by selection. One large offspring is probably optimal for all large primates, and the size–number trade-off is not expressed.

Summary of the evidence

In the flowering plants as a whole, seed size is more strongly associated with way of life and lineage than with size of parent. In the primates, the correlation of offspring size with size of parent is very tight, and it is different for the two main groups of primates. At a given size, haplorhines have larger offspring then strepsirhines, and only small primates normally have the multiple litters in which the trade-off could be observed. The offspring number–offspring size trade-off is lineage-dependent.

Within a species, there is often no trade-off between offspring size and number, but there are notable exceptions (Appendix 2(*e*)). Of 15 species, the trade-off was observed in eight: a bean, a grasshopper, a copepod, a cladoceran, three small birds, and man. When populations of a single species of mosquito fish were compared, the evidence was mixed—the trade-off was strong among populations in the field and disappeared in the laboratory. In Pacific salmon there is a latitudinal cline in the trade-off. Across closely related species of goldenrods and salamanders, the trade-off is strong.

Seed weight is generally much less variable among individual plants within species than are seed number or other reproductive traits. When wheat was grown at different densities, the number of ears per plant varied 56-fold, the number of seeds per plant 833-fold, and the number of grains per ear 1.43-fold, but the mean weight per grain varied only 1.04-fold (Puckridge and Donald 1967). Within species in amphibians (Salthe 1969; Salthe and Duellman 1973) and sticklebacks (Wootton 1973) egg size is also relatively canalized.

In exceptional cases, seed size does vary within and among individuals, as indicated by 15-fold variation in seed weight among 17 *Lomatium* individuals (Thompson and Pellmyr 1989), by 17-fold variation in the seeds from a single plant of *Trifolium subterraneum* (Black 1957), and by the complex seed size polymorphism in *Heterosperma pinnatum* (Venable and Burquez 1989—a good entry to the literature on seed size polymorphism).

There is no general pattern of negative phenotypic or genetic correlations for the major life history trade-offs, including those between reproduction and survival and between early and late reproduction. The evidence for negative phenotypic correlations between reproduction and growth is good, but some disappointing results may not have been reported. Check whether a trade-off between two traits exists. Do not simply assume it is there.

DISCUSSION

How should one measure a reproductive effort?

The only general answer implies a lot of work—carefully, with several different methods, and taking the normal range of environmental and genetic variability into account. The best physiological measure is the proportion of surplus power devoted to reproduction, but this measure is not meaningful until individual variation in foraging efficiency has been measured. Variation has both genetic and environmental components, and it is hard to do physiological measurements on a sample of individuals large enough to give insight into the effects of genetic variation in reaction norms on reproductive effort.

In plants, results have been mixed when hypotheses on reproductive effort have been tested, perhaps as a result of having used the wrong units. Samson and Werk (1986) have shown that over a fairly broad range of plant sizes, reproductive effort is fairly constant per leaf or per branch within a species. Plants respond in part to environmental heterogeneity by varying the number of modules out of which they are built; within each module, the level of reproductive effort stays fairly constant. In stressful environments, the number of modules is low and the influence of capital expenditures on structures shared by all modules (tap root, main stem) appears.

Promising developments include using doubly-labelled water to measure energy expenditures under natural conditions and following marked individuals for long periods to measure individual variation. The method of doubly-labelled water, with an application to a free-flying bird, can be found in Hails and Bryant (1979). The work on kestrels and geese done in Holland exemplifies the advantages.

Measures of reproductive effort should be dynamic, should measure calories allocated, should measure proportions and not absolute amounts, and should take individual variation in activity and foraging efficiency into account.

Signs and shapes of trade-off functions

The simplest negative relationships between two traits are linear, convex, or concave. Such differences in shapes strongly influence predictions (Schaffer 1974*a*). Thus there are three questions about the relationship between two traits: (1) Is it positive or negative? (2) What is its shape? (3) What are its interactions with other traits and with environmental factors? The first two questions would be answered if we knew the signs of the first and second partial derivatives of the second trait with respect to the first.

Consider a matrix where rows and columns are indexed by traits ($1 \cdots n$) and the entries represent tradeoffs. The matrix of first partial derivatives defines the *linear trade-off structure*. It tells us whether the relationship is positive or negative. This is all that we usually know. The matrix of second partial derivatives defines the *simply curvilinear trade-off structure*. It tells us whether the curve relating the two traits bends up or down, important for certain predictions. One of the few experimental demonstrations of a significant curvature either up or down is that for percentage survival as a function of eggs laid per host by the parasitoid wasp *Trichogamma* (Charnov and Skinner 1984). The matrix containing a graph in each cell defines the *complexly curvilinear trade-off structure*. The curve relating the two traits could be a sigmoid or something more complexly nonlinear.

Interaction effects among trade-offs would go beyond that and would have to be represented by graphs in three or more dimensions (Pease and Bull 1988). The graph in each cell for the complete trade-off structure might describe a complex surface among three or more traits, rather than a line relating two traits, to show interactions where the trade-off between two traits depends on the value of other traits. Such effects are as yet poorly described.

Schaffer's (1974*a*) and Bell's (1980) analyses of the trade-offs between reproduction and survival and between current and future reproduction suggested that the curvature of trade-offs is critical. Whether an organism should invest everything in current reproduction, to the detriment of growth and survival, or choose an intermediate level of reproductive investment, reserving something for the future, depends on whether those trade-offs are convex or concave. This is a qualitative condition—we just want to know whether the lines bend up or down—but it has been elusive. Bell and Koufopanou (1986) summarized the few attempts to measure that curvature, and in every case concluded that the pattern was not detectably different from a straight line.

CHAPTER SUMMARY

Physiology contributes to life history evolution the analysis of the trade-offs that constrain adaptation. Physiological trade-offs within single individuals mean that giving more resource to one trait implies that less will be allocated to the other. They are boundary conditions on the optimization of life history traits.

The amount of energy acquired per unit time depends on foraging behaviour, which is itself physiologically constrained. The scope for activity of many organisms lies in the range of 3–4 times the basal metabolic rate.

When the energy acquired and proportion allocated to two functions vary among individuals, the sign of the phenotypic correlation will be positive when the amount of energy acquired varies a lot and allocation varies little, and negative when the amount of energy acquired varies little and allocation varies a lot.

Microevolutionary trade-offs are defined by the response of a population to selection. A proper measurement of a trade-off would take genetic variation among individuals and genotype by environment interactions into account. By this criterion, no trade-off has yet been well measured.

The phenotypic evidence for a trade-off between reproduction and parental survival and between current and future reproduction is mixed. There is good evidence for a trade-off between growth and reproduction, but it is mixed for a trade-off between offspring size and number. Seed and egg sizes are often less variable than other reproductive traits, but there is remarkable variation within some individual plants and among individual grasshoppers.

RECOMMENDED READING

Bell, G. and Koufopanou, V. (1986). The cost of reproduction. *Oxford Surv. Evol. Biol.* **3**, 83–131.

Clutton-Brock, T. H., Guinness, F. E., and Albon, S. D. (1982). *Red deer: behavior and ecology of two sexes.* (University of Chicago Press, Chicago).

Partridge, L. (1991). An experimentalist's approach to the role of costs of reproduction in the evolution of life-histories. In *Towards a more exact ecology.* BES Symposium (eds P. J. Grubb and I. Whittaker). (Blackwell Scientific, Oxford).

Sibly, R. M. and Calow, P. (1986). *Physiological ecology of animals.* (Blackwell Scientific, Oxford).

PROBLEMS

1. What conditions would select for an optimal range of variation in offspring size—5-fold rather than 15-fold, for example?
2. The size of an egg or a seed often varies less within a population than other life history traits. Which life history traits might one expect to be canalized against both genetic mutations and environmental influences?
3. In a large hypothetical data set of life history traits for several hundred species of gastropods, the correlation between reproductive effort and lifespan at the level of the family is strong and negative. Among genera within families, it is less strong but still negative. Among species within genera, there is no significant correlation. Among individuals within species, the correlation is significantly positive. How do you interpret the trade-off in this lineage?
4. What is the relationship between a trade-off and a constraint?
5. Is the cause of a trade-off in the selection pressure or in the system that reacts to it? Can we separate selection from the organisms under selection?
6. Why might physiological functions differ in their sensitivity to the amount of energy allocated to them?
7. Choose an organism in which the trade-off between reproduction and survival might be measured. Make a computer model of the organism in which feeding and energy allocation are represented so that the parameters vary

among individuals. Design thought experiments in which phenotypic correlations are measured in environments heterogeneous for (a) food, affecting energy intake, and (b) temperature, affecting the amount of energy needed for maintenance. Carry them out on the computer.

8. Generalize the model to represent genetic variation with several genes with pleiotropic effects mediating the trade-off between reproduction and survival. The genes obey Mendel's rules. Keep the physiological allocation model, and let the genetics influence the allocation decisions. Run a simulation in a heterogeneous environment, as before, but this time calculate genetic and phenotypic correlations. When do the genetic and phenotypic correlations have the same sign, and when do they not?

5

LINEAGE-SPECIFIC EFFECTS

It is generally acknowledged that all organic beings have been formed on two great laws—Unity of Type, and the Conditions of Existence. By unity of type is meant that fundamental agreement in structure which we see in organic beings of the same class, and which is quite independent of their habits of life. On my theory, unity of type is explained by unity of descent. . . . Hence, in fact, [natural selection] is the higher law; as it includes, through the inheritance of former variations and adaptations, that of Unity of Type

Charles Darwin, 1859

CHAPTER OVERVIEW

This chapter begins with examples of lineage-specific effects on life histories. Some of these depend allometrically on body weight; others describe patterns associated with higher taxonomic groups after the effects of weight have been removed. It then discusses what a lineage is and how to define it using *phylogenetic systematics*, a method used in historical explanations of life history patterns. Those using multivariate statistics to analyse extant variation within and among lineages concentrate on patterns of covariation among traits and between traits and environmental factors. Some of them interpret the patterns as adaptive. Those using functional morphology and developmental biology locate the irreversible evolutionary transitions that have created major new opportunities and constraints. Some of them interpret the patterns as evidence for constraint. Both view patterns of variation among species, genera, and higher taxa as influenced by history. The approaches answer different questions. Comparative evidence brings generality, suggests hypotheses, and places intraspecific patterns into context.

INTRODUCTION

An adaptationist claims that life history traits are adapted to each other and to local environmental conditions and that optimality models predict the states of the traits. This assumes that evolution occurs rapidly enough to bring populations into equilibrium with the environmental conditions in which we find them and that enough genetic variation of the right sort is available. To be fair to this claim, life history traits can evolve quickly (Discussion, Part I), and plenty of genetic variation is available (Chapter 3). However, two observations suggest that the strong adaptationist interpretation is incomplete.

First, some life history traits are fixed at high taxonomic levels, not varying within populations at all. All species in the Order Procellariformes (tubenose seabirds) have a clutch size of one egg; all six species of Pacific salmon are semelparous; no birds are viviparous; no bats lay eggs. Such traits offer no genetic variation on which micro-evolution could work. While fixed traits are not being further adapted, they may still have an adaptive function. All bats have exactly two wings, and they are adaptive. However, not all fixed traits have clear functions, and some fixed traits constrain the evolution of others. Their effects need further study.

Secondly, to solve an optimality model, one needs boundary conditions. Optimality theorists build lineage-specific effects into their boundary conditions. For example (Chapter 6), to predict optimal age and size at maturity, one assumes relations between fecundity and weight and between juvenile survival rates and age at maturity of the mother. These relations differ among higher taxonomic groups. Optimality models, like quantitative genetics, explore local adaptation within a framework of lineage-specific effects.

Demographic theory and quantitative genetics do not tell us how lineage-specific effects evolve, at which taxonomic level they arise, or among which groups they differ. These questions belong to comparative biology. That evolutionary explanation has two components—adaptation and lineage-specific effects—was Darwin's (1859) standard view. He explained Unity of Type through descent with modification of related species from a common ancestor and adaptation as modification of the distribution of types within a population by natural selection. A balanced interpretation of an evolutionary pattern requires both explanations; recently techniques have been developed to measure their relative contribution.

These new techniques have prompted a surge of interest in historical explanation. Basic to all such work is a reliable phylogeny, and most now use the techniques of phylogenetic systematics to determine relationships among species and the sequence in which traits evolved (Hennig 1950; Wiley 1981; Ridley 1983; Maddison 1990; Lessios 1990). Some combine this information with statistics to discover at what taxonomic level traits vary, at what level which patterns arise, and whether traits covary with environmental factors at any taxonomic level (Pagel and Harvey 1988; Harvey and Pagel 1991). Others use the evolutionary implications of physiological, developmental, and morphological constraints to analyse the causes and consequences of irreversible change (e.g. Roth and Wake 1985; Lauder 1986; Wake and Larson 1987; Müller 1989; Wake and Roth 1989*a*, *b*). The two approaches complement each other in explaining comparative patterns.

EXAMPLES OF LINEAGE-SPECIFIC EFFECTS

Whereas microevolution concentrates on traits that vary within populations, comparative physiology emphasizes biology characteristic of species or larger groups. Some important lineage-specific physiological effects and the levels at which they originate are listed in Table 5.1.

Table 5.1 Some important lineage-specific physiological distinctions: illustrative, not exhaustive

Distinction or effect	Characteristic of
(1) Allometric variation	All levels
(2) Homeotherms vs. poikilotherms	Classes, Phyla
(3) Capital vs. income breeders	Families, Orders
(4) Determinate vs. indeterminate growth	Classes, Phyla

Allometric variation

Many life history traits have a nonlinear relation to body size. Power functions describe such *allometric relations*:

$$Y = aW^b \tag{5.1}$$

where Y is the value of a trait, a and b are constants, and W is body weight (see Appendix 3 for a review).

Unicells, homeotherms, and poikilotherms

Unicells, homeotherms, and poikilotherms have different metabolic rates, growth rates, and operating costs (Fig. 5.1). Because of the way they were estimated, the exponents of the allometric equations are the same for all three groups, but the coefficients differ. It was assumed that the homeotherms operated at 39°C and the others at 20°C (Peters 1983).

The homeotherm relation implies that 'a half-ton moose degrades at least 440 J of energy to heat every second; 1 half-ton of 20-g mice would degrade at least 5 400 J in the same time'. The differences between homeotherms and poikilotherms imply that 'the same rate of energy supply could support about thirty times more poikilotherms than homeotherms because of the difference in metabolic rates' (Peters 1983, pp. 31–2). One ton homeotherms and 100 mg poikilotherms must eat about the same proportion of their body weights each day.

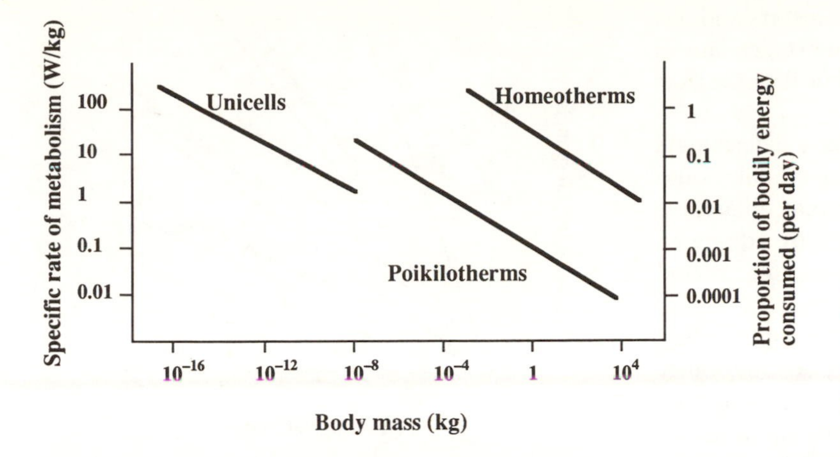

Figure 5.1 Specific metabolic rates: the equations relating specific power production (R) to body mass are: $R_{\text{homeotherms}} = 4.1W^{-0.249}$; $R_{\text{poikilotherms}} = 0.14W^{-0.249}$; $R_{\text{unicells}} = 0.018W^{-0.249}$ (after Peters 1983).

Within the homeotherms and poikilotherms, fish and reptiles, marsupials and placental mammals, and altricial and precocial birds also differ (Fig. 5.2). Growth rates per unit weight change during growth. To standardize comparisons, these rates were all measured during the phase of rapid growth.

Major groups grow at different rates. Natural selection cannot arbitrarily modify the constraints that size and phylogeny place on growth rates. Growth is involved in the cost of reproduction, the determination of age at maturity, and the evolution of reproductive effort.

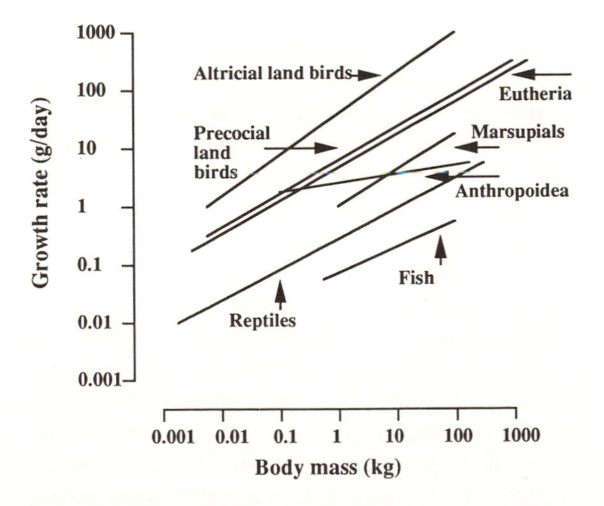

Figure 5.2 The growth rates of different vertebrate groups differ. There are major differences among fish, reptiles, marsupials, eutherians, and altricial land birds. Anthropoid apes grow slowly for mammals—large apes and man grow more slowly than marsupials of the same size. Altricial land birds grow fastest of all major groups (after Case 1978).

When the temperature drops, all physiological processes in poikilotherms slow down. For example, in plants cool nights imply small metabolic costs; warm nights imply large metabolic costs. When nights are cool and days are warm, plants have more surplus energy to put into growth and reproduction. In most organisms, body temperature varies with environmental temperature, and the distinction between physiological time and calendar time is useful. We need both calendar time and physiological time for life history evolution, for mortality rates and birth rates depend on seasonal events and on the life cycles of homeotherms (Taylor 1981).

Mode of growth

Organisms with indeterminate growth continue to grow after they mature: most perennial plants, fish, amphibians, reptiles, crustaceans, worms, and molluscs. Organisms with determinate growth stop growing at or before maturation: most insects, birds, mammals, and annual plants. Organisms with indeterminate growth allocate energy between growth and reproduction many times throughout adult life. This allocation decision is made in organisms with determinate growth only once, at maturity.

These categories contain important variety. In the fruit fly *Drosophila*, a determinate grower, the first two larval instars last until a certain minimum weight is reached, and their duration varies with conditions. The third instar prepares the larva for pupation, and its period—plus pupation—is relatively inflexible. Food scarcity early in development can be compensated for by longer instars and has little effect on adult size. Food shortages late in development produce smaller adult flies (Bakker 1959; Gebhardt and Stearns 1988).

Hemimetabolous insects also have determinate growth, achieving adulthood after a final moult, and their mode of growth means that differences in age and size at maturity evolve by adding or dropping an instar from the juvenile period (Hassall and Grayson 1988).

Among indeterminate growers, the continuously growing fish differ from the discontinuously moulting crustaceans. Moulting requires resorbing and storing the exoskeleton, taking up water to increase volume as much as possible, then feeding, storing energy, and perhaps reproducing before moulting again. When reproduction is coupled to moulting, the volume of the clutch is constrained by the size of the carapace. The life cycle is subdivided into intermoult periods within each of which allocation decisions are made.

Lineage-specific modes of growth have many consequences for life history evolution, including (1) determining when a trait is particularly sensitive to environmental influences, (2) how surplus power is allocated to growth, storage, or reproduction, and (3) the timing of growth and reproduction.

Life history ratios are lineage-specific

Charnov and Berrigan (1990) suggest that the ratio of adult lifespan to age at maturity is fixed within lineages but differs among them. They present regressions of average adult lifespan on age at maturity for species from different major groups (Fig. 5.3).

Although these remarkable relations imply that age at maturity and lifespan evolve in fixed ratio, variation is concealed within these major groups (Charnov and Berrigan, pers. comm.). Length of life as a multiple of age at maturity increases from fish, through snakes and lizards (which do not differ) to mammals, with birds living longest of all for a given age at maturity. The range of values for the ratio is about 0.1–0.4 for fish, 0.6–2.2 for mammals, and 1.1–3.5 for birds. Homeotherms live longer after maturation than poikilotherms, and within the vertebrates birds live longest of all. These analyses, which would be easier to interpret

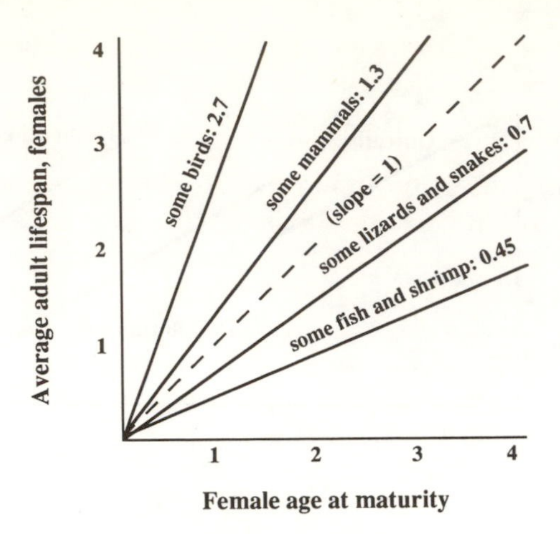

Figure 5.3 Relation of adult lifespan to age at maturity for different taxa (after Charnov and Berrigan 1991).

with data corrected for body size, suggest that the basic design common to a higher taxon shapes the rough boundaries within which the life cycles of all its constituent species evolve. To make such a claim solid, we at least need to know how to define a higher taxon, and that comes next.

WHAT A LINEAGE IS AND HOW TO DEFINE IT

All methods in comparative biology rely on knowing the relationships among species. A lineage—or clade—consists of all the species that share a common ancestor. Recent progress in systematics has greatly increased the clarity with which those relationships can be established. This has led to some major taxonomic revisions.

Two key advances were Hennig's (1950) introduction of phylogenetic systematics (also called cladistics) and molecular systematics, where the non-transcribed part of the genome provides data independent of the confusing changes that evolution can produce in phenotypes. In testing historical hypotheses on traits thought to be adapted to local conditions, one can avoid circularity by using a phylogenetic hypothesis constructed with traits (e.g. molecular ones) not involved in the adaptation. Nei (1987) reviews molecular evolution, and Felsenstein (1988) reviews methods for reliably inferring phylogenies from molecular data. Wiley (1981) discusses phylogenetic systematics in depth.

Phylogenetic systematics aims to define the relationships among species to reflect evolutionary history precisely. If we want to define the relationships among species in the dog family (Canidae), traits shared by all dogs, such as warm-bloodedness, hair, and having two eyes, are of no use, for they are shared with all mammals. Instead we must focus on *traits shared by some dogs but not by others and not shared with the nearest relatives that are not dogs.* Such traits have arisen within dogs after they split off from other lineages and, within dogs, have arisen only in the common ancestor of certain species. These are called shared derived traits, or *synapomorphies.*

To work with shared derived traits, one must be able to identify a trait and to see that the trait is homologous across two or more species. Traits are pragmatically defined as things that can be measured or classified: e.g. number of fin rays or pattern of dentition. To see when traits are homologous, we cannot use an hypothesis of phylogeny, for 'if phylogeny determines homologies, and homologies are required to reconstruct phylogeny, one must know the phylogeny in order to obtain the homologies necessary to build the phylogeny' (Brooks, pers. comm.). To avoid the circularity, we define homology with non-phylogenetic criteria, such as shared developmental pathways (Remane 1952). The general homology problem is discussed in Wagner (1989) and Roth (1991).

The cladistic approach can be summarized in four steps. First, use non-phylogenetic criteria to establish homologies, then determine similarities in traits among species. Second, by comparing with out-groups, establish shared general, shared derived, and unique traits. Third, use only the shared derived traits to form groups, depict their relationships as an evolutionary tree called a cladogram, and indicate all transitions of traits on it. Fourth, interpret inconsistencies as cases of parallel and convergent evolution.

One decides that a certain group shares a derived trait (a synapomorphy) by comparing that group with another that does not have the character. To make an *out-group comparison,* one must choose an out-group. The most appropriate out-group is the one that is more closely related than any other to the group being analysed; it is called the *sister-group.* (The choice of a sister-group assumes that an independently derived, higher level phylogenetic hypothesis exists.) Once a pattern of shared derived traits is defined by reference to sister-groups, one can erect a number of phylogenetic hypotheses. In the absence of informed judgment to the contrary, one accepts the most parsimonious—the one requiring the least change—as the best working hypothesis.

This application of the Principle of Parsimony, the idea that the best explanation is the least complex, remains controversial. Some argue that it is an arbitrary logic-chopping assumption, that evolution can be very complex and we have no reason to assume otherwise. Others argue that because the resultant classifications are simply working hypotheses, there is no reason to make them more complex than we have to. I lean toward the latter view. Keep in mind that phylogenies change as more information comes in.

We illustrate the method with the classification of the tetrapods (Maddison 1986) and start with a table defining the characters and their states (Table 5.2). For each group, one then codes the character state and forms a data matrix (Table 5.3) from which a phylogeny can be constructed (Fig. 5.4) with a variety of methods.

The ray-finned fish are taken as the sister-group. For this simple example, one could construct a phylogeny by hand, but with many groups and many characters it is more efficient to use a computer program that allows one to explore the consequences of various assumptions. One such program produces a phylogeny (Fig. 5.4) suggesting that 'reptiles' are not a natural group and that homeothermy evolved twice. Note the insights suggested by placing the changes in character state on the cladogram, in this case by using darker lines for the lineages that are homeothermic.

APPLICATIONS OF CLADISTICS TO ECOLOGY

Phylogenetic systematics contributes alternative, historical hypotheses to biological fields with an adaptationist orientation. In the first two of the examples that follow, an adaptationist hypothesis

Table 5.2 Characters and character states for the tetrapods. From Maddison (1986)

Character	State	Relevant group
1. amnion:	absent, present	
2. appendages:	fins, legs only, legs and wings	tetrapods
3. body covering:	derm. scales, smooth, epid. scales, feathers, hair	
4. thermoregulation:	poikilotherm, homeotherm	birds and mammals
5. int. nostrils:	absent, present	sarcopterygians
6. atrial septum:	absent, present	sarcopterygians
7. temp. fenestrae:	none, one, two	diapsids
8. hemipenes:	absent, present	squamates
9. suspensorium:	not strep., streptosylous	squamates
10. gizzard:	absent, present	archosaurs
11. antorb fenes:	absent, present	archosaurs
12. lat. sphen:	ossified, not	archosaurs
13. teeth:	not ped., pedicillate	lissamphibians

Table 5.3 Data matrix for the tetrapods. From Maddison (1986)

Group	Codes for characters 1–13
Ray-finned fish	00000 000?0 010
Frogs	01101 10000 011
Turtles	11201 10000 010
Lungfish	00001 100?0 010
Salamanders	01101 10000 011
Crocodiles	11201 12001 100
Lizards	11201 12110 010
Birds	12311 12001 100
Mammals	11411 11000 010
Snakes	1?201 12110 010

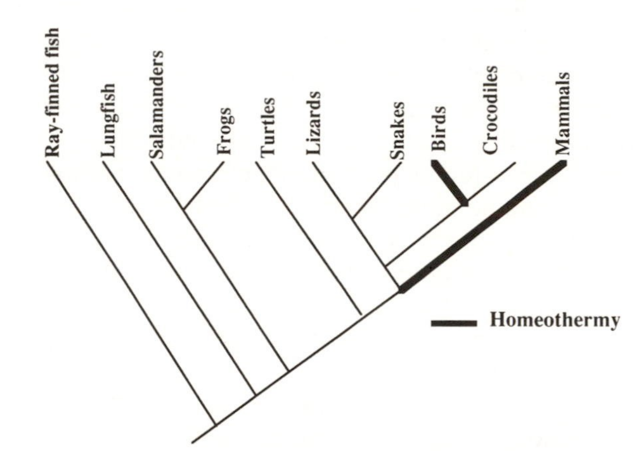

Figure 5.4 A provisional phylogeny of tetrapods and lungfish that indicates: (1) the ray-finned fish are the sister-group. (2) The salamanders and frogs, the lizards and snakes, and the birds and the crocodiles each share a common ancestor not shared by the others. (3) The 'reptiles' are not a natural group, for the birds are more closely related to the crocodiles than they are to the mammals, and the crocodiles are more closely related to the birds than they are to the snakes and lizards. (4) Homeothermy evolved twice, once in the ancestor leading to the mammals and once in the ancestor leading to the birds (after Maddison 1986).

is contrasted with an historical alternative. In the third, a cladistic analysis is used to support an adaptive association between two traits.

Leaf retention in oaks

Young European oak and beech trees retain their dry leaves over the winter, and it has been suggested that leaf retention is an adaptation for more efficient nutrient cycling (Otto and Nilsson 1981). Wanntorp (1983) placed this trait on to a cladogram of the Family Fagaceae, to which beech and oak belong. Most of the 1000+ species in the family are evergreen trees living in rain forests in south-east Asia or in Mediterranean climates.

The simplest historical hypothesis is that leaf retention is ancestral and that leaf shedding is the derived feature needing explanation. Taking the strictly evergreen, mostly tropical oak genus *Lithocarpus* as the sister-group of *Quercus* gives a

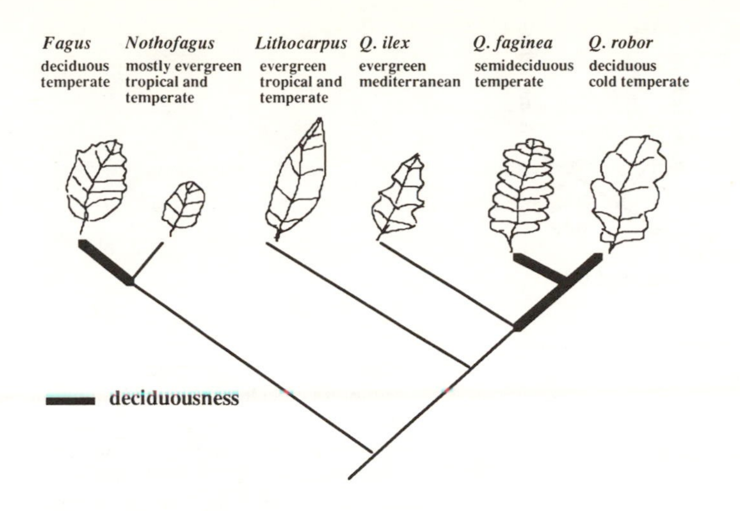

Figure 5.5 A simplified cladogram of the Fagaceae that suggests the deciduous state evolved twice, once in the ancestor of the beeches and once in the ancestor of the temperate oaks (after Wanntorp 1983).

cladogram in which states of leaf shedding run from evergreen species in tropical and Mediterranean areas through semi-deciduous species in intermediate habitats to deciduous species that reach Scandinavia (Fig. 5.5). This changes what needs to be explained from leaf retention to leaf shedding. Leaf shedding has evolved at least twice within the family, both times in temperate species. Now the questions are 'What promotes leaf shedding?' and 'What maintains leaf retention?'.

Stickleback behaviour

Sticklebacks (Gasterosteidae) are small fishes widely used in behavioural and ecological research. It has been suggested that breeding systems, territorial defence, the opportunity for sexual selection, feeding behaviour, anti-predator behaviour, and parental investment strategies have been adapted to local environments. McLennan *et al.* (1988) constructed a stickleback cladogram based only behavioural traits (Fig. 5.6) that closely resembles cladograms based only on morphological traits.

Together with the observation that some stickleback species have the same behaviour patterns under different ecological conditions, this pattern suggests that history and phylogeny have played as large a role in determining patterns of stickleback behaviour as has adaptation to local conditions. By placing the characters on the cladogram in the sequence in which they are thought to have evolved, one can see which characters are candidates for strict phylogenetic adaptation—special derived traits like building a nest with no tunnel exit in *Culaea*—and those which are not, like digging a nest pit in *Gasterosteus*. The nest-building behaviour of the three-spined stickleback is a mosaic of ancient and recent traits. Some traits are ancient and either fixed within the lineage or maintained by selection pressures shared by all species in it (e.g. building a nest that has a tunnel exit and entrance). Other traits specific to particular species may have something to do with its particular breeding biology (e.g. rolling on side during the nesting display).

Breeding season and group composition in primates

A phylogenetic hypothesis of adaptation suggests that an association between a trait and an environmental factor has arisen repeatedly and independently. Such convergences are strong evidence that the trait represents an adaptation to the environmental factor. When the association is between two traits, their repeated independent evolution is evidence that they are co-adapted.

Ridley (1983) developed a method to address the problem of non-independence in species that share ancestors. He used it to test the hypothesis that primates with shorter breeding seasons have multi-male troops and those with longer breeding seasons have harems (Ridley 1986). When short breeding seasons force females to come into oestrus in near synchrony, no single male could

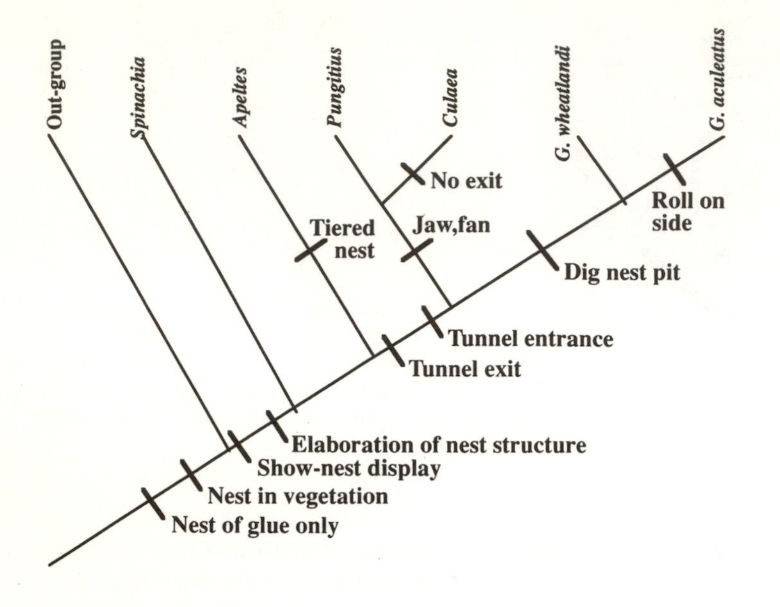

Figure 5.6 When one constructs a cladogram of sticklebacks based only on behavioural traits, the phylogeny closely resembles that obtained with morphological traits and the evolutionary sequence of character change becomes obvious (after McClennan *et al.* 1988).

mate with all the receptive females while defending them from matings by other males. So many matings would be lost to other males that it would not pay a dominant male to defend a harem. In a long breeding season, oestrus is not so tightly synchronized, and males should defend harems because they can mate with each receptive female in sequence.

To test this idea, he constructed a cladogram for 33 species of primates for which length of breeding season and social system were known. The association between length of breeding season (short vs. long) and group structure (single male vs. multi-male) could then be tested by constructing a 2 × 2 table and counting the number of cases of each type (long breeding season, multi-male, etc.). Now comes the critical point. Should one count the number of species, the number of genera, or the number of times the association has arisen? If two species inherited the same condition from a common ancestor, then the occurrence of the condition in one of them is not independent of its occurrence in the other. 'Evolutionary relationships should be weighted not by the number of species that eventually inherit them, but by the number of times they have arisen ... [Furthermore,] a broad array of taxa, not just taxa that provide large numbers of (species) data points, should be sought out to reduce the probability that some alternative factor explains the result' (Pagel and Harvey 1988).

Table 5.4 Relation between group composition and breeding season in primates calculated from three different contingency tables. From Ridley (1986)

	Breeding season		
	Long	Short	*P*
Species			
Multimale	4	18	0.0000033
Single male	12	0	
Genus			
Multimale	2	9	0.0003
Single male	9	0	
Independent events			
Multimale	2	8	0.007
Single male	5	0	

Ridley performed the test three ways—with species, with genera, and with independent evolutionary events (Table 5.4). To count the independent evolutionary events, he noted whenever a descendent taxon had changed in either trait from its ancestral state and counted the state of the two traits in the descendant as one event. All three methods of analysis support the hypothesis, but counting species or genera rather than independent events considerably overestimates the significance of the pattern.

The phylogenetic assumptions implicit in comparative methods

Several methods of analysing phylogenies implicitly assume that it is equally likely that a given amount of evolutionary change might occur on any branch and in either direction. This is the case with Ridley's method. However, we know from molecular or fossil data that branches are often not of equal length.

Phylogenetic analyses (1) identify ancestral and derived conditions and therefore what is to be explained by adaptation. (2) They establish the sequence in which characters evolved, thus permitting explicit tests of hypotheses for which sequence is critical. (3) By placing traits on a cladogram in the order in which they evolved, one can also estimate the number of convergences in a lineage and evaluate any adaptationist hypothesis based on such convergences.

STATISTICAL APPROACHES TO THE COMPARATIVE METHOD

Several approaches have been used for the comparative analysis of quantitative traits. Each is better at answering some questions than others; none is wholly without problems. Taken together, they give considerable insight into evolutionary patterns. The first three methods, like Ridley's count of independent events, measure how two traits covary independently of other factors. Method 1 isolates a set of statistically independent differences; method 2 aims to define the taxonomic level at which the mean value of the trait can be treated as statistically independent of other such values; method 3 deals with allometric covariation where body weight is the confounding factor. Pagel and Harvey (1988) call the combination of Methods 2 and 3 the ANOVA/allometry approach.

Method 1: Phylogenetically independent differences

Suppose we want to test the hypothesis that two traits share a particular relationship—for example, that age at maturity and lifespan are positively correlated in the mammals. We cannot use species as independent data points, for species share ancestors, and two related species may both have delayed maturity and a long life because a common ancestor did. Felsenstein (1985) suggested a solution. Consider a phylogeny of four species (Fig. 5.7). We measure two traits, X and Y, and get values $X_1 \cdots X_4$ and $Y_1 \cdots Y_4$ for the species' means. While Y_1 and Y_2 are not independent of one another, the difference $(Y_1 - Y_2)$ is independent of the difference $(Y_3 - Y_4)$, for each difference occurred after the independent speciation events E_1 and E_2. While the absolute values of traits are influenced by common ancestry, appropriately chosen *differences* between pairs of trait values are independent of one another and can be used for statistical tests of evolutionary hypotheses. Harvey and Pagel (1991) discuss significance tests.

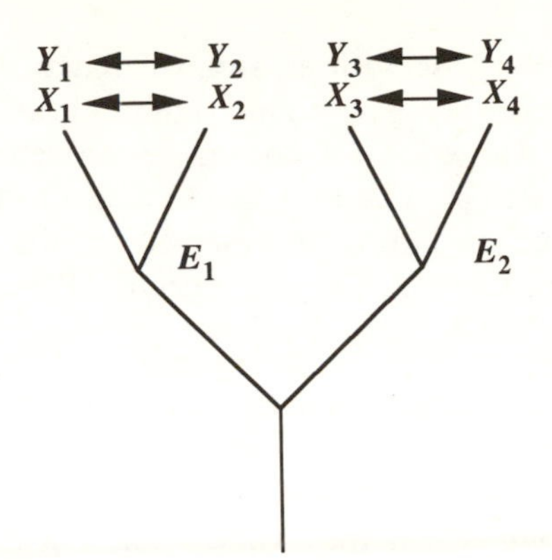

Figure 5.7 A phylogeny of four species on which two traits, X and Y, are measured (after Felsenstein 1985).

Method 2: Finding the right taxonomic level for comparison

Phylogenetic relationships can confound comparative trends just as weight does.

> Suppose we find that macaques . . . have larger brains for their body sizes than langurs . . . Macaques are terrestrial while langurs live in trees. . . . The assertion that brain size varies with habitat might be tested by means of a two-by-two contingency table: large or small brain size for a given body size versus terrestrial or arboreal habitat. If we have data on seven species within each genus, then the contingency table would contain two diagonal cells each

with seven observations, and two diagonal cells that are empty . . . a highly significant association . . . However it is not unreasonable in this instance to assume that there are at most two independent points represented in the contingency table: one for each genus. (Pagel and Harvey 1988).

To solve this problem, one must find the appropriate unit for comparative analysis—a taxonomic unit within which the traits vary little but among which they vary a lot. Lack (1968) suggested that this unit was the subfamily for birds. This has since been confirmed in several studies (e.g. Harvey and Clutton-Brock 1985), which is why the names of subfamilies are on most of the points in Fig. 5.9. However, there is nothing special about subfamilies, and it is only a coincidence that the appropriate level of analysis turned out to be subfamilies in both birds and primates. The tastes of systematists vary, and different groups have been classified, especially at higher levels, by different criteria. Even after groups have been classified in the same way—a distant dream—a quantitative justification will have to be given for the decision to use a given taxonomic level in a given analysis.

Life history variation in the mammals

A straightforward way to identify the taxonomic level at which traits vary is to do a nested analysis of variance and assign the total variance to orders, families, genera, and species. For life history traits in mammals, most variation arises at the level of orders and families (Table 5.5). Clutton-Brock and Harvey (1977) and Harvey and Mace (1982) interpret a pattern like that in Table 5.5 as estimating the degree of statistical independence to be found at different taxonomic levels. In this case, the means of each trait for species and genera are not independent of the families to which they belong. Families vary a bit within orders, and orders vary a lot within the class. Making comparisons among orders rather than species reduces sample size but comes closer to estimating the number of independent evolutionary events in the lineage. In practice, one picks a taxonomic level that offers a reasonable sample size and represents most of the variation. In Table 5.5, that level would be the order. Harvey and Clutton-Brock (1985) used this technique to isolate the subfamily as the level of choice for their study reported above (cf. Fig. 5.9).

This choice is a pragmatic compromise, not a conclusion forced by logic. Of several traits, some will vary more among families, others more among orders. The pattern of variation itself depends on what is included in the data set. With orders included within classes, variation among orders will usually explain much of the variation. With subfamilies included within one family, the subfamilies will explain most of the variation.

Table 5.5 Percentages of variance at different taxonomic levels for selected life history and size variables in mammals

Trait	Species within Genera	Genera within Families	Families within Orders	Orders within the Class
Gestation length	2.4	5.8	21.1	70.7
Age at weaning	8.4	11.1	18.9	61.6
Age at maturity	10.7	7.2	26.7	55.4
Inter-litter interval	6.6	13.5	16.1	63.8
Maximum lifespan	9.7	10.1	12.4	67.8
Neonatal weight	2.9	5.5	26.6	64.9
Adult weight	2.9	7.5	21.0	68.5

Nested ANOVAS were done on logarithmically transformed data. About 80 to 90 per cent of the variation in each variable is accounted for by differences among orders and families. Genera and species do not vary much independently of phylogeny. From Read and Harvey (1989).

Repeating the analysis within taxa is a good check on conclusions drawn at higher levels (Dunham and Miles 1985; Pagel and Harvey 1988).

One should use a nested analysis of variance to identify the level that explains most of the variation before proceeding to the next step, Method 3, for in an analysis of allometric residuals one has to decide whether to use residuals from regression lines calculated with order means, family means, genus means, or species means. Each residual has a different meaning, and one should use the same type of residuals for all traits used in the analysis. If most variation for most traits is among families, then the residuals of the regressions of traits on weight should be calculated as the deviations of the family means from the regression of family means on weight.

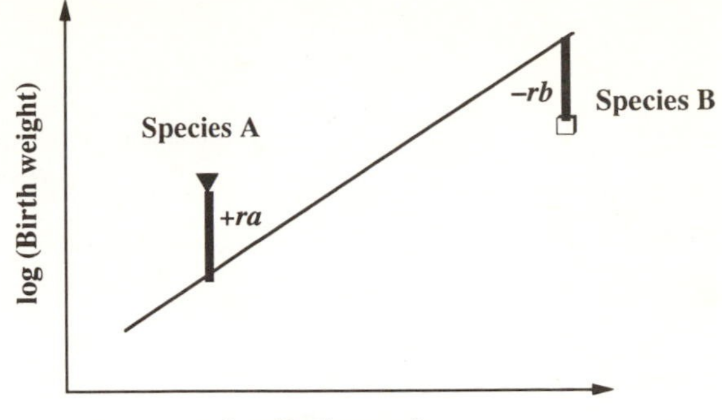

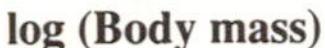

Figure 5.8 To compare the birth weights of two species of different body masses, one calculates the regression between birth weight and body mass across the taxon containing both species, then measures the birth weights of each as deviations from the value expected for an organism of that size in that taxon. Here the smaller species A has a larger than expected birth weight (the residual is positive), and the larger species B has a smaller than expected birth weight (the residual is negative). The line is estimated from many species, but the residuals are depicted only for two.

Method 3: Correcting for weight through allometric residuals

In comparing the birth weight of a shrew with that of an elephant, one must control for body mass. Otherwise differences due to other causes would be totally obscured by the enormous difference in weight. To avoid this problem, one can calculate the deviations of birth weights from the line that would be expected for all mammals on a log–log plot (Fig. 5.8). When residuals are needed, normal linear regression is preferred, for it is the only regression model whose residuals are not correlated with the x-variable (Pagel and Harvey 1988).

The method of allometric residuals functions best when applied to two or more traits correlated with size. One calculates the deviation from the expected regression line for each trait, then plots the residuals against each other. Any relationship in the residuals exists despite the correlation of both traits with size. This amounts to a multiple regression analysis in which the correlations of two traits are calculated after the influence of size on both is accounted for.

Life history variation among primate subfamilies

Harvey and Clutton-Brock (1985) used this method to analyse life history variation among primate subfamilies. Note the proximity of groups with different body weights, e.g. man and tarsiers (Fig. 5.9). Independent of body size, relative neonatal weight increases with relative age at maturity of females. Gibbons (Hylobatidae) have relatively small offspring for a primate that matures that late, and *Callimico*, the only marmoset that typically bears one offspring rather than twins, has a relatively large offspring for a primate that matures so early.

From several such analyses, Harvey and Clutton-Brock concluded that 'primate families that produce relatively large neonates also have relatively long gestation lengths and weaning periods, advanced ages at maturity, and longer lifespans. They also have relatively large neonatal and adult brain sizes' (Harvey and Clutton-Brock 1985, p. 573).

Life expectancy and age at maturity in mammals

Harvey and Zammuto (1985) showed that mammals with a relatively long lifespan for their body size also start reproducing relatively late (Fig. 5.10). Some species with very different body weights again lie close to each other, e.g. elephants and pikas or mice and warthogs. Harvey and Zammuto also checked the effects of gestation length, litter size, neonatal weight, and litter weight by removing those effects with partial correlations. In no case 'did the correlation

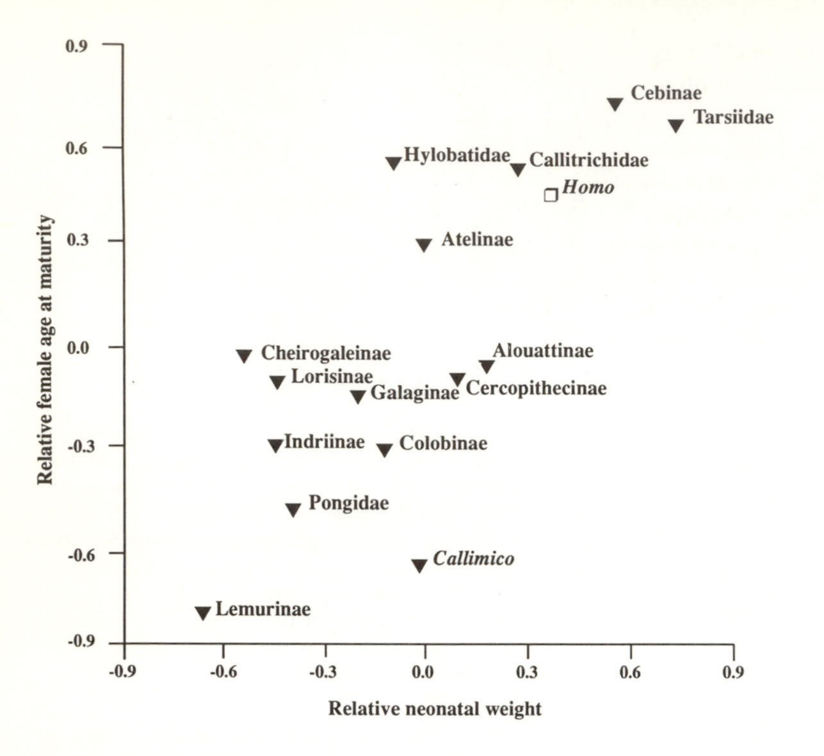

Figure 5.9 An example of the analysis of residuals from trait–weight (allometric) relationships (after Harvey and Clutton-Brock 1985).

between age at first reproduction and life expectancy fall below 0.9' (Harvey and Zammuto 1985, p. 320). They suggest that the mechanism underlying the correlation involves the cost of reproduction: 'Perhaps the most straightforward explanation is that each bout of reproduction reduces subsequent reproductive success . . . and that older mothers produce larger litters (and/or more viable offspring) than do younger mothers. As life expectancy at birth decreases under such circumstances, selection will favour maturation at earlier ages'.

Causal hypotheses can only be tested with other methods (e.g. modelling and manipulation experiments), but there is no denying the pattern. Note the connection to Charnov's ratio of age at maturity to lifespan (Fig. 5.3).

Read and Harvey (1989) used the method of residuals to identify patterns in the mean values of life history traits in 17 orders of mammals. For their sizes, bats have the longest and rabbits the shortest gestation (Fig. 5.11), and gestation length increases with age at maturity. For their size, bats also have the lowest and rabbits the highest annual fecundity, bats the longest and rabbits the shortest period of maternal investment (Fig. 5.12). Bats have the fewest offspring per litter and the largest offspring, for their size, while rabbits have the largest litters (Fig. 5.13). Although the offspring of whales are very large in absolute terms, they are small in relative terms.

Figures 5.10–5.13 demonstrate the power of the comparative method to document broad patterns. The radiation of mammals was an evolutionary experiment in which different combinations of age at maturity, gestation length, fecundity, weight at birth, and parental investment were tried out. Only certain combinations were successful.

Read and Harvey tested several non-adaptationist hypotheses for these associations. Body size cannot explain the patterns, for its effects were removed from the data before they were plotted. Metabolic rate had no effect on inter-order differences, which were also not affected by brain rate or growth rate of neuronal tissue. Read and Harvey conclude that the patterns represent co-adaptation of life history traits to one another and to mortality rates. Gittleman (1986) did a similar analysis on families of carnivores.

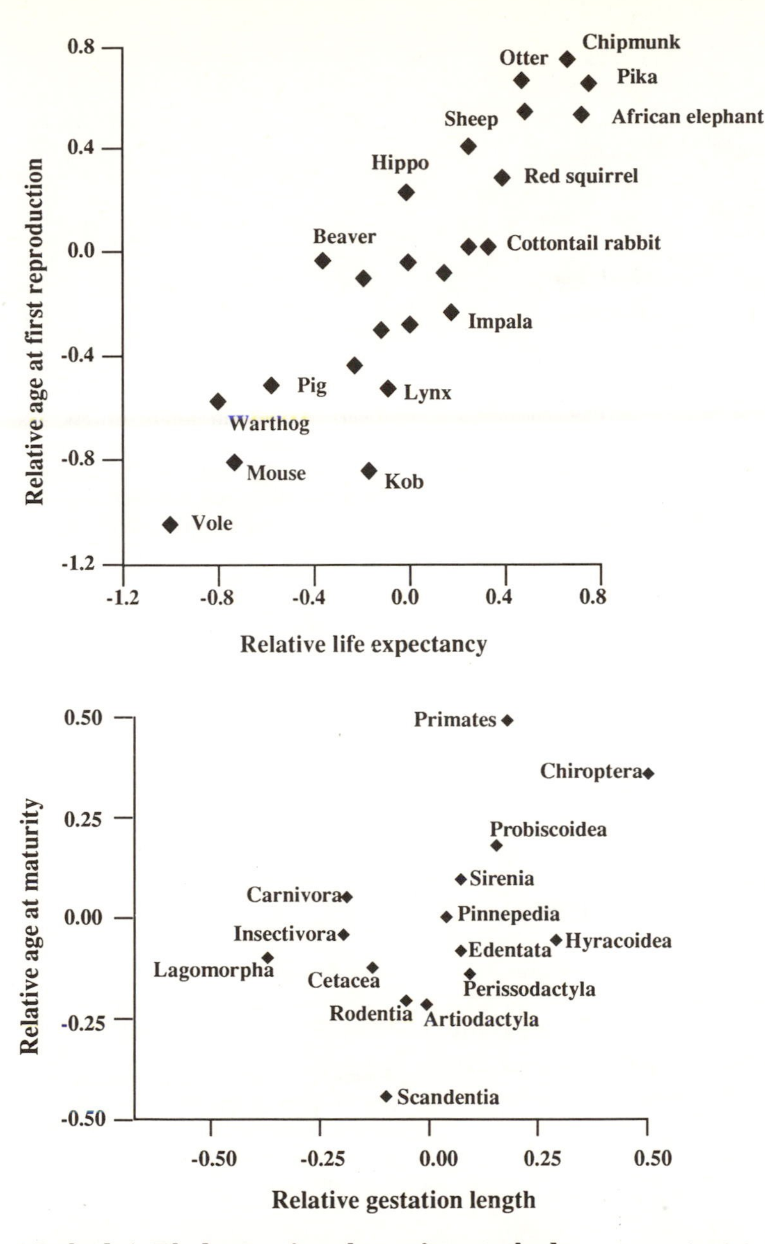

Figure 5.10 Relative age at first reproduction plotted against relative life expectancy at birth for 24 species of mammals (after Harvey and Zammuto 1985).

Figure 5.11 Relationship between gestation period and age at maturity for eutherian orders with effects of body size removed. $r = 0.60$ (after Read and Harvey 1989).

Method 4: Phylogenetic-subtraction methods

The *phylogenetic-subtraction method* approaches the problem from the opposite direction. Harvey and his co-workers identify a higher taxonomic level at which statistical independence can be claimed and compare units at that level. The variation among those units is seen as adaptive, variation within those units is a distraction. The phylogenetic-subtraction method assumes that the variation among families and orders represents an historical, phylogenetic effect concealing the meaningful variation among genera within families and among species within genera. Here the variation among higher levels is a confounding influence that must be removed before patterns in the lower levels can be seen.

A researcher using the phylogenetic-subtraction method might criticize a nested-ANOVA/allometry result as indicating little more than taxonomic differences associated with phylogeny.

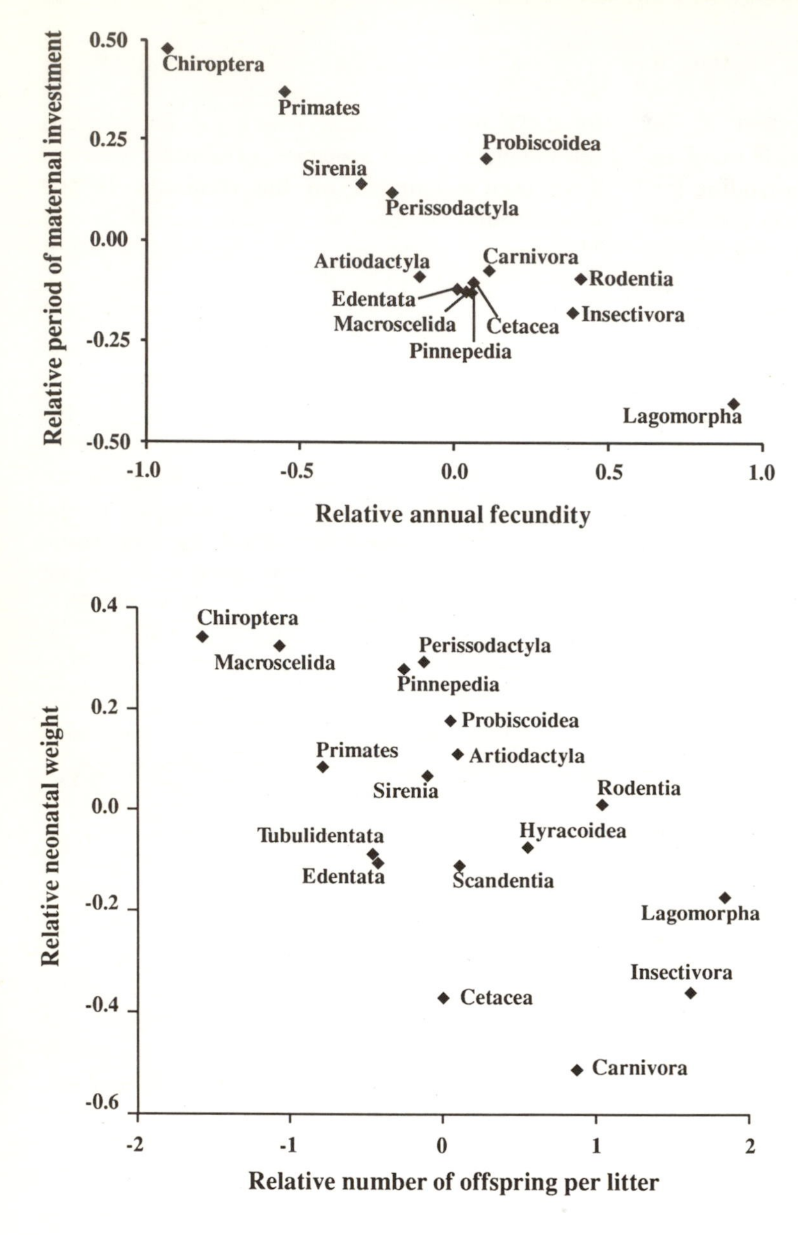

Figure 5.12 Relationship between annual fecundity and length of maternal investment for eutherian orders with effects of body size removed. $r = 0.89$ (after Read and Harvey 1989).

Figure 5.13 Relationship between number of offspring per litter and neonatal weight for eutherian orders with effects of body size removed. $r = 0.89$ (after Read and Harvey 1989).

The nested-ANOVA/allometry practitioner might criticize the phylogenetic-subtraction result for missing most of the variation and for using non-independent species' residuals in the final analysis. Both would be partially correct. The nested-ANOVA/allometry result must be shown to hold independently of taxonomy, that is, the result should hold separately within a number of the higher-level taxa . . . Conversely, if the phylogenetic-subtraction approach shows no effect among the variation remaining at lower levels, but an effect exists at a higher level, an explanation should be given for why it is believed that the higher level effect cannot be separated from taxonomy, and why variation associated with taxonomy cannot be used to address functional questions. (Pagel and Harvey 1988).

The last two points are discussed at the end of the chapter.

The phylogenetic-subtraction method is an analysis of residuals applied to correlation matrices of many traits. It is equivalent to controlling for phylogenetic effects by using categorical variables in multiple regression, with phylogenetic relationships represented as a series of categorical variables in the regression equation (Pagel and Harvey 1988; Harvey and Pagel 1991. They refer to Draper and Smith 1981, p. 241, for statistical details.) The method was first used to answer the question 'Do life history tactics exist at the level of the species, or are they a pattern only detected at higher taxonomic levels?'.

An early product of life history theory was *r* and *K*-selection. It was claimed (MacArthur and Wilson 1967; Pianka 1970) that life history variation in several traits could be ordered on a single continuum. On one end lay the '*r*-selected' species, which matured early, had many small offspring, made larger reproductive efforts, and had a short life. On the other end lay the '*K*-selected' species, which matured late, had a few large offspring, made smaller reproductive efforts, and had a long life. Although the hypothesis was explicitly microevolutionary, broad comparative data were used to support the claim—insects were compared with mammals.

It was noticed (Stearns 1980) that people working comparatively with families, orders, and even phyla were more likely to have found life history 'tactics' than people studying variation within populations. The tendency for a number of life history traits to covary might have been a phylogenetic artefact found in higher-level comparisons and related more to ancient lineage-specific effects than to the local adaptation of populations within species. Subsequent work (Stearns 1983*d*; Brown 1983; Dunham and Miles 1985; Wootton 1987) has supported this claim.

Phylogenetic effects on covariance patterns in the life histories of mammals and birds

The technique was applied to patterns of covariation in the life history traits of mammals (Stearns 1983*d*). Since body weight influences most life history traits, but may be considered characteristic of orders and families as well, I did the analysis with and without having first removed the effects of weight. Means for each trait were calculated for each order (or family) then subtracted from the means for each species. Correlation matrices were then calculated on the residuals. In the unmanipulated data, the pattern called '*r* and *K*-selection' was strong. It accounted for about 68 per cent of the covariation in 10 life history traits. The strength of the pattern was reduced by removing the effects of weight (to about 42 per cent), of orders (to about 32 per cent), and of families (to about 33 per cent). Thus much of the tendency to see the pattern was accounted for by comparisons among higher taxonomic levels. When the effects of weight and phylogeny to the level of family were removed from the data, some covariance of the sort described by '*r* and *K*-selection' did remain.

Gaillard *et al.* (1989) used a variant of this technique on a sample of 80 mammal and 114 bird species for which body weight, fecundity, age at first reproduction, and adult life expectancy were available. In their sample, adult body weight accounted for 40 per cent of the demographic variation in mammals and 60 per cent in birds. They attribute the difference to the constraints of flight. When the effects of body weight were removed, the species could be ranked along a 'fast–slow' gradient analogous to '*r* and *K*-selection', which accounted for 45 per cent of the residual demographic variation in mammals and 35 per cent in birds. This ranking was more closely associated with phylogeny in birds and with ecology in mammals.

Similar effects in squamate reptiles . . .

Dunham and Miles (1985) used the technique, as well as nested analysis of covariance (see below), to perform a similar analysis on the snakes and lizards. Their conclusions differed in some important details. Phylogenetic effects at the level of orders were not strong, but there were important family-level effects on clutch size, mode of reproduction, brood frequency, and age at maturity. Their within-family patterns suggested 'that local adaptation, plasticity of response to local environmental heterogeneity, and physiological constraints . . . are likely to be important determinants of life history variation' (Dunham and Miles 1985,

p. 250). They did not explain why a pattern within family rather than within species supports local adaptation.

... and for age at maturity in mammals
Wootton (1987) examined the relations of body mass, phylogeny, habitat, and trophic level to one trait, age at first reproduction, in 547 mammal species. He concluded,

> Body mass correlated positively with age at first reproduction, explaining 56% of the variance. Habitat and trophic groups deviated significantly from the allometric curve in a pattern generally consistent with predictions from r/K selection ... However, mammalian orders also deviated significantly from the allometric curve, and different habitat and diet groups contained different ratios of mammalian orders. When the effects of orders were removed, residual deviations did not differ among ecological groups. Adjusting for ecological differences did not eliminate the differences between orders. These results suggest that body mass (or some correlated factor) and phylogeny strongly constrain age at first reproduction. Ecological factors appear to have little effect on the evolution of age at first reproduction (Wootton 1987, p. 732).

Compare that statement with the strong adaptationist interpretation of life history evolution given in the first paragraph of the introduction to this chapter. The difference is remarkable. An adaptationist could justly reply that neither habitat nor trophic classification tell us much about those aspects of ecology relevant to life history evolution. Those are captured in how mortality and fecundity rates change as other traits or environmental factors change. Wootton's paper tells us not that life histories are unrelated to ecological factors, but that the relationship is not captured in his ecological categories.

Method 5: The nested analysis of covariance
One uses the nested analysis of covariance (ANCOVA) to test for differences in the *relationship* between two traits, such as weight and clutch size, among species within genera, among genera within families, and so forth. Because the results within each lower-level group are relatively independent of taxonomy, each result can be treated as a statistically independent point bearing on the idea being tested (Pagel and Harvey 1988). This has the effect of increasing the number of times one idea can be tested on a given set of hierarchically organized data.

Bell (1989) suggests that nested ANCOVA can be used to partition the covariation of two traits across genera into components reflecting the between- and within-genera covariances. He argues that the within-genus covariance represents functional covariation, and the between-genus covariance represents the effects of phylogeny. In this interpretation, 'phylogenetic' means 'whatever went on at or above this level in the past', and 'functional' means 'whatever went on below this level since then'.

One's ability to interpret a nested ANCOVA also depends on whether the regression lines for the taxonomic groups cross each other. If they cross frequently, then covariates interact strongly with taxonomic groups to determine the pattern. One way to interpret such covariance is to assign it to phylogenetic effects, since the slopes of the lines depend strongly on the groups for which they are calculated. In such cases, there is no clear way to remove, for example, the effects of weight, and one might get just as much insight by doing a nested ANOVA, letting the weight effects be incorporated into the phylogenetic effects.

Nested ANCOVA applied to squamate reptiles
Dunham and Miles (1985) used nested analysis of covariance to examine patterns of covariation in the life history traits of snakes and lizards. For each of four traits they partitioned the variation into components attributable to body size, families within orders, and orders (Table 5.6). Order and family effects were usually stronger than length effects, suggesting that differences in body plan are more important than differences in body size within this group. The large residual variation suggests that life history traits in squamate reptiles differ most at the level of genera and species. This analysis shows how one can account for the effects of size and phylogeny while isolating the variation not explained by those effects.

Table 5.6 Proportion of variance accounted for in a nested ANCOVA with length as a covariate for 110 species from five families of lizards and three families of snakes (after Dunham and Miles 1985)

Trait	Length	Order	Family	Residual
Clutch size	11	17	24	47
Age at maturity	15	19	19	48
Mode of reproduction	1	18	20	60
Broods per year	6	28	9	57

Method 6: Phylogenetic autocorrelation

Cheverud *et al.* (1985) apply techniques originally developed for spatial autocorrelation analysis to measure the importance of ancestry in comparative data. Their analysis partitions variance into a portion that can be attributed to ancestry and a portion that remains variable within the lowest taxonomic level analysed. Specifically:

$$V_y = V_p + V_s + 2\,\mathrm{Cov}_{ps} \qquad (5.2)$$

where V_y represents the total variance in trait y, V_p is the portion attributable to phylogeny, V_s the portion attributable to species, and Cov_{ps} the covariance between phylogenetic and specific effects. The covariance component would be large if, for example, species whose ancestors had larger clutches tended to evolve larger *changes* in clutch size. Cheverud *et al.* attribute the covariance component to the effects of phylogeny. To estimate the phylogenetic effects, they express the trait as a function of its phylogenetic and nonphylogenetic components:

$$\mathbf{y} = p\mathbf{W}\mathbf{y} + \mathbf{e} \qquad (5.3)$$

where $\mathbf{y}$ is a vector of length n containing the species values of the trait, p is a scalar 'phylogenetic autocorrelation coefficient', $\mathbf{W}$ is an $n \times n$ 'phylogenetic connectivity matrix', and $\mathbf{e}$ is a vector of species values free of phylogenetic effects.

One critical step is the construction of $\mathbf{W}$, the matrix containing estimates of the relationships among the species. More closely related species should receive large coefficients of relationship, and more distantly related ones lower coefficients, but they did not state how the coefficients should be weighted.

Phylogenetic autocorrelations for sexual dimorphism in primates

They found that 50 per cent of the variation in sexual dimorphism in primates could be attributed to ancestry. Most of the impact of ancestry was mediated by size. Forty-two per cent of the variation was accounted for by differences among species, and again, most of this variation was mediated by size differences among species. What distinguishes their technique is its ability to say that 28 per cent of the variation attributable to size results from common ancestry and 34 per cent from other differences among species. This method also avoids the problem of interpreting a nested analysis of covariance in which the slopes have different signs.

Some problems with phylogenetic autocorrelations

Pagel and Harvey (1988) see two problems with phylogenetic autocorrelation. First, the results are quite sensitive to the coefficients of relationship. No agreed-upon methodology exists to determine phylogenetic distances. In those relatively few cases where we have information on genetic distances from studies of electrophoresis, DNA hybridization, or DNA sequencing, they could be used to construct the coefficients of relationship in $\mathbf{W}$. Phylogenetic systematists might prefer a matrix that counted the number of speciation events (including those in hypothetical ancestors) lying between two species on a cladogram.

Secondly, Pagel and Harvey take issue with the idea of including the covariance between phylogenetic and specific effects in the phylogenetic term. They suggest that the covariance could represent functional, adaptive effects in the lineage, an interpretation consistent with their general adaptationist stance on higher-level taxonomic comparisons. This is a problem of interpretation. Whether interaction effects are adaptive or represent historical constraints is an issue that cannot be settled with these kinds of analysis.

The different methods make different assumptions about the role of lineage-specific effects in biology. Methods 1 and 2 were developed to isolate patterns of variation at higher taxonomic levels that suggest adaptations. Methods 4 and 6 were developed to remove those patterns of variation at higher levels so that the adaptive variation at lower levels could be studied. The methods are, however, independent of the reasons for their invention and could be used to either end. They cannot themselves decide whether the effects of size and phylogeny should be treated as adaptations or as constraints.

This survey has not come close to exhausting the analyses that could be done with comparative data. Felsenstein's (1985) method of phylogenetically independent contrasts best controls for the confounding effects of relationship.

Some rules of thumb for comparative analysis

1. Do not begin with multivariate methods. First analyse the data set one trait at a time, then with bivariate techniques, before proceeding to multivariate analyses. Only then can one interpret the multivariate results properly.
2. Try to get a cladogram for the organisms in your data set, for the phylogenetic control offered gives insights into relationship and sequence of origin of traits that cannot be attained by other methods.
3. If you do not have a cladogram, proceed anyway, for most phylogenetic analyses bring some insight, whether the tree used was built cladistically or not.
4. Apply several different techniques to the data set to get complementary insights. For example, use the nested ANOVA/allometry approach, the phylogenetic subtraction approach, and the ANCOVA approach on the same data and compare results.
5. Comparative methods can rule out some hypotheses, but they cannot establish causes because they all rely on correlations rather than experimental manipulations. For many questions, this is the best that can be done because experiments would be impractical.
6. Comparative analyses are the only way to establish the generality of experimental results. There is no escape from the trade-off between experimental rigour and phylogenetic generality.

THE ORIGIN OF PHYLOGENETIC EFFECTS

In the comparative analyses discussed above, it was assumed that common ancestry affects patterns of variation in different ways in different lineages, but no explanation was given for how those effects arose. They were just assumed to be there. An hypothesis for the origin of phylogenetic effects (Stearns 1989*a*) builds on the traditions of comparative morphology and developmental biology. The approach is exemplified with a research programme that places functional morphology into an explicit phylogenetic framework, revealing the constraints, opportunities, and irreversibilities introduced into evolution by key transitions in life history traits. Such research programmes could get at phylogenetic constraints on life history evolution directly, through experimentation in a comparative context.

The origin of phylogenetic constraint

Whether we imagine that genes existed first and evolved organisms, or that organisms existed first and evolved genes, is a distinction with consequences for causation. I think they evolved together, but no matter what the sequence, the genes have not had point-for-point control over the organism since they *were* the organism in the primordial soup. The reasons include the following.

First, building complex phenotypes was advantageous, and evolution hit early on modular design as a method of rapidly building up complex structures. Organisms are often built out of repeating parts.

Second, those parts, especially organelles and cells, had intrinsic properties arising from their construction materials and their interactions.

> Genes do not make structures directly. They code for molecules which either regulate the expression of other genes or confer characteristic properties on cells. Differential gene activity endows cells with certain developmental properties. These include . . . : rates of cell division; cell death; timing of differentiation; cytoskeletal properties (e.g., cell motility, including the capacity to change cell shape, to move and be guided by certain external cues . . . , and the

capacity to exert mechanical tractions on neighboring cells and the extracellular matrix); cell adhesion and cell–cell recognition properties . . . ; and secretion of extracellular matrix materials, which changes the chemical constitution of the cell's environment . . . So equipped, the cells then construct organs and structures in accordance with the laws of physics and chemistry (Oster *et al.* 1988).

The genes did not store information about the properties intrinsic to the structures whose production they controlled, but they could modify those properties within limits.

Third, genes mutate, and most mutations are disadvantageous. Canalized developmental systems buffered against genetic change evolved for traits whose integrity was crucial for survival and reproduction.

Some new parts were added sequentially as repetitions of existing modules (e.g. segments, limbs, digits). The new parts were then modified. Sometimes old parts were discarded or incorporated in the new parts. Evolution did not always proceed by modular addition, and although modular design broke down to some degree, enough remained for its traces to be detected in developmental regeneration experiments and in atavistic mutations. In both cases, functioning subsystems appear as more-or-less intact modules at a place and a time in development that correspond to the historical sequence of evolutionary transformations that can be inferred on other grounds (cf. Müller 1989).

To survive and reproduce, an organism must function in an integrated fashion, not as a collection of independent parts. Therefore, the other parts of an organism are the most important elements in the selection regime of any trait. Each trait must interact successfully with the others to produce a functioning whole. As an organism evolves, some of its traits are fixed as its design is improved to produce efficient and reliable function. The fixation of function and form is carried out by the developmental processes that buffer phenotypic design against genetic change. This canalization is rarely perfect, but it does make change more difficult.

When form and function have been fixed for one trait or module, they become reference points for selection on the other parts of the organism. They are taken for granted, and other changes are made on the assumption that things are going to stay as they are. Once this integration has started, it results in constraints, for some traits can only function if other traits remain fixed. Irreversibility enters evolution. The modules acquired earliest have been fixed in their essential features for over a billion years: meiosis, membrane structure, glycolysis, many macromolecules, coenzymes, organelles. Those added most recently remain more flexible, but any feature shared by all species in a phylum or class has been fixed for hundreds of millions of years and is now deeply buried in a framework of integration.

This model explains the evolution of lineage-specific constraints through the fixation of form and function in the traits that characterize the lineage: the backbone and internal skeleton of vertebrates, the segmented limbs and exoskeleton of arthropods, and so forth. At some point in the past, deeply embedded features that now characterize phyla were the relatively labile differences between recently separated sister species. Only after they became differently incorporated in complex subsystems did they constrain the further evolution of each lineage in different ways.

The fixation of key traits leads to organismal integration that progressively constrains evolution into the lineage-specific patterns measured by the comparative method as phylogenetic effects. The hypothesis is consistent with some interpretations of regeneration experiments and atavistic mutations, but it is by no means fully corroborated.

Bull and Charnov (1985) discuss several examples of irreversibility, including all-female parthenogenesis, polyploidy, heteromorphic sex chromosomes, haplo-diploidy, selfing, and dioecy.

A qualification—blocks of traits can evolve independently

This is Günter Wagner's idea (pers. comm.). If we have measured the genetic variance and covariance of many traits, we can ask, what permutation of the rows and columns of the matrix will produce the most block-like structure (Fig. 5.14)? For purposes of illustration, define the covariances

$$\begin{bmatrix} sn\ sn\ nn \\ ns\ nn\ nn \\ sn\ sn\ nn \\ ns\ ns\ ns \\ nn\ nn\ sn \\ nn\ ns\ ns \end{bmatrix} \qquad \begin{bmatrix} ss\ nn\ nn \\ ss\ nn\ nn \\ nn\ ss\ nn \\ nn\ ss\ nn \\ nn\ nn\ ss \\ nn\ nn\ ss \end{bmatrix}$$

Figure 5.14 Genetic variance–covariance matrix permuted to define blocks of covarying characters.

simply as significant (*s*) or non-significant (*n*). The traits in the matrix on the left are arranged into rows and columns in a way that appears to be biologically arbitrary. Those on the right are arranged so that traits appear as blocks within which connections are significant but among which they are weak. The transformation from left to right was accomplished by switching trait 2 with trait 3 and trait 4 with trait 5.

If traits can in fact be arranged in submatrices within which interactions are important, but among which traits vary more freely, then we need to know the following: (1) Do the submatrices correspond to functional units, e.g. a limb, an organ system, all traits coded by an imaginal disc? (2) Is this a method of defining homologues (traits)? A trait would the correspond to a submatrix. (3) How does the evolution of such independent units occur? It must require the uncoupling of developmental and physiological processes. With this sort of structure, trait fixations would constrain traits within submatrices more than among them.

Constraint and opportunity in plethodontid salamanders

Wake and Larson (1987) describe the details of this case, presented here in condensed form. The ancestor of the plethodontid salamanders had aquatic larvae, metamorphosis, and semi-terrestrial adults. This life cycle can still be found in the family, but some plethodontids now mature sexually as larvae while others have direct developing embryos. The direct developers include both subterranean and arboreal species. Two transitions in plethodontid history involved fixation of key traits that led to irreversible changes: the evolution of lunglessness with a switch to dermal respiration (a feature that distinguishes the whole family) and the evolution of directly developing embryos.

Wake and Larson discuss three structures, the premaxilla (part of the jaw), the autopodium (the foot and wrist), and the feeding system. Irreversible change is seen most clearly in the feeding system. Plethodontids project their tongues to capture prey. Both the loss of lungs and the switch to direct development have created opportunities for improvements in tongue-projection that have led to irreversible change. The bones and muscles used for breathing in lunged ancestors have specialized in the throat and mouth to form part of the tongue projection system. There are two ways to fold the skeleton during projection of the tongue. Force can be transmitted either through the first or through the second ceratobranchial element, which in turn determines the folding pattern.

The history of the lineage determines the option taken. Plethodontids with aquatic larvae (e.g. *Eurycea*) used the first option because the first ceratobranchial plays a role in larval feeding that causes it to be larger than the second ceratobranchial. The larval stage constrains options in the adult because metamorphosis is not sufficiently profound to reverse the relationship. Where the larval stage has been lost, as in the Bolitoglossini, the development of the throat is restructured, the first ceratobranchial remains small, and the second develops early into a stout structure. 'Lunglessness and direct development are necessary preconditions for the evolution of option 2 tongues, and a larval phase constitutes a developmental constraint that imposes option 1 on the evolution of freely projectile tongues by plethodontids that lack the derived life history trait' (Wake and Larson 1987, p. 45).

The change from larval to direct development released the development of the jaw from constraints imposed in the larval stage. Subsequent selection for efficiency in adult feeding eliminated development structures previously used in larval feeding. A return to a larval stage in the Bolitoglossini is probably not possible. In this case the mechanical causes of organismal constraint on life history evolution are clear. These constraints apply to the range of habitats occupied by each type of salamander.

> The adaptive zone inferred to be ancestral for plethodontids featured habitation of mountain streams by both larval and metamorphosed adult salamanders. The replacement of lungs with cutaneous respiration is believed to have been a key feature permitting utilisation of this environment; reduction and loss of lungs are observed in other lineages of salamanders that use stream habitats.... The directly developing plethodontids (Bolitoglossini and Plethodontini) never use stream habitats and may be viewed as occupying a novel, terrestrial adaptive zone. Subsets of both terrestrial lineages evolved adaptations for arboreality and subsets of the bolitoglossines have also evolved adaptations for fossorial existence. In each case, these adaptive transitions featured the fixation of novel features ... that altered the boundaries on organismal-environmental interaction and stimulated subsequent speciation and morphological diversification (Wake and Larson 1987, p. 47).

Key innovations imply irreversibility because they result in organismal integration and lead to dramatic changes in habitat and selection pressures. These cause further changes, including changes in habitat selection, making unlikely a return to the environment in which the prior state of the key innovation was favoured.

Phylogenetic constraints on parthenogenesis

Parthenogenesis never occurs naturally in frogs or mammals (Elinson 1989), but the reasons differ in the two groups. In frogs, the centrosome that replicates to form the poles of the first mitotic spindle is contributed by the sperm. Without sperm-contributed centrosomes, egg division aborts. Gynogenetic frogs use the sperm centrosomes but avoid nuclear fusion to produce functionally parthenogenetic offspring. In mammals, the egg supplies the centrosome, but normal development requires one egg- and one sperm-derived nucleus. 'A diploid mouse egg with two haploid sets of egg-derived chromosomes does not develop. ... The marking of the sperm- and egg-derived nuclei is known as imprinting and may involve DNA methylation ...' (Ellinson 1989, p. 259). This prevents the evolution of parthenogenetic mammals. Alberch and Gale (1985), Oster *et al.* (1988), and Müller (1989) discuss other persuasive examples of developmental constraints.

DISCUSSION: CAN COMPARATIVE RESULTS DEMONSTRATE ADAPTATION?

Problems with the adaptationist interpretation

Pagel and Harvey (1988) raise two important points in saying 'an explanation should be given for why it is believed that the higher level effect cannot be separated from taxonomy, and why variation associated with taxonomy cannot be used to address functional questions'. Often, there is not enough evidence available to provide a convincing answer one way or the other, but in the plethodontid salamanders the evidence is rich. There a key trait, direct development, is fixed at the level of the tribe and is responsible for many of the differences in life histories found among tribes. Since it and the changes consequent to it are among the traits used to define taxa, higher-level effects on life histories cannot be separated from taxonomy, and we can regard those effects as phylogenetically constrained. Such well-worked examples are rare, however, and not all higher-level variation is confounded by phylogeny.

On the other hand, the same changes that constrained life histories produced opportunities exploited in improving the efficiency of the feeding structure. Those changes, associated with phylogeny and identified through careful studies of development and functional morphology, were used to address functional questions. If the freely projectile tongues of bolitoglossines are not an adaptation, nothing is. Taxonomic variation can be used to address functional questions, but it cannot be used convincingly without additional information from approaches—like these—that separate the effects of adaptation and constraint.

Suppose that one has done an ANOVA/allometric analysis for birth weight and habitat within the Order Rodentia. One finds that most of the variation in birth weight is explained by variation at the level of the family and above, so one chooses the family as the appropriate level of analysis. This means that variation among genera and species is regarded as an irrelevant distraction; it will be absorbed into the family means. One then looks for patterns in the association of families with certain habitats or lifestyles. One is careful to acknowledge that the species within a family may occupy several habitats and defines the 'habitat occupied' by a family by working only with families in which at least 80 per cent of the species occupy the same habitat. One might then find that the weight-corrected birth weights of rodents living in grasslands are smaller than those of forest-dwelling rodents.

What are we to make of this result? The analysis has identified the taxonomic level at which an evolutionary comparison can be made with points that are statistically independent. At that level, one finds a correlation between relatively heavy birth weights and living in forests rather than grasslands. Is a heavier birth weight therefore an *adaptation* to forest living? That conclusion can only be tested experimentally, but the analysis has made it more plausible and therefore more worth testing.

One problem is that many factors could confound the comparison. Experimental micro-evolutionary studies might later show that birth weight has little to do with the capacity to survive and reproduce in one habitat or the other, but that it is phylogenetically correlated with physiological processes that are decisive. Second, the comparison was carried out on families, and rodent families are between 10 and 50 million years old. In what sense can a pattern that is 10–50 million years old represent an adaptation to local conditions? The assertion implicitly assumes that species have been living in the same habitats since the origin of the families. This is conceivable but neither convincing nor easy to check. Third, the families of rodents are more heterogeneous than the analysis indicates. If it were to be repeated within each family, the amount of variation in each trait accounted for by genera and species would be significant for some traits in some families. The relative unimportance of that variation was a statistical artefact caused by the inclusion of order and family effects.

Not all arguments count against adaptation at higher taxonomic levels. Harvey *et al.* (1989) reject the notion that life history variation in the mammals is primarily a consequence of body size, brain size, or metabolic rate, all of which could be seen as constraints. They argue instead that life history variation results from differences among fertility and mortality schedules that vary with habitat. In support of this view, they present data taken from Promislow and Harvey (1990) on 48 mammal species belonging to 21 families in 10 orders where life tables were available. They found significant correlations across families between life history traits and mortality rates when the effects of body size had been removed (Table 5.7). Two of these traits have been particularly thoroughly investigated with optimality models that assume adaptation within populations, not among families, and for those two traits—age at maturity and number of offspring per litter—the correlations are precisely what one would expect. Here the phylogenetic effects on a broad, comparative data set were apparently not strong enough to mask the effects of microevolutionary adaptation.

Table 5.7 Correlations across families of placental mammals between life history traits and mortality rates with the effects of body size removed. From Harvey *et al.* (1989)

	r	n	p
Gestation period	−0.44	21	0.048
Age at weaning	−0.50	18	0.040
Time from weaning to maturity	−0.52	18	0.026
Age at maturity	−0.61	20	0.004
Inter-litter interval	−0.47	20	0.037
Maximum recorded lifespan	−0.03	17	0.240
Neonate weight	−0.39	20	0.120
Litter weight	−0.26	20	0.260
Number of offspring per litter	+0.59	21	0.005

Problems with the interpretation in terms of constraints

Claims that certain traits or patterns are phylogenetically, morphologically, or developmentally constrained have been made more frequently than they have been well supported. Lauder and Liem

(1989) propose a method for bringing rigour to the study of constraints. While their method was developed for morphological traits, it can be applied, with appropriate modifications, to life history traits. In brief, they suggest that one should take the following steps.

First, define the novelty—for example, semelparity, viviparity, parthenogenesis. This is best done with a cladistic analysis. Second, formulate a causal model that connects the novel feature with other parts of the life history. Third, sample in-group and out-group taxa and quantify the relationships involved in the causal model. This should uncover exceptions to the rule and make some implicit assumptions explicit by showing where they break down. If the pattern of evidence continues to support the interpretation of constraint, then the claim has been strengthened by having survived systematic scrutiny. The strength of the claim of constraint must rest on the plausibility of the causal model, the depth to which it is analysed, and the quality of the evidence used to test it.

One should not claim that a trait is constrained just because it does not vary within a lineage. That pattern only suggests a hypothesis that must then be tested both with a phylogenetic analysis and with a causal model.

CHAPTER SUMMARY

All comparative work relies on a sound phylogeny. Comparative studies can be used to establish generality, reject certain hypotheses, and suggest others. Important lineage-specific physiological constraints include those characterizing: homeothermy vs. poikilothermy and altricial vs. precocial birds and mammals. Allometric relations are those in which two traits are related by a power function. They usually describe the way a trait varies with size, or how two traits both of which have allometric relations to size vary with respect to each other. Allometric effects on growth and metabolic rates are well documented and differ among lineages and physiological categories.

The lineage-specific mode of growth affects the growth and reproduction of individuals. The major divisions are between determinate and indeterminate and between continuous and discontinuous growth. Determinate growth stops at maturity; indeterminate growth continues. Discontinuous growth characterizes organisms with exoskeletons; continuous growth, those with endoskeletons.

Phylogenetic systematics reconstructs the relationships of species as natural groups. The method has four steps. First, non-phylogenetic criteria are used to find similarities in homologous traits among species. Then out-group comparisons distinguish shared general, shared special, and unique homologies. Groups are formed according to shared special traits; all traits are placed on the resulting tree. Finally, inconsistencies are interpreted as parallel and convergent evolution.

Cladograms can be used to determine ancestral and derived conditions; it is the derived conditions that most clearly need explanation as adaptations. Cladograms reveal the sequence in which traits evolved and therefore test any hypothesis dependent on that sequence. One can also count the number of times an evolutionary transition occurred independently on a cladogram and use these in statistical tests in which the points are independent of one another.

The different statistical approaches to comparative analysis have complementary strengths and weaknesses; several of them should normally be applied to the same data set. Comparative data should be taken seriously. No other approach establishes the taxonomic breadth of a pattern so well. Few other approaches generate so many hypotheses so quickly.

The origin of phylogenetic constraints lies in the fixation of key traits followed by organismal integration with consequent irreversibility of the fixation. The plethodontid salamanders illustrate irreversibility for the key transition from larval to direct development. Changes in feeding structures made a return to indirect development impossible.

A claim that a trait is constrained because it does not vary within a lineage is a hypothesis that must be tested with phylogenetic analysis and a causal model.

RECOMMENDED READING

Felsenstein, J. (1988). Phylogenies from molecular sequences: inference and reliability. *Ann. Rev. Genet.* **22**, 521–66.

Harvey, P. and Pagel, M. (1991). *The comparative method in evolutionary biology*. (Oxford University Press, Oxford).

Maynard-Smith, J., Burian, R., Kauffman, S., Alberch, P., Campbell, J., Goodwin, B. *et al.* (1985). Developmental constraints and evolution. *Q. Rev. Biol.* **60**, 265–87.

Pagel, M. and Harvey, P. (1988). Recent developments in the analysis of comparative data. *Q. Rev. Biol.* **63**, 413–40.

Vogel, S. (1988). *Life's devices: the physical world of animals and plants*. (Princeton University Press, Princeton).

Wake, D. B. and Roth, G. (eds). (1989*a*). *Complex organismal functions: integration and evolution in vertebrates*. Dahlem Workshop Report. (John Wiley & Sons, New York).

PROBLEMS

1. Using the *Zoological Record* or other sources, get the taxonomic descriptions of all species in a small genus that interests you. Do the same for its sister genus. Code the traits and, using software like *MacClade*, generate and evaluate cladograms until you are satisfied with the result. Compare your result with the standard taxonomy in the literature. Discuss.
2. Paterson has suggested that in most speciation events, the mate recognition mechanisms evolve first and lead to reproductive isolation, not the other way round as Mayr suggests. How would you test this hypothesis on a recent adaptive radiation?
3. In Fig. 5.10, a common allometric line was calculated for all mammals, and the deviations of elephants and shrews were obtained by subtracting the value for an average mammal of that weight from the values of the two species. When should one use a line for all members of a class, and when should one use separate lines for each order, family, or genus? How do the residuals differ in meaning? See Dunham and Miles (1985) and Harvey and Pagel (1991).
4. When is a correlation found at the level of the subfamily between one trait and an environmental variable an adaptation? Consider variation within subfamilies and the time elapsed since the subfamilies were separated.
5. How might one improve and test the idea that phylogenetic constraint originates in the fixation and subsequent integration of key traits?
6. In Chapter 6, age at first reproduction is analysed with optimality models that assume a microevolutionary equilibrium driven by demographic forces incorporating ecological factors. We will find a remarkable agreement between the predictions of optimality models and observations on natural populations. Wootton (1987) interpreted his results as demonstrating strong phylogenetic constraints on age at maturity. Can both his interpretation and the optimality results be correct?
7. Get hold of a database for a fairly large group—such as the mammals, the birds, or one of their orders—and apply first the ANOVA/allometry method, then the method of phylogenetic subtraction, then the nested analysis of covariance. Contrast the insights gained.

DISCUSSION OF PART I

DISCUSSION OVERVIEW

The goal of this section is to tie together Chapters 2–5 before proceeding to the analysis of life history traits in Part II. To understand a pattern of variation in life history traits, whether interspecific or intraspecific, we need information on phylogenetic effects, phenotypic variation, genetic variation, trade-off structure, and the selection pressures generated by demography. Such a multidimensional analysis of life history evolution has not yet been achieved for any group of organisms. Its strength would lie in the capacity of each approach to generate an alternative view of the same pattern and, through their combination, to weight the effects of adaptation and historical constraint. Despite considerable evidence for phylogenetic effects on life history traits, their evolution can be rapid enough to allow local adaptation. This is no paradox. Some are fixed within lineages and evolve very slowly. Others vary within populations and evolve rapidly. Which traits are 'slow' or 'fast' depends on the lineage.

CONSTRAINTS

Traits that are fixed within lineages constrain within certain bounds those that continue to vary. Only those that vary are involved in trade-offs. Each lineage, with a different pattern of fixations, is constrained in a different way. Each lineage, with a different set of traits that vary, has a different trade-off structure. A phylogenetic analysis is necessary to determine the pattern of fixations. These fixations can then be taken as a hypothesis for the origin of constraints on the trade-off structure.

The longer a trait has been fixed, the more deeply embedded it is in the structure of the organism. Traits fixed early are probably more important constraints than those fixed later, but their antiquity makes experiments difficult. Disturbances of ancient patterns often lead to death early in development. However, combining comparative phylogenetic analysis with functional morphology and development can clarify many effects of fixations (cf. Wake and Larson 1987). We need to demonstrate the mechanisms connecting lineage-specific fixations with the broad-sense trade-offs that remain. That would reduce the arbitrary assumptions that have to be made in an optimality analysis.

THE RATE OF EVOLUTION OF LIFE HISTORY TRAITS

Without a rapid response to selection, environmental conditions will change before the population can adapt. Only traits varying within species can respond rapidly to selection. One measure of evolutionary speed is the darwin, defined (Haldane 1949) as a change by a factor *e* (the base of the natural logarithms) per million years, or about 0.1 per cent of the value of a character in 1000 years. It is measured as the difference in the logs of the initial and final values of the trait divided by the time interval in millions of years. The unit was invented for the fossil record, where generation times are not easy to determine.

The fastest change found in the fossil record is for dwarfing in elephants that colonized islands in the Mediterranean during the Pleistocene. They shrank at a rate of 12–16 darwins to half their former size within 50 000 years. That rate may seem rapid, but in the laboratory and in colonization events, evolution can run much faster (Gingerich 1983—Table D1.1). The rate of evolution of bristle number in *Drosophila* or body size in mice can reach hundreds of thousands of darwins but can be maintained only until genetic variation is exhausted. As Yoo (1980) showed, in experiments in which the rate of change of bristle number was about 517 000 darwins and in which the response to selection had not stopped after 85

Table D1.1 Rates of morphological evolution (in darwins) for selection experiments, colonization episodes, and fossil data. Note the consistent negative correlation of rate with the interval over which it was measured. From Gingerich (1983) with Yoo's (1980), Stearns' (1983*a*, *c*), and Reznick *et al.*'s (1990) data added.

Domain and correlation	Sample size	Time interval Range	Time interval Mean*	Evolution rate (d) Range	Evolution rate (d) Mean*
I Selection experiments >0.60	9	1.5–10 y	3.7 y	12 000–517 000	65 100
II Field manipulations§	1	—	13 y	6 000–24 000	10 700
III Colonization 0.78	104	70–300 y	170 y	0–353 000	>370
IV Post-Pleistocene mammals 0.22	46	1 000–10 000 y	8 200 y	0.11–32.0	3.7
V Fossil invertebrates 0.52	135	0.3–350 my	7.9 my	0–3.7	0.07
VI Fossil vertebrates	228	0.008–98 my	1.6 my	0–26.2	0.08

* Geometric means.
§ Reznick *et al.* (1990).

generations (3.27 years), that can take quite a while. Often after artificial selection is relaxed the trait returns part of the way towards the value it had before selection began.

Evolution is also rapid during colonization episodes. Rates are much lower in fossils than in living organisms, but much of this difference results from the different time intervals involved. Evolutionary rate is negatively correlated with the time interval over which it is measured; thus over short intervals faster rates are measured than over larger intervals. Also note the broad ranges. Even during colonization, evolutionary rates can be zero, and when measured in detail in natural populations, they vary markedly from year to year.

For example, long-term studies of great tits in Holland have shown that the most productive clutch size varies from year to year, that clutch size is heritable, and that clutch size responds genetically in the short term to selection (van Noordwijk *et al.* 1981). The most productive clutch size varies from 6 to 12, and if one looked only at the pattern over a 30 year interval, one would probably conclude that all clutch sizes between 6 and 12 were 'equally fit'. Finer analysis shows, however, that fluctuating selection pressures

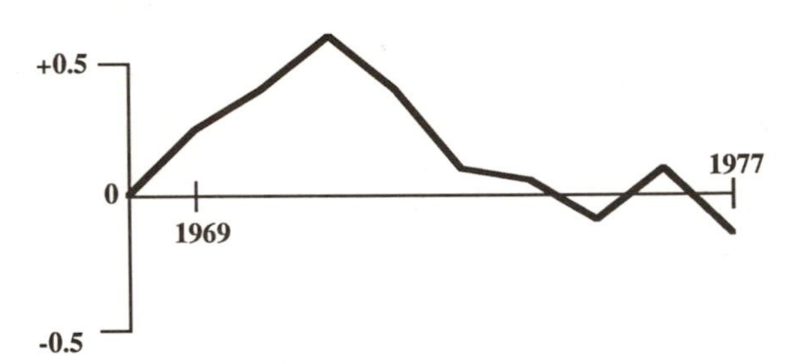

Figure D1.1 Cumulative selection differential on clutch size of great tits on an island off the Dutch coast (after van Noordwijk *et al.* 1981*b*).

(Fig. D1.1) are causing heritable changes in clutch sizes from year to year. In the long term, selection is stabilizing—it holds clutch sizes within broad limits; in the short term, selection is directional and the response to it is rapid, on the order of thousands of darwins.

A similar pattern over a longer interval can be seen in the fossil record of a Miocene stickleback (Bell *et al.* 1985). Here samples were collected over short time intervals. If we only had samples at the beginning and the end of the sequence, we would conclude that relatively little change had occurred. Because the intermediates are available, a completely different picture emerges. Population

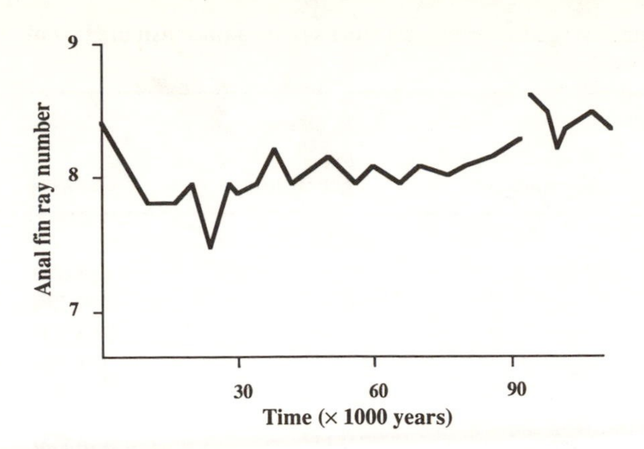

Figure D1.2 Changes in anal fin ray number in Miocene sticklebacks, *Gasterosteus doryssus.* The change from start to finish corresponds to 224 darwins over 110 000 years. The most rapid short-term change corresponds to 4960 darwins over 5000 years (after Bell *et al.* 1985).

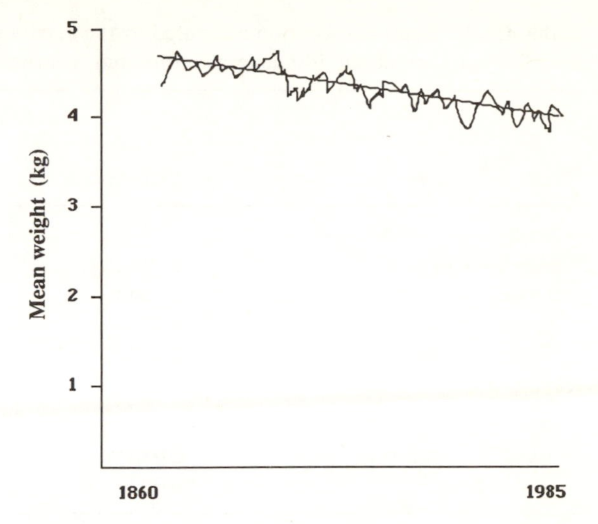

Figure D1.3 Mean weights of two-sea-year salmon in the Godbout River, Quebec, from 1864 to 1983 (after Bielak and Power 1986).

means fluctuate, sometimes around a long-term trend, sometimes with no trend at all, but with significant short-term changes that cancel each other out in the long run (Fig. D1.2). As with the great tits, selection is directional in the short-term but stabilizing in the long-term.

Field studies also show that life history traits can respond rapidly and directionally to selection. Fishing exerts a selection pressure on commercially exploited stocks. Where nets are used, selection usually favours fish too small to be caught in the nets, and the population evolves towards smaller mean size. Bielak and Power (1986) have shown that since 1864 the Atlantic salmon caught in the Godbout River (Quebec, Canada) have become significantly younger and smaller (Fig. D1.3). The decrease in size amounts to a change of 0.5 to 0.9 per cent per year, depending on the age class. Salmon that spent two years at sea decreased from an average of 4.8 kg to 4.0 kg in 119 years—a rate of 3065 darwins.

Mosquito fish introduced to Hawaii in 1905 were put into sugar plantation reservoirs with a broad range of selection regimes with few predators or competitors (Table D1.2). About 140 generations later, the maximum divergence among six stocks raised under constant conditions in the laboratory amounted to 0.08 to 0.5 per cent per generation. (That the changes took place gradually and continuously is supported by the similarity of the rates for Trinidad guppies—see below—which were measured over a 13 year period.) They range from 138 000 to 353 000 darwins and approximate the rates seen in selection experiments.

In a field manipulation experiment in Trinidad, Reznick *et al.* (1990) changed the age and size specific selection pressures on guppies by changing the predators that the fish encountered (Table D1.3). After 13 years they found several significant changes, especially in early life history traits. The changes were rapid, occurred in response to known, manipulated selection pressures, and took the direction predicted by theory (cf. Chapter 7).

Although the fossil record indicates that the rate of change of quantitative characters is slow, these examples show that some life history traits can evolve rapidly. The conditions for rapid evolution are an ample supply of genetic variation and strong selection. Heritabilities for many life history traits are substantial. Colonizations and the impact of human activities both provide strong selection pressures. The result is evolution so rapid that evolutionary and ecological time overlap. Local adaptation then produces a large change in the direction of the optimum predicted by equilibrium theory.

Limits on the evolution of life history traits in

Table D1.2 Rates of evolution measured by the maximum divergence among six stocks of mosquito fish in Hawaii over 70 years, or about 140 generations. From Stearns (1983*c*)

Trait	Difference	Significance	% change per generation	darwins
Females				
Age at maturity	18.9 days	0.001	0.16	138 600
Length at maturity	1.5 mm	0.001	0.06	139 700
Growth rate	0.0048	0.001	0.50	265 700
Weight at birth	0.23 mg	0.001	0.15	335 000
Males				
Age at maturity	16.1 days	0.05	0.11	344 900
Length at maturity	2.0 mm	0.001	0.08	346 500
Growth rate	0.0079	0.05	0.30	352 900

Table D1.3 Rates of evolution measured by the divergence of controls and experimentals in mosquito fish in Trinidad over 13 years, or about 30 generations. From Reznick *et al.* (1990)

Trait	Difference	Significance	% change per generation	darwins
Females				
Age at maturity	9.7 days	0.05	0.12	5 707
Weight at maturity	24.1 mg-wet	0.01	0.23	10 700
Weight at birth 1st brood	0.08 mg-dry	0.10	0.15	6 770
Weight at birth 2nd brood	0.12 mg-dry	0.05	0.22	9 630
Size of 1st brood	1.2 young	0.05	0.23	23 860
Males				
Age at maturity	9.7 days	0.05	0.30	14 020
Weight at maturity	8.6 mg-wet	0.001	0.20	9 230

large, natural populations are probably not set by the exhaustion of genetic variation for any one trait, but by trade-offs among traits with significant phenotypic and genetic covariance.

THE INTERPRETATION OF OPTIMALITY MODELS

From Chapter 5, it should be clear that there can be no such thing as one general theoretical model for life history evolution. Demographic theory (Chapter 2) tells us what questions to ask. It focuses our attention. Comparative analysis defines the boundary conditions on life history evolution by giving us information on the lineage-specific pattern of fixations and trade-offs (Chapter 5). Using that information, we can build a model for one species. Because many traits are involved and the answers for one trait often depend on the values of the others (Chapter 2), the predictions we make for an optimal life history are specific to an organism. A model built for a fish cannot be applied to a tree until we can find 'Source Laws' for trade-offs. Until then, the models will remain local, but the framework and approach are general. We always need to ask, what traits are fixed and how do the variable traits trade off? Then we can apply demographic theory to explore the selection pressures on the phenotype (Part II).

We now know that there is abundant genetic variation for many life history traits that respond rapidly to selection in both field and laboratory, but that they do not normally evolve rapidly in nature. Only two explanations are consistent with this pattern: symmetrically varying directional selection or stabilizing selection, but not steady directional selection or disruptive selection. Symmetrically varying directional selection is unlikely; stabilizing selection for an intermediate optimum that varies from place to place and year to year is more plausible. This is one justification for shifting, in Part II, to an emphasis on optimality models of evolutionary equilibria—such an approach assumes a situation that appears to occur frequently.

PROBLEMS

1. What is an organism? What is an individual? Are birds, trees, corals, and slime moulds all organisms in the same sense? How does the answer to this question affect the definition of a life history and the model of an organism used to concentrate attention on certain traits?
2. What is a trait? Organisms are integrated. What are the implications of the decision to partition an organism in a given way? Is there always a preferred partitioning for a well-defined problem? Is there a natural partitioning given by genetics? Is there a natural partitioning given by evolutionary morphology? Are they the same partitioning?

PART II

THE EVOLUTION OF THE MAJOR LIFE HISTORY TRAITS

6

AGE AND SIZE AT MATURITY

Small organisms are usually small not because smallness improves fecundity or lowers mortality. They are small because it takes time to grow large, and with heavy mortality the investment in growth would never be paid back as increased fecundity. So optimal size depends strongly on mortality, but mortality is often size-dependent. This reciprocal relationship is surely one source of the great variability of life histories found in nature

Jan Kozlowski, 1991

CHAPTER OVERVIEW

This chapter analyses the evolution of age and size at maturity. Maturation divides a life into preparation and fulfilment. Age at maturity is pivotal, for fitness is often more sensitive to changes in this trait than to changes in any other. With maturation, selection pressures and trade-offs change dramatically. The microevolutionary selection pressures and trade-offs are sketched in the introduction. The demographic pressure to mature early must be balanced by trade-offs with other fitness components to explain delayed maturity.

The second section documents variation in age and size at maturity among and within vertebrate classes, among and within populations of single species, and between the two sexes. Species that delay maturity tend to be large, long-lived, and, if they are birds or mammals, to have a few, large offspring, or, if they are reptiles, to have many, small offspring. Much of this pattern is associated with higher taxonomic levels; it weakens but does not disappear within populations.

An optimal age and size at maturity are attained where the benefits and costs of maturation at different ages and sizes balance at a stable equilibrium point or along a reaction norm. Two approaches are discussed in some detail. The first concentrates on two trade-offs, one between early maturation and fecundity, the other between early maturation and offspring survival. It takes r as its fitness definition. The second approach concentrates on growth and fecundity and takes lifetime production of offspring, $R_O = \sum l_x m_x$, as its fitness definition. Both approaches predict optimal reaction norms for age and size at maturity.

I then discuss the relation between optimality models and genetic variation, consider the meaning of apparently successful predictions, and conclude that confirmations of the models tell us more about assumptions than predictions.

INTRODUCTION

In life history theory, age at maturity is *defined* as age at first birth in animals and age at which seeds ripen in plants, not as the age at which some morphological or physiological criterion of maturation is met. The evolutionary equilibrium of any trait can be analysed as the sum of *costs* that reduce fitness and *benefits* that increase fitness. The costs of earlier maturation are the benefits of later maturation (Fig. 6.1).

Benefits of earlier maturity

The principle benefit to early maturation is demographic. Simply because they spend less time

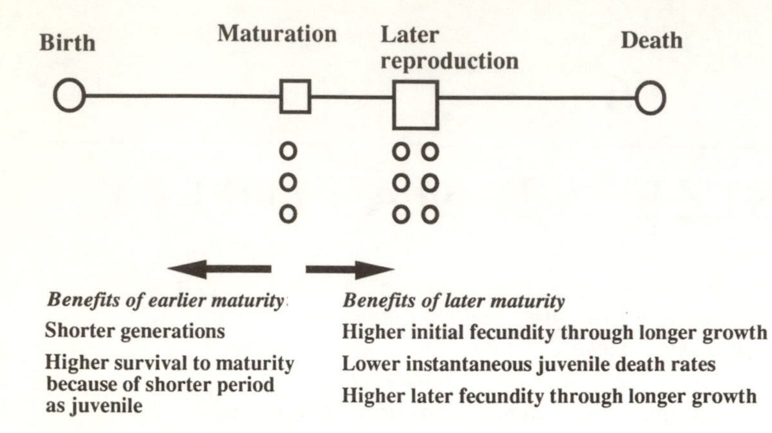

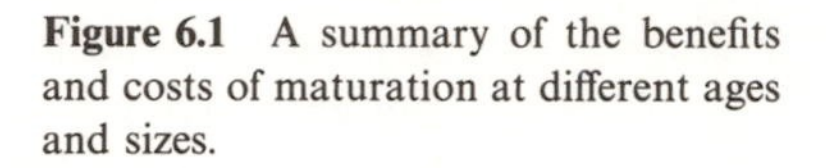
Figure 6.1 A summary of the benefits and costs of maturation at different ages and sizes.

as juveniles, early maturing organisms have a higher probability of surviving to maturity (Bell 1980). Organisms that mature earlier also have higher fitness because their offspring are born earlier and start reproducing sooner (Cole 1954; Lewontin 1965; Hamilton 1966). The magnitude of this benefit depends on the type of life history. It is weaker in organisms that mature early and stronger in organisms that delay maturity (when the change is measured as a percentage—cf. Table 2.8). While its magnitude varies, the benefit is always present. Compared to other life history traits, changes in age at maturity and in juvenile survival have large impacts on fitness across a wide range of types of life histories.

Costs of early maturity

Of the many factors that could act to delay maturity, two are generally important.

1. If delaying maturity permits further growth and fecundity increases with size, then delayed maturity leads to higher initial fecundity. This effect can outweigh the advantages of earlier maturation. Maturity will be delayed for this reason until the fitness gained through increased fecundity is balanced by the fitness lost through longer generation time and lower survival to maturity.

2. If delaying maturity means that the quality of the offspring produced or of the parental care provided will be higher, then delaying maturity reduces the instantaneous juvenile mortality rate of the offspring. This effect can also outweigh the benefits of earlier maturation. Maturity will be delayed for this reason until the fitness gained through the increase in juvenile survival rate due to the mother's greater age at first birth balances the fitness lost through a longer generation time and that portion of survival to maturity that is attributed simply to the delay and not to any change in instantaneous juvenile mortality rate.

Although the simplest way to improve survival to maturity is to mature earlier (Bell 1980), maturity will be delayed if it improves the instantaneous juvenile survival rate sufficiently. These statements are consistent because they deal with two separate parts of survival to maturity. We can represent survival to maturity, $l(\alpha, x)$, as a decreasing exponential function of the instantaneous juvenile death rate, $d(\alpha, x)$, as follows:

$$l(\alpha, x) = \exp\left(-\int_0^\alpha d(\alpha, x)\,\mathrm{d}x\right) \qquad (6.1)$$

Any decrease in age at maturity has its impact on survival to maturity through a change in the upper limit to the integral, α. If a delay in maturity is to improve juvenile survival, it must have its impact through a decrease in the instantaneous juvenile mortality rate, $d(\alpha)$, that is large enough to compensate for the longer period over which integration occurs.

For example, if the instantaneous juvenile mortality rate is a constant $c = 0.1$, then the probability of survival to age 4 is $\exp(-0.1 \times 4) = 0.67$ and to age 5 is 0.61. Now if instantaneous juvenile mortality increases as age at maturity decreases, how strong would such an effect have to be for survival to age 5 also to be 0.67? If $d(\alpha) = c\alpha^{-\gamma}$, then the exponent γ, which can be manipulated to cause the desired effect, should be about 0.14. Thus instantaneous juvenile mortality must fall off exponentially with age at maturity at least as fast as $\alpha^{-0.14}$ if the effect of delayed maturity on juvenile survival is to delay maturity from 4 to 5 years. That is a weak effect about equal to the inverse of the seventh root of age at maturity. Thus the quality of offspring must not improve very rapidly as age at maturity is delayed to more than balance the risk of having to survive a longer period. This does not depend on the constant background mortality rate *c*.

3. If delayed maturity means that the organism will live longer, grow larger, have more reproductive events, have higher fecundity later in life, or otherwise gain in lifetime reproductive success, then maturity may be further delayed. Bell (1980) called the reduction in future fecundity entailed by earlier maturation the 'potential fecundity cost'. Such delayed effects figure in Kozlowski's models of delayed maturity, described below.

Frequency-dependent effects on maturation patterns

When social interactions affect reproductive success, then frequency dependence can maintain more than one maturation phenotype in the population, and ESS models are more appropriate for the analysis than optimality models. Good examples of such systems are the bluegill sunfish studied in Canada by Gross (e.g. Gross and Charnov 1980) and the blue-headed wrasses studied in Panama by Warner (e.g. Warner 1984). In both systems, two types of males can be present: large, dominant males that defend nests or mating territories and attract females through elaborate courtship displays, and small, cryptic, female-mimicking males that achieve their reproductive success by sneaking into the spawnings of the large, dominant males. These two systems have spawned many ideas on the evolutionary maintenance of alternative reproductive strategies and illustrate strong interactions between the evolution of life histories and behaviour.

PATTERNS OF MATURATION

Now we look briefly at taxonomic variation in age at maturity. Early maturation is easily understood. It is delayed maturity that calls for explanation.

Variation in age at maturity

Bell (1980) compiled frequency distributions of 'typical' ages at maturity in several groups (Fig. 6.2). The trend towards early reproduction is stronger in the placental mammals than in squamate reptiles, amphibians, or fish, but the

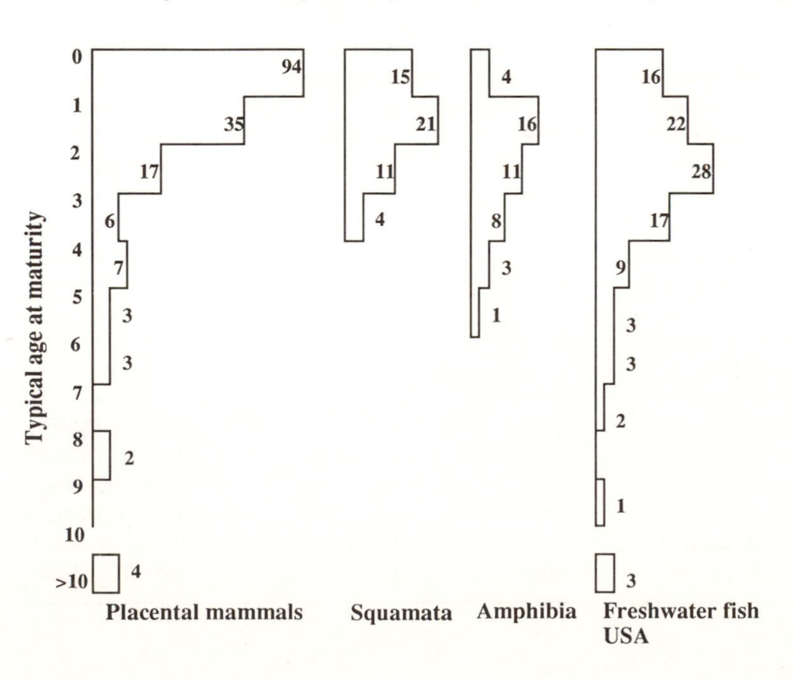

Figure 6.2 Frequency distributions of 'typical' ages at maturity in species drawn from five groups of vertebrates. 'Typical' age at maturity is the midpoint of a range where a range was given, or mean value when that was given or could be calculated. A great deal of intraspecific variation, both genetic and phenotypic, is not represented here (after Bell 1980).

Table 6.1 Mean age at maturity in years of shad (*Alosa sapidissima*) from five Atlantic Coast populations. Populations with similar ages at maturity have the same letter in the significance column (from Leggett and Carscadden 1978)

Males			Females		
Age	Significance	River system	Age	Significance	River system
3.8	a	St Johns (FL)	4.2	a	St Johns (FL)
4.1	b	Connecticut	4.5	b	St John (NB)
4.2	b	York (VA)	4.6	b	Miramichi
4.2	b	St John (NB)	4.7	c	York
4.2	b	Miramichi (Quebec)	4.8	c	Connecticut

range of ages at maturity is as large in the mammals as it is in the fish. Not all species mature as early as would be 'physiologically possible' for their lineage. Roff (1984) documents similar patterns in 69 species of north temperate marine fish.

Mean age at maturity can vary considerably among populations within species. For example, age at maturity in American shad increases from Florida to Canada (Table 6.1). Genetic differentiation among populations was distinguished from environmental effects on six populations of mosquito fish derived from common ancestors introduced to Hawaii in 1905 (Table 6.2). Differences expressed in the laboratory indicated that the populations of mosquito fish had diverged

Table 6.2 Mean age and length at maturity (in days) of mosquito fish from six Hawaiian populations. The fish were raised under similar conditions in the laboratory. Populations with similar values have the same letter in the significance column. From Stearns (1983*c*)

Males			Females		
Age (days)	Significance	Population	Age (days)	Significance	Population
71.3	a	Twin (Hawaii)	78.0	a	Kay (Hawaii)
74.2	ab	Kay (Hawaii)	81.9	a	Res. 33 (Maui)
79.8	ab	Res. 81 (Maui)	85.6	ab	Twin (Hawaii)
84.4	bc	Res. 33 (Maui)	86.9	ab	Res. 81 (Maui)
85.2	c	Res. 31 (Maui)	90.1	ab	Res. 31 (Maui)
86.4	c	Res. 40 (Maui)	96.9	b	Res. 40 (Maui)

Males			Females		
Length (mm)	Significance	Population	Length (mm)	Significance	Population
16.6	a	Kay (Hawaii)	18.3	a	Res. 81 (Maui)
16.1	a	Res. 81 (Maui)	18.5	a	Res. 33 (Maui)
16.3	a	Twin (Hawaii)	19.0	ab	Kay (Hawaii)
18.5	b	Res. 33 (Maui)	19.4	ab	Res. 40 (Maui)
18.5	b	Res. 40 (Maui)	19.8	b	Twin (Hawaii)
18.6	b	Res. 31 (Maui)	19.9	b	Res. 31 (Maui)

genetically in age and size at maturity over the previous 75 years. As in the shad, the patterns for males and females were different. The mean values given in the table hide considerable individual variation within populations. The fastest developing fish, for example, reached maturity at 35–40 days, the latest at 140–160 days. Within a single population the range was as high as 35–100 days.

Even when one rears organisms in similar environments, one cannot be sure that the differences expressed reflect the natural situation. Because each population has probably evolved a different reaction norm, the move from the field into the laboratory may have changed the rank order of the populations for each trait (cf. Chapter 5). Lotz (1990) avoided that problem in a study of 5 populations of *Plantago major* in which seeds from every population were reared in every environment. He found genetic differences among populations corresponding to the selection pressures in each environment.

Table 6.3 Below the diagonal: correlations between life expectancy (LE), age at maturity (α) and fecundity (Fec) before effects of adult weight are removed. Above the diagonal: after effects of weight are removed. From Gaillard *et al.* (1989)

	Birds			Mammals		
	LE	α	Fec	LE	α	Fec
LE	1	0.78	−0.81	1	0.70	−0.55
α	0.85	1	−0.74	0.79	1	−0.56
Fec	−0.87	−0.83	1	−0.77	−0.80	1

The effects of size and phylogeny

As noted in Chapter 5, age at maturity increases with adult body size. This relationship must exist when growth slows or stops after maturation, for one must grow for a longer time to get larger. Age at maturity was strongly correlated with adult body size in 65 mammal species ($r = 0.75$, *slope* = 0.26; Stearns 1983*d*) and in the squamate reptiles ($r = 0.53$, *slope* = 0.32; Dunham and Miles 1985). Following Harvey and Zammuto (1985), Gaillard *et al.* (1989) calculated the correlations among age at maturity, adult life expectancy, and fecundity before and after removing the effects of adult weight for 80 mammal species and 114 bird species (Table 6.3).

The correlations for birds and mammals resemble each other and do not change much when the effects of weight are removed. In both, delayed maturity is associated with long life and low fecundity. Heavier birds and mammals mature later, live longer, and have lower fecundity than lighter ones.

Part of the overall correlation of age at maturity with body size is confounded by taxonomic differences. Whales are larger than rodents, deer are larger than rabbits, and not all the differences in age at maturity among those groups can be attributed to body size alone—the body plan, the shared ecology of the group, and a shared evolutionary history contribute to the variation. In both the mammal and the squamate data, for example, there were strong interactions among age at maturity, adult body size, and taxonomic group. Gaillard *et al.* (1989) found that whales matured relatively early and had high reproductive rates, while bats matured late and had low reproductive rates *for their weights*. The separate effects of size and phylogeny can be weighed with analysis of covariance and autocorrelation analysis. Whenever this has been done, the conclusion has been that both size and phylogeny covary with age at maturity. Body plan makes a difference (Stearns 1983*d*; Dunham and Miles 1985; Harvey and Zammuto 1985; Saether 1987; Gaillard *et al.* 1989).

Social effects: bimaturism

The breeding structure of a species influences the *difference* in age at maturity between males and females. In polygynous species where males compete directly with one another for females, males tend to delay maturity, grow larger, and gain more experience before attempting to reproduce, a pattern that Wiley (1974*b*) called *bimaturism*. Such effects are well known in seals, sea lions, porpoises, horses, goats, sheep, deer, grouse, wrasses, and primates (Fig. 6.3(*b*), Table 6.4).

On the other hand, in species with promiscuous mating, external fertilization and indeterminate growth—where males do not control access to females—one expects the opposite. Because females gain fecundity with size at a higher rate than males,

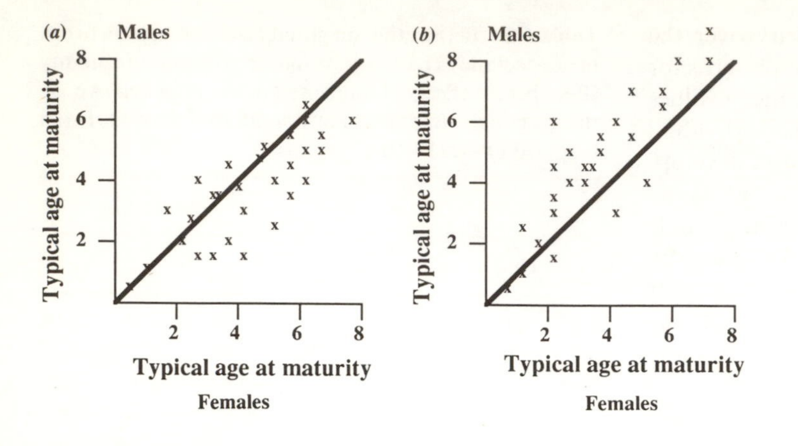

Figure 6.3 In fishes (*a*), females tend to mature later and larger than males, whereas in birds and mammals (*b*), the trend is in the opposite direction. The fish represented are a summary of North American freshwater fish; the birds and mammals are a sample containing some dramatically polygynous species (horses, dolphins, and elephant seals, for example) (adapted from Bell 1980).

one expects the males to be smaller and younger at maturity than the females. That is the case in many fish (Fig. 6.3(*a*)).

Wiley's (1974*a*) review of the interaction of mating systems with life history patterns in grouse is particularly illuminating because of the strong contrasts in mating patterns within the family and the phylogenetic control over the comparisons. Wiley refers to three categories of social structure:

Category I	Promiscuous species that form leks
Category II	Promiscuous species with dispersed males
Category III	Species that form pair-bonds

Categories I and II involve polygyny of various types. The essential point is that in leks (I) males compete directly with each other, by fighting, for the mating sites preferred by females. When the males are dispersed (II), the competition is less direct and less intense. In pair-forming species (III), there is little or no competition among males for mates following pair-formation. In all three categories, females mature at one year, but male maturation is delayed in social systems that promote competition among males for mates (Table 6.5).

Implications of the patterns

That age and size at maturity vary among closely related species, among populations within species, and among individuals within populations suggests that they can respond rapidly to natural selection (cf. Fig. D1.4, Table D1.2). Age at maturity in flour beetles responds to artificial selection (Soliman 1982), and $h^2 > 0$ for age and size at maturity in many species (Mousseau and Roff 1987). This suggests that age and size at maturity can be adjusted by natural selection to local conditions within populations. On the other hand, body size and phylogenetic relationship have strong effects on age and size at maturity and their covariation with other life history traits, suggesting that constraints on age and size at maturity imposed by history and design are also important. Analyses of the selection pressures on these traits within populations should accommodate lineage-specific effects.

OPTIMALITY MODELS OF AGE AND SIZE AT MATURITY

The basic idea: a balance of costs and benefits

Optimization assumes that a balance of costs and benefits determines the value of a trait within the range of available variation (Gadgil and Bossert 1970). In life history theory, the costs and benefits are measured in the currency of fitness. If we plot the summed costs and benefits against age, we can determine the optimal age at maturity graphically (Fig. 6.4). Such diagrams bury complex biology beneath simple functions. We can be certain that one cost of early maturation, the instantaneous mortality rate of the offspring, does climb rapidly when females mature earlier. The capacity of younger and smaller females to produce offspring that can survive must decrease with age; at some point the female will be so small and poorly

Table 6.4 Average age at maturity in males and females of various species of hoofed mammals and primates where the social structure favours delayed maturity in males more strongly than in females. From Schaffer and Reed (1972), Clutton-Brock *et al.* (1982), Harvey and Clutton-Brock (1985)

Taxonomic group	Males	Females	Common name
Family Bovidae			
Rupicapra rupicapra	3.0	2.0	Chamois
Hemitragus jemlahicus	4.0	2.5	Thar
Ovis canadensis	6.0	2.5	Bighorn sheep
Ovis musimon	4.0	2.0	Sheep
Family Cervidae			
Cervus elaphus	6.0	3.0	Red deer
Family Callitrichidae			
Lenotopithecus rosalia	2.4	1.5	Lion tamarin
Family Callimiconidae			
Callimico goeldii	1.4	0.7	Goeldi's marmoset
Family Cercopithecidae			
Papio cynocephalus	6.2	4.3	Baboon
Papio ursinus	5.0	3.2	Baboon
Cercopithecus neglectus	6.0	4.0	Guenon monkey
Miopithecus talapoin	9.5	4.0	Talapoin monkey
Erythrocebus patas	3.5	2.8	Patas monkey
Family Hylobatidae			
Hylobates lar	6.5	9.0	Gibbon
Family Pongidae			
Pongo pygmaeus	9.6	6.0	Orang-utan
Pan troglodytes	13.0	9.8	Chimpanzee
Gorilla gorilla	10.0	6.5	Gorilla

Table 6.5 Maturation of male grouse in different social systems (after Wiley 1974*a*, © The Stony Brook Foundation, reproduced with the permission of the author and the University of Chicago Press)

Type of social system	Incidence of breeding by first-year males: Never/Rarely	Occasionally	Usually
Category I	Sage grouse Prairie chicken Sharp-tailed grouse Black grouse Capercaillie		
Category II		Blue grouse Ruffed grouse	
Category III		White-tailed ptarmigan	Rock ptarmigan Willow ptarmigan

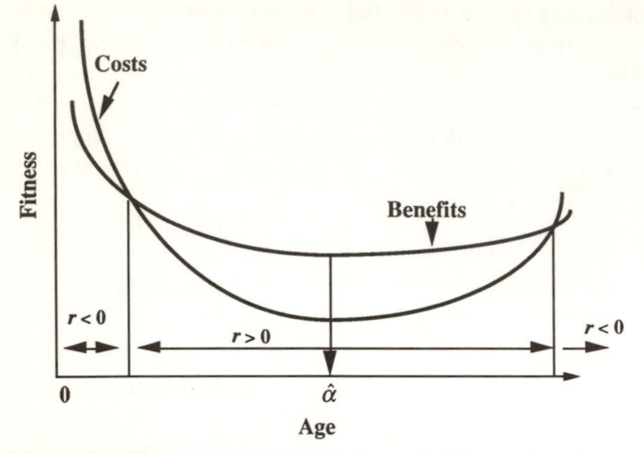

Figure 6.4 The reasoning behind the optimality approach to predicting age at maturity. Where the benefits of a given age at maturity exceed the costs by the widest margin, the organism should mature (after Schaffer and Reed 1972).

developed herself that she cannot produce any offspring at all. For example, in humans the infant mortality rate rises rapidly as the mother's age at first birth decreases (Fig. 6.5). On the other hand, past a certain point delaying maturity can increase juvenile mortality rates. In humans, the probability of birth defects and difficult births starts to rise at about age 35.

The assumption that fecundity increases with size has been abundantly documented, primarily in fish, amphibians and reptiles (e.g. Fig. 6.6). Note that here fecundity increases with length within species but not among species.

Earlier maturing forms have shorter generation times, higher r, and a decreased period of exposure to mortality before maturity; later maturing forms may have increased fecundity and/or lower instantaneous juvenile mortality rates (cf. Fig. 6.1). At an intermediate point, an optimum should exist. The challenge is to estimate the relevant costs and benefits, model their interactions, then predict the optimum and see whether it matches the data.

Implicit assumptions

To visualize the implicit assumptions of optimization, conceive of the population as a set of haploid clones in which reproduction and survival are influenced by age and size at maturity. Endow each clone with a different age and size at maturity and with the corresponding age-specific schedule of fecundity and mortality. Optimization then tells us which clone will have the highest lifetime reproductive success and will be best represented in future generations. To make the calculation, it is convenient to assume stable age distributions and to define fitness as the rate at which the clones are growing. Any one clone could have a rate of increase which is positive, negative, or zero while the size of the whole population remains constant (Stearns and Crandall 1981*a*).

We can also think about optimization in terms of a mutation that causes variation in age at maturity, survival, and reproduction. When it is

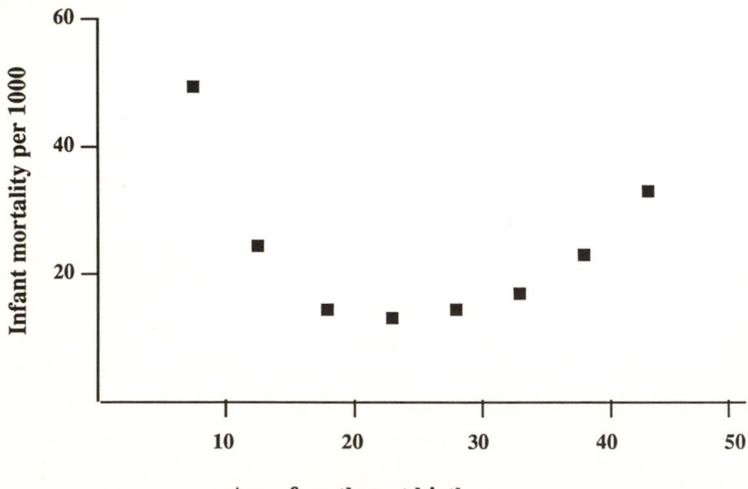

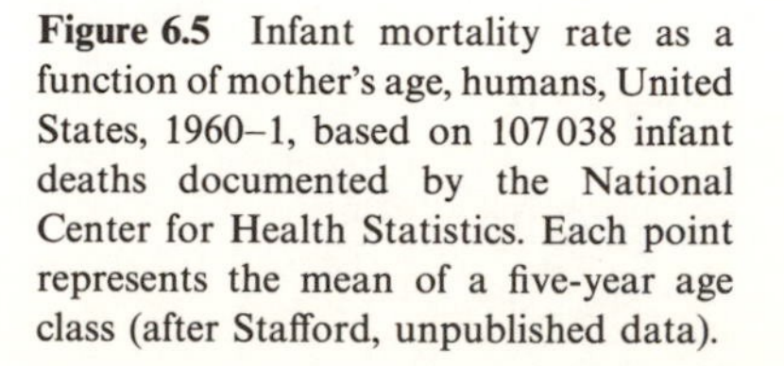

Figure 6.5 Infant mortality rate as a function of mother's age, humans, United States, 1960–1, based on 107 038 infant deaths documented by the National Center for Health Statistics. Each point represents the mean of a five-year age class (after Stafford, unpublished data).

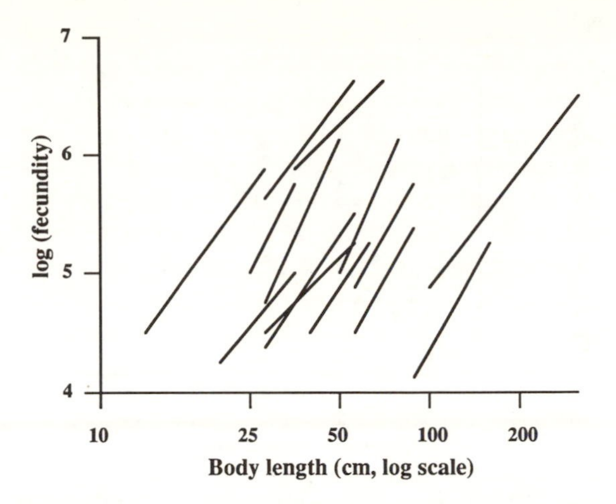

Figure 6.6 Fecundity vs. length for 14 species of flatfish (after Roff 1982).

moved by recombination through a large sample of organisms, it will have an average effect on their life history. Given many such mutations, which will be fixed and which will be eliminated? Charlesworth (1973) showed that a rare, incompletely recessive gene will spread if it affects life history traits in such a way that heterozygotes have a higher *r* than homozygotes. If the effects of frequency dependence, density dependence, or interactions among loci are not decisive, then those genes will be fixed that move the phenotype closer to the values predicted by optimality models (cf. Charlesworth and Williamson 1975; Charlesworth 1980).

In other words, optimality models assume simple, unrealistic genetics, but that assumption does not appear to make a crucial difference to the design of the phenotype. Many genetically unrealistic assumptions of optimization models affect not the equilibrium expected but the rate at which it is approached. One should not expect optimality models to predict a precise match of the organism to its changing and heterogeneous environment, for the genetic system may not be able to come to equilibrium rapidly enough to track the environment. Where the problem is to predict the adjustment of one life history trait to another, the picture is quite different. Because such interactions occur reliably in every generation, in the development and reproduction of every individual, over extensive periods of evolutionary time, we can expect the interactions of traits that determine fitness within a single organism to be fine-tuned by selection. Here optimality models should work.

The assumption of a stable age distribution (cf. Chapter 2) can also be questioned. Natural populations are rarely in stable age distribution, but when they are not, they are usually returning to one. As a reference point representing the average situation, the stable age distribution is a good choice. However, attempts have been made to study models of populations that are not in stable age distributions, where mortality and fecundity rates are changing in time and space, to see if that makes any qualitative difference to the predictions. It can: not necessarily for age and size at maturity, but certainly for the number of reproductive events per lifetime and the evolution of the lifespan (Orzack and Tuljapurkar 1989). Much remains to be done on testing the complex hypotheses of nonequilibrium life history theory.

A final class of assumptions deals with the manner in which various life history traits change with age, for example, that fecundity increases with size, or that the instantaneous juvenile mortality rates of the offspring decrease with the age at maturity of the mother. We cannot check such assumptions prior to maturation. In practice, one estimates the relation from a sample of just-mature females, then extrapolates into the juvenile age and size range.

A brief survey of modelling approaches

Analytical work on the evolution of age and size at maturity, and of reproductive effort in general, began with Gadgil and Bossert (1970). Their ideas are adumbrated in Cole (1954), Lewontin (1965), MacArthur and Wilson (1967), and in Hutchinson's lecture notes, which were discussed in the 1950s but only published in 1978. Gadgil and Bossert's was the first important attempt to analyse the forces acting on an entire life history and had wide influence. Their assumptions and hypotheses have since been worked out in more detail, but few approaches to the problem have differed fundamentally from theirs. Their prediction was that 'sexual maturity, that is, the stage at which optimal reproductive effort is greater than zero, will not be postponed beyond the size at which the curve [of fitness] vs. [size] has its peak' (p. 16). Thus maturation should occur at or before the peak in the $l_x m_x$ curve, a qualitative conclusion that has been discussed by Charlesworth (1980) and Kozlowski and Wiegert (1987) and tested by Roff (1984) for age at maturity and Roff (1981*b*, 1986) for size at maturity.

To test such a prediction quantitatively, one must estimate functions relating (a) age, size, and growth rates and (b) fecundity and survival. These have been estimated for many fish because of their commercial importance. Roff used this database to make his test. He started with the general allometric relationship between fecundity at a given age x, m_x, and length at that age, L_x:

$$m_x = aL_x^3 \tag{6.2}$$

The exponent is not always precisely 3 but is usually close to it, reflecting the linear relation of fecundity to weight and the cubic relation of weight to length. He used the von Bertalanffy function for growth,

$$L_x = L_\infty(1 - e^{-kx}) \tag{6.3}$$

where L_∞ is asymptotic length and k is the constant growth rate. To model survival, he assumed that there was a constant larval survival p, followed by a constant instantaneous mortality rate d:

$$l_x = p\, e^{-dx} \tag{6.4}$$

These expressions can be combined to give an analytic function for $l_x m_x$:

$$l_x m_x = p\, e^{-dx} a L_\infty^3 (1 - e^{-kx})^3 \tag{6.5}$$

If we assume that growth starts rapidly, then decreases steadily towards an asymptotically limiting length, that fecundity is proportional to weight, and that a constant proportion of the population dies, after the larval period, in each time unit, then the expected contribution of a given age class to reproduction is expressed in the function above. To find its peak, we take the derivative $(\partial l_x m_x / \partial x)$, set it equal to zero, and solve for x:

$$x = \frac{1}{k} \ln\left(\frac{3k + d}{d}\right) \tag{6.6}$$

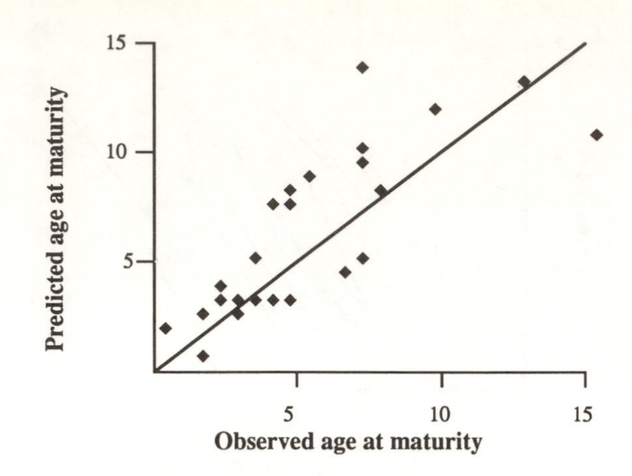

Figure 6.7 Observed and predicted ages at maturity for 24 fish species or populations of the same species, on the assumption that maturation occurs at the peak in the $l_x m_x$ curve (after Roff 1984).

Such a prediction does not imply that x is the *optimal* age at maturity, for the costs and benefits in fitness of different ages at maturity have not been used.

Roff's prediction works well (Fig. 6.7). The correlation between predicted and observed age at maturity is significant ($r = 0.78$, $P < 0.001$), and the slope of the least-squares regression line does not differ significantly from 1 ($P > 0.10$). The model accounts for 61 per cent of the observed variation in age at maturity; the null expectation would be 13 per cent. In these data, Gadgil and Bossert's hypothesis has withstood a test.

The next step is to introduce the effects of maturation on fecundity and survival into the Euler–Lotka equation and find the age at maturity that maximizes fitness (r or R_O). This approach has been taken by Charnov (1979), Bell (1980), Roff (1981*b*, 1984), Stearns and Crandall (1981*a*, 1984), Crandall and Stearns (1982), Stearns and Koella (1986), and Sibly and Calow (1989), who predicted optimal reaction norms as responses to physiological stress.

Bell and Roff used the fecundity–size relation to explain delayed maturity and predict maturation events. Stearns, Crandall, and Koella used the relations between age at maturity and fecundity and between age at maturity and instantaneous juvenile mortality rates to predict both maturation events and norms of reaction for a range of growth rates.

A second approach has been taken by Kozlowski and his co-workers (Kozlowski and Wiegert 1987; Kozlowski and Uchmanski 1987) and by Roff (1984). Like Bell, they concentrated on the connection between fecundity and size, but unlike Bell or Stearns, they maximized lifetime reproductive success, $\sum l_x m_x$, instead of r. Using Koslowski's and Roff's approach, one can predict not only age at maturity, but the age distribution of reproductive effort, the optimal growth rate, and the optimal size of the organism as well (e.g. Roff 1986).

PREDICTING OPTIMAL AGE AND SIZE AT MATURITY: MAXIMIZING r

This section introduces two reasons for delaying maturity. Can an increase in fecundity with size explain the evolution of delayed maturity without any change in instantaneous juvenile mortality rates? Can a decrease in instantaneous juvenile mortality rate with age at maturity explain delayed maturation without any effect of size on fecundity? The answer is yes in both cases (Stearns and Crandall 1981*a*). Both models are tested against the same data to compare their explanatory power. The symbols used are listed in Box 6.1.

Box 6.1 Symbols used in the analysis of optimal age and size at maturity in this and the next section

Symbol	Meaning
x	age
α	age at maturity
r	fitness as defined by the Euler–Lotka equation
r_c	critical r: the solution to the equation $\partial r/\partial\alpha = 0$
s	size
W	weight
l	survival
k	growth rate
A	asymptotic size
B	difference between A and size at birth, expressed as a proportion of A
m	fecundity
F	rate at which fecundity increases with size or age prior to maturation
G	rate at which fecundity increases with age after maturation
H	intercept of line relating fecundity to size or age
a	instantaneous adult mortality rate; can be affected by the growth rate
j	instantaneous juvenile mortality rate; can be affected by mother's age at maturity and by the individual growth rate
τ	power to which inverse of growth rate is raised; changing τ changes the strength of the association between growth rates and adult mortality rates
λ	a constant that can be changed to increase or decrease juvenile mortality rate
σ	the power to which the inverse of the growth rate is raised; changing σ changes the strength of association between growth rates and instantaneous juvenile mortality rates
γ	the power to which the inverse of age at maturity is raised; changing γ changes the strength of the relationship between mother's age at maturity and the instantaneous juvenile mortality rates of her offspring.

When fecundity increases with age

Suppose that age has *two* effects on fecundity. First fecundity increases linearly with age at maturity, then after maturity it could increase, remain constant, or decrease linearly. We refer to this as the Linear Fecundity Model.

Let the mortality rate be constant, ($a = j =$ constant) and let fecundity vary linearly both with age at maturity and with post-maturational age:

$$m(\alpha, x) = F\alpha + Gx + H \qquad \text{for } x > \alpha$$
$$= F\alpha + H \qquad \text{for } x \leqslant \alpha. \qquad (6.7)$$

Note that here the model assumes a *potential* fecundity that the organism would have had if it had matured earlier. For a given age at maturity α to maximize fitness r, the following conditions must be satisfied: (1) When we substitute functions for fecundity (m_x) and survival (l_x) into the Euler–Lotka equation,

$$1 = \int_0^\infty e^{-rx} l_x m_x \, dx \tag{6.8}$$

we require that r solves the equation. (2) The partial derivative of r with respect to α must equal 0,

$$\partial r / \partial \alpha = 0. \tag{6.9}$$

When these two conditions are met and (3) the second partial derivative is negative, age at maturity is optimized in a population in stable age distribution.

Under the assumption that $l_x = e^{-ax}$ and $m_x = F\alpha + Gx + H$, the Euler–Lotka condition becomes

$$1 = \int_\alpha^\infty (F\alpha + Gx + H)\, e^{-ax} e^{-rx} \, dx, \tag{6.10}$$

which implies that

$$1 = \frac{e^{-(r+a)\alpha}}{(r+a)}(F\alpha + G\alpha + H) + \frac{e^{-(r+a)\alpha}}{(r+a)^2} G. \tag{6.11}$$

Taking the derivative and setting it equal to zero, we find

$$r = \frac{F}{F\alpha + G\alpha + H} - a. \tag{6.12}$$

If we now substitute this value of r back into eqn (6.11), we get

$$1 = \frac{(F\alpha + G\alpha + H)^2}{F}\left(1 + \frac{G}{F}\right) e^{-\{\alpha F/(F\alpha + G\alpha + H)\}}. \tag{6.13}$$

When eqn (6.13) is satisfied and $\partial^2 r / \partial \alpha^2 < 0$, r is maximal.

This model has two interesting properties. If $H = 0$, maturity will be delayed whenever $F > 0$ and $(F + G) > 0$. Thus an increase in fecundity with age is itself sufficient to delay maturity. Secondly, this increase will be effective so long as the fecundity gained by delaying maturity (slope F) exceeds the rate at which fecundity may decline after maturity (slope $G < 0$). One could measure a negative post-maturational relationship between age and fecundity although delayed maturity had evolved because of a positive pre-maturational relationship.

When instantaneous juvenile mortality rate decreases as maturity is delayed

Now let us assume that delayed maturity has no effect on fecundity at all, but that instantaneous juvenile mortality rate decreases as the mother's age at maturity increases. We refer to this model as the Juvenile Mortality Model. Thus $F, G = 0$ and $m_x = H$ for $x \geqslant \alpha$. In other words, fecundity is constant. The critical assumption is that juvenile mortality is a function of the mother's age at maturity:

$$\begin{aligned} d(\alpha, x) &= j(\alpha) \qquad \text{for } x < \alpha \\ &= a \qquad \text{for } x \geqslant \alpha. \end{aligned} \tag{6.14}$$

so that

$$l(\alpha, x) = \exp\left(-\int_0^x d(\alpha, y)\, dy\right). \tag{6.15}$$

The instantaneous juvenile mortality rate is given by

$$j(\alpha) = \frac{\lambda}{\alpha^\gamma} + a. \tag{6.16}$$

This equation models juvenile mortality as a function of age at maturity with a negative exponential curve asymptotic to a constant adult mortality rate a (Fig. 6.8). We then ask that the Euler–Lotka equation be solved and that $\partial r / \partial \alpha = 0$. This occurs when

$$1 = \frac{H\, e^{-\gamma \lambda \alpha^{-\gamma+1}}}{\lambda(\gamma - 1)\alpha^{-\gamma}}, \tag{6.17}$$

and r is then given by

$$r = \frac{\lambda(\gamma - 1)}{\alpha^\gamma} - a. \tag{6.18}$$

There will only be a solution for finite positive r and α if $\gamma > 1$. Thus juvenile mortality must drop off faster than $1/\alpha$ if effects of juvenile mortality alone are to delay maturity. Recall that the

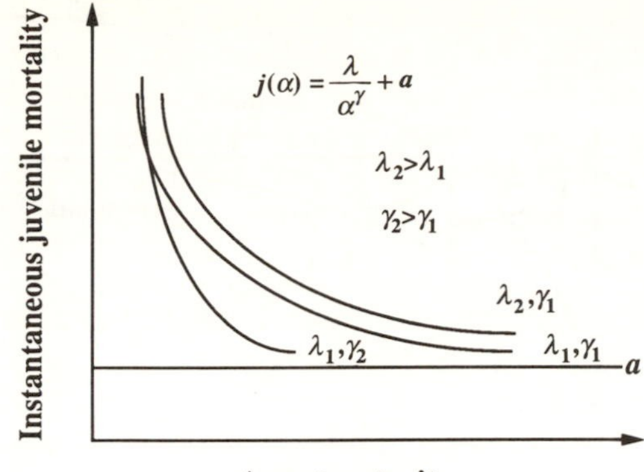

Figure 6.8 In the Juvenile Mortality Model, fecundity is held constant but the juvenile mortality rate decreases with the mother's age at maturity. Increasing the parameter λ shifts the curve up by a constant amount. Increasing the exponent γ bends the curve towards the origin.

relationship can be as weak as the inverse of the seventh root of age at maturity and still be strong enough to delay maturity from 4 to 5 years of age. These statements are not contradictory. The first indicates how strong the dependence of instantaneous juvenile mortality rate on age at maturity must be to cause a delay from age 0. If maturity is already delayed to 4 years, a much weaker effect can cause a further delay to 5 years.

Behaviour of the models

Take a concrete example, the Easter fence lizard, *Sceloporus undulatus*. For the South Carolina population, whose average age at maturity in the field is 10 months, we can estimate from Tinkle and Ballinger's (1972) data that $F = 28.8$, $H = -13.8$, $\lambda = 0.28$, and $a = 0.50$. If in addition we assume that $G = 0$ and $\gamma = 2$, then we can calculate fitness profiles for the two models (Fig. 6.9).

The fitness profiles are steeper to the left of the optima than to the right, suggesting that selection against earlier maturation may be stronger than selection against later maturation. The models were tested with data for nine populations of five species of lizards and salamanders (Stearns and Crandall 1981*a*). In the six cases where the Linear Fecundity Model (eqn 6.13) could be solved, the correlation of predicted with observed ages at maturity was $r = 0.90$ ($P < 0.05$). The Juvenile Mortality Model (eqn 6.17) could be solved for all nine cases, with $r = 0.93$ ($P < 0.01$). The models accounted for 80–86 per cent of the variation: age at maturity may be adjusted to an intermediate optimum (cf. Fig. 6.1).

Comment

This exercise in applied mathematics encounters a common dilemma. To solve the equations, one needs functions with convenient properties, but there is no guarantee that they are biologically realistic. The critical assumptions are that fecundity increases linearly with age and that instantaneous juvenile mortality rate declines with the inverse square of age at maturity. Neither is generally true.

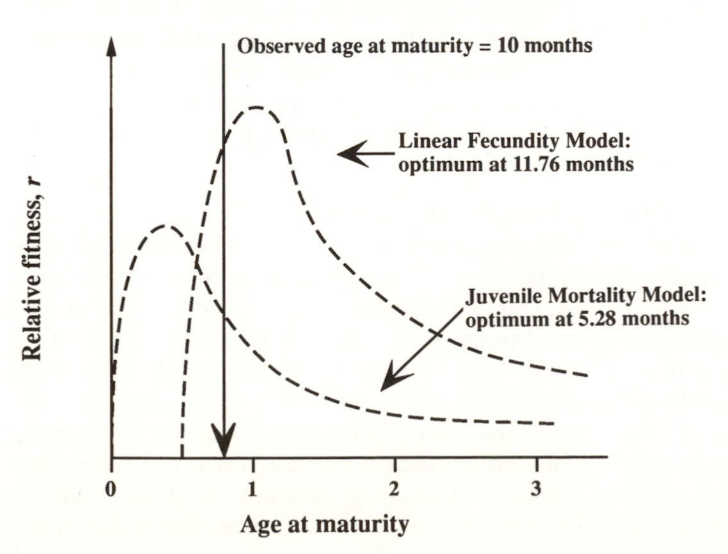

Figure 6.9 The solutions to eqns (6.11) and (6.15) for the case of the Eastern fence lizard in South Carolina. Both models show optimal ages at maturity in the neighbourhood of the one actually observed, but neither model fits the datum precisely (after Stearns and Crandall 1981*a*).

The next section develops a more general framework with a range of fecundity and mortality relationships.

EXTENDING THE MODEL FROM POPULATION MEANS TO REACTION NORMS

The approach taken by Stearns and Crandall (1981*a*) did not take two important factors into account. The first, growth, was emphasized by Gadgil and Bossert (1970) and Roff (1984). Roff could make more predictions because he modelled the size–fecundity and size–age relationships explicitly, rather than assuming a linear relationship between fecundity and age. The second factor, phenotypic plasticity in age and size at maturity, had not been approached with optimality models. However, data from several studies (e.g. Alm 1959; Grainger 1953; Robertson 1957; Frisch 1978; Stearns 1983*b*) indicated that phenotypic plasticity accounted for much of the variation in age and size at maturity. It therefore made sense to combine variation in growth rates with phenotypic plasticity in age and size at maturity, an approach sketched in Stearns (1983*b*) and analysed in greater depth in Stearns and Crandall (1984) and Stearns and Koella (1986). This version is based largely on Stearns and Koella (1986).

Consider a graph in which both growth curves and a reaction norm appear (Stearns 1983*b*; Fig. 6.10). This one illustrates rapid and slow growth for a fish population that under the best conditions can mature at 2 years of age and 8 kg. The question is, what sort of rule should the fish follow when forced to grow slowly? If it adopts Rule 1, 'Always mature at the same size', then the problem it faces is mortality. It needs 5.5 more years at the slower growth rate to attain 8 kg: during that period it could die for any reason. Rule 1 is risky. Or it could adopt Rule 2, 'Always mature at the same age'. Here the problem is a loss of fecundity. If the fish always matures at 2 years of age, then when it is growing slowly it will be 5 kg instead of 8 kg at maturity. Smaller fish produce fewer eggs. Rule 2 is costly. Between fecundity cost and mortality risk a compromise should be struck; organisms should have evolved a reaction norm for age and size at maturity that lies between the simpler options of maturing at a fixed size or at a fixed age. This compromise is indicated as Rule 3.

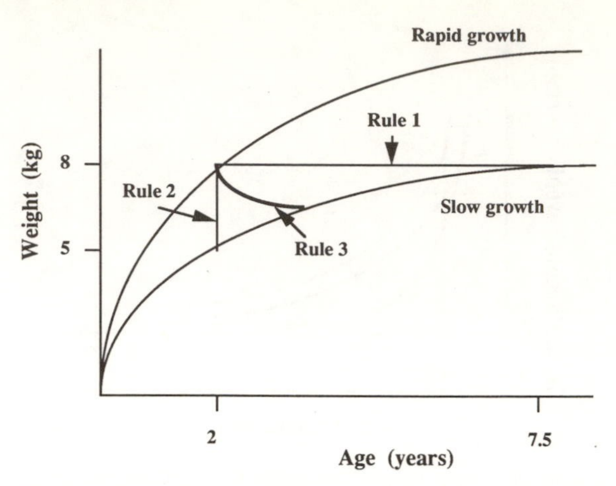

Figure 6.10 A graphical interpretation of three simple rules of thumb for the maturation event. Rule 1: always mature at the same size. Here the problem is mortality risk. Rule 2: always mature at the same age. Here the problem is fecundity cost. Rule 3: strike a compromise between mortality risk and fecundity cost and produce a reaction norm where each maturation event is optimal for the given growth conditions (after Stearns 1983*a*).

To calculate the optimal reaction norm for age and size at maturity, we extended the approach sketched above. We built a model in which fecundity is related to size, in which the relationship of size and age are related by growth, and with an explicit relationship of size or age at maturity to instantaneous juvenile mortality rate. By calculating the optimal age and size at maturity for different growth rates, we generated a continuous reaction norm that describes the optimal response across the range of growth conditions.

The optimality conditions

The first problem is to know where the optimum is reached. Crandall and Stearns (1982) and Stearns and Koella (1986) found the optimality conditions for Euler–Lotka models where a distinction is made between juvenile and adult mortality rates. Following Table 6.6, we can write x for age, k for growth rate, α for age at maturity, d for instantaneous mortality rate, a for instantaneous adult mortality rate, and j for instantaneous juvenile mortality rate. The mortality functions

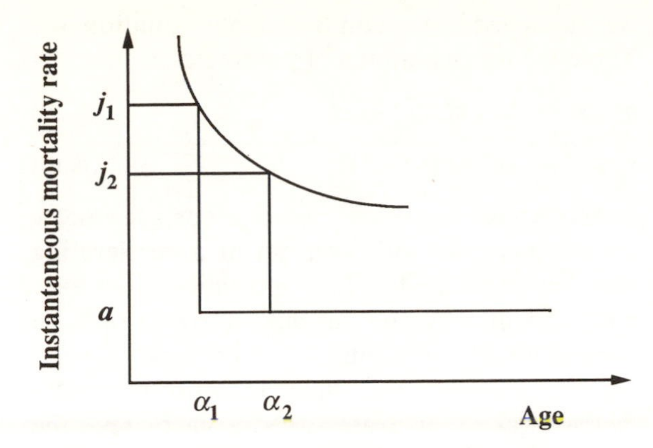

Figure 6.11 Instantaneous juvenile mortality is assumed to decrease as age at maturity increases. Adult mortality is not affected by age at maturity (after Stearns and Koella 1986).

(Fig. 6.11) are

$$d(x, \alpha, k) = a(k) \qquad \text{for } x \geqslant \alpha$$
$$= j(\alpha, k) \qquad \text{for } x < \alpha \tag{6.19}$$

and the fecundity functions (Fig. 6.12) are

$$m(x, k) = b\{s(x, k)\} \qquad \text{for } x \geqslant \alpha$$
$$= 0 \qquad \text{for } x < \alpha. \tag{6.20}$$

Here, b stands for birth rate and $s(x, k)$ determines the size of the organism.

If fitness is defined by the Euler–Lotka equation and we indicate the dependence of fecundity and survival on factors in addition to age by $b(x, \cdot)$ and $l(x, \cdot)$, then

$$\int_{\alpha}^{\infty} b(x, \cdot) l(x, \cdot) \, e^{-rx} \, dx = 1 \tag{6.21}$$

where survivorship is

$$l(x, \cdot) = e^{-j(\alpha, k)x} \qquad \text{for } x < \alpha$$
and
$$l(x, \cdot) = e^{-j(\alpha, k)\alpha} \, e^{a(k)(x-\alpha)} \qquad \text{for } x \geqslant \alpha. \tag{6.22}$$

Equation (6.21) defines the stable age-distribution condition. We also require that

$$\partial r / \partial \alpha = 0 \tag{6.23}$$

and that

$$\partial^2 r / \partial \alpha^2 < 0. \tag{6.24}$$

As is shown in Stearns and Koella (1986, Appendix), the solution to eqn (6.21) is a curve that defines values of r, denoted r_c, corresponding to values of α that satisfy the Euler–Lotka equation:

$$r_c = \frac{1}{\alpha} \ln\left(\frac{b(\alpha) l(\alpha)}{[\partial \alpha (a - j) / \partial \alpha]} \right). \tag{6.25}$$

Moreover, when the condition expressed by eqn (6.25) is satisfied, it can be shown (Stearns and Koella 1986, Appendix) that

$$\partial r_c / \partial \alpha > 0. \tag{6.26}$$

These equations define sufficient conditions to find the optimum α. One finds the simultaneous solution to eqns (6.21) and (6.25) after substituting eqn (6.25) into eqn (6.21). For most mortality and fecundity functions with any realism, this must be done numerically. One then checks to see whether the α found satisfies eqn (6.26).

In brief, one requires that the population be in stable age distribution and that neither later nor earlier maturation would produce higher lifetime reproductive success under the prevailing constraints on fecundity and mortality.

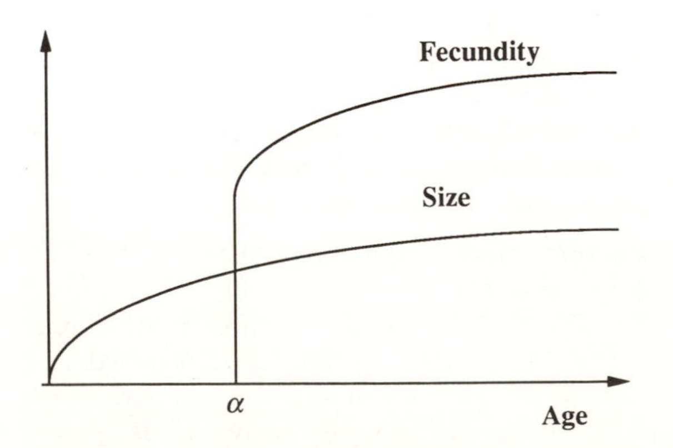

Figure 6.12 An example where fecundity = 2 × size (after Stearns and Koella 1986).

The choice of functions to represent fecundity, mortality, and growth

To find an optimal α, one needs a growth function relating size to age, a relationship between fecundity and size (and to age through the growth function), and a relationship between instantaneous mortality rates and age, age at maturity, or growth rates. We chose parameters for the functions and the functions themselves with two criteria in mind: the questions guiding the choice had to be interesting, and the resultant equations had to be tractable.

We chose the functions and parameters so that we could analyse the consequences of different environmental cues. Suppose that decreases in growth rates are reliably associated with increases in juvenile mortality rates. That would hold when the impact of food shortages is felt more by juveniles than adults, as is true for humans. Here the growing organism could use its growth rate as a cue containing information about changes in the demographic environment. Such an organism should evolve quite a different reaction norm from that of organisms in which decreases in growth rates signified larger increases in adult than in juvenile mortality rates—e.g., triclad flatworms (Calow and Woollhead 1977). How should the reaction norms of humans differ from those of triclad flatworms? We need a framework in which such comparisons can be generated by appropriate changes of parameters and functions.

We expressed instantaneous adult mortality rate a_0 as an inverse function of growth rate k raised to a power τ. When $\tau = 0$ or when $k = 1$, then adult mortality rate is a constant, and by varying τ one can produce a range of dependencies of adult mortality on growth rate and thus build the notion of an environmental cue into the model:

$$a(k) = \frac{a_0}{k^\tau}. \tag{6.27}$$

We expressed instantaneous juvenile mortality rate as an inverse function of both growth rate k and age at maturity α raised to powers σ and γ, respectively. When the exponents $= 0$, there is no dependency on those factors, but by varying the exponents we could generate a range of dependencies of juvenile mortality on growth rate and age at maturity:

$$j(\alpha, k) = \frac{\lambda}{k^\sigma \alpha^\gamma} + a_0. \tag{6.28}$$

Fecundity b was expressed as a linear function of size, measured either as length or as weight, whichever was appropriate, and growth in size was modelled by the von Bertalanffy equation (see Table 6.6 for definitions of parameters)

$$b\{s(x, k)\} = Fs(x, k) + H \tag{6.29}$$

$$s(x, k) = A(1 - B\,\mathrm{e}^{-kx}). \tag{6.30}$$

According to these assumptions, fecundity always increases with age, but at a decelerating rate; juvenile and adult mortality rates are independent of age, but juvenile mortality rate can depend on age at maturity and both can depend on the growth rate. Juvenile mortality rates increase as λ increases. As γ increases, the hyperbola relating juvenile mortality rates of offspring to age at maturity of parents bends in towards the axes. σ represents the strength of the relation between decreasing growth and increasing juvenile mortality, and τ represents that between decreasing growth and increasing adult mortality.

These assumptions only apply in the neighbourhood of the actual maturation event: the procedure finds a local optimum among phenotypic variants located nearby. Thus the fecundity assumption is not as unrealistic as it might seem if one thought it was supposed to apply to the entire life history.

The model reduced to a single equation

If one inserts these functions into the Euler–Lotka equation (6.21) and applies the condition $r = r_c$, one finds that the optimal age at maturity is the solution α_0 to the equation

$$\frac{(a + r_c)b(\alpha_0, k) + kb(\infty, k)}{(a + r_c)(a + r_c + k)}\,\mathrm{e}^{-\alpha_0(j + r_c)} = 1. \tag{6.31}$$

The solution α_0, which can be found numerically, is optimal if eqn (6.23) is also satisfied. When solutions are found for a range of growth rates, a reaction norm for age and size at maturity is generated. Every point on it represents the highest possible fitness that can be achieved under the given conditions (Fig. 6.13).

Factors affecting the shape of the reaction norm

The shape of the optimal reaction norm is quite sensitive to the relationships between growth rate and mortality rates. For all the sets of parameters explored, age at maturity increased as growth rates decreased, but whether size at maturity decreased or increased depended on the parameters

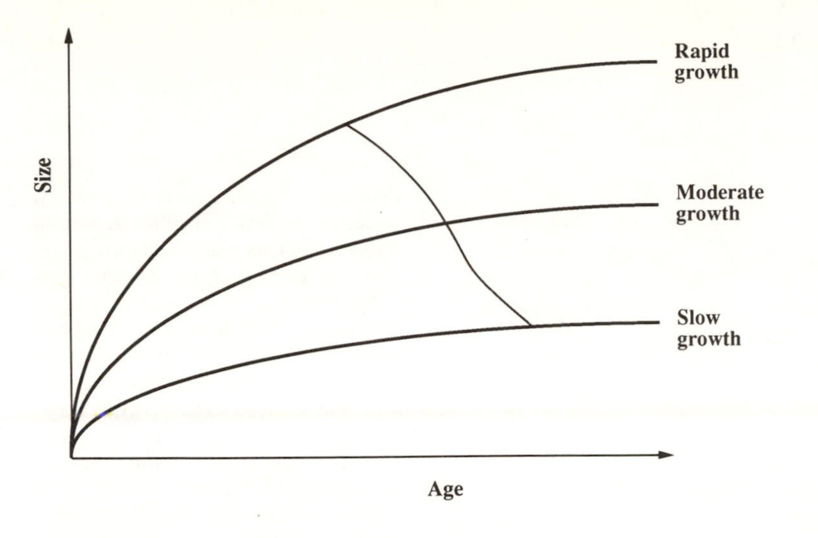

Figure 6.13 Across the entire range of growth rates from rapid to slow, an optimal age at maturity can be calculated by solving eqn (6.21). The norm of reaction connects those points. The figure shows three growth curves and one reaction norm (after Stearns and Koella 1986).

Table 6.6 Four submodels of reaction norms for age and size at maturity that differ in the way mortality depends on growth rate. Each submodel is associated with one or two distinct reaction norms depicted in Fig. 6.14. From Stearns and Koella (1986)

Submodel	Reaction norm
1. No dependence of mortality rate on growth rate	L-shaped
2. Juvenile mortality increases as growth rate decreases	Sigmoid-shaped for weak dependence, paraboloid for strong dependence
3. Adult mortality increases as growth rate decreases	L-shaped
4. Juvenile and adult mortality both increase as growth rate decreases	Keel-shaped

used. Table 6.6 summarizes the submodels and Fig. 6.14 illustrates the types of reaction norms predicted for the different situations. The commonest reaction norms in the literature are either L-shaped or sigmoid, i.e. as growth decreases, age at maturity increases and size at maturity decreases. There are a few cases in which both age and size at maturity may increase as growth rates decrease; these should correspond to the assumptions that produce the paraboloid or keel-shaped reaction norms in Fig. 6.14.

All the reaction norms depicted tend to flatten out to the right. As growth decreases, age at maturity continues to increase, but size at maturity changes less and less, for the effects on fecundity of a given change in growth rate are different for rapid and for slow growth (Fig. 6.15).

These models use growth rate as a potential cue to demographic conditions. The commonest

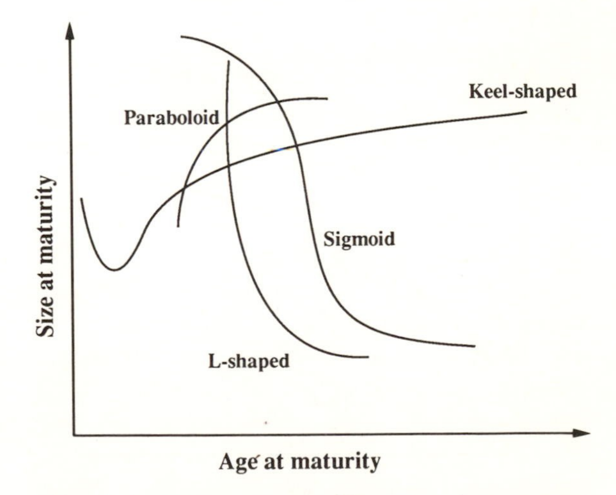

Figure 6.14 All four curves are different reaction norms, not growth curves, and they illustrate how the behaviour of the model changes qualitatively when the growth–mortality relationships change as described in Table 6.6 (after Stearns and Koella 1986).

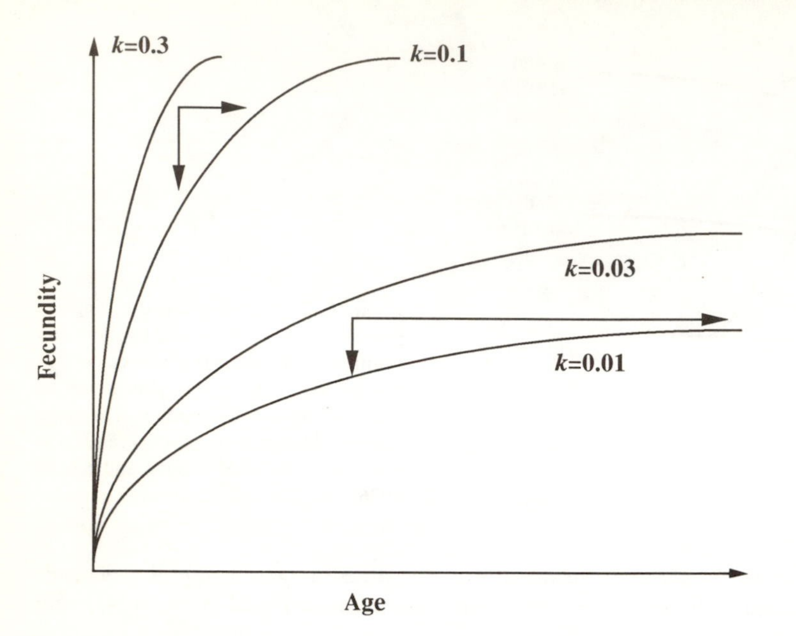

Figure 6.15 When growth is rapid (two curves on the left), a one-third reduction in growth rate decreases fecundity markedly at a given age but the same level of fecundity can be restored by a small increase in age (horizontal arrow). When growth is already slow (two curves on the right) a further one-third reduction in growth rate reduces fecundity by about the same amount, but it takes a much greater increase in age to restore the same level of fecundity (horizontal arrow). This explains the tendency of all reaction norms to flatten out with decreasing growth rates (after Stearns and Koella 1986).

correlation in nature is probably between decreasing growth rates and increasing instantaneous juvenile mortality rates. As the strength of this relation increases, resulting in larger increases in juvenile mortality when growth rate decreases, then the model predicts that the reaction norms should rotate upward to the right (Fig. 6.16). This predicts that as conditions deteriorate for juveniles, maturity should be delayed, thus increasing the period during which the juveniles are at risk!

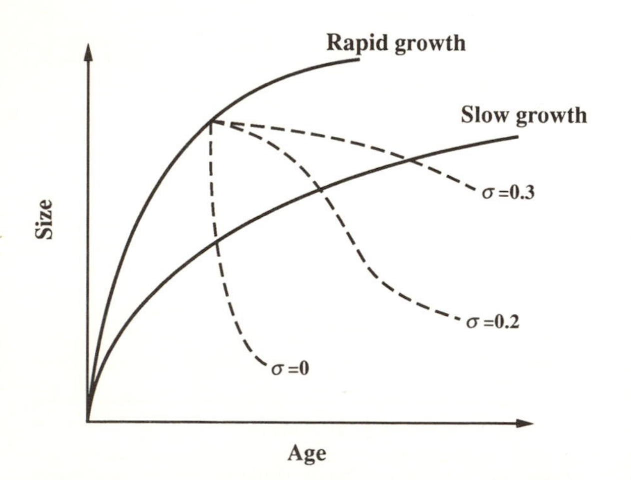

Figure 6.16 As the correlation between slower growth and higher juvenile mortality increases, here through the increase in σ from 0 to 0.3, the reaction norms should rotate up and to the right.

A deeper look shows that the response makes sense after all. Recall the discussion of eqn (6.1). So long as the fitness gained by increasing fecundity and reducing instantaneous juvenile mortality rates *though delayed maturity* can compensate the increases in instantaneous juvenile mortality rate *associated with slower growth*, it will pay to delay maturity. Two factors affect juvenile mortality here, growth rates and age at maturity, and they do so in a compensating fashion. The qualitative behaviour of the model depends on the quantitative interplay of these two factors in influencing juvenile mortality rates.

Factors affecting the position of the reaction norm

Changes in certain parameters change not the shape of the optimal reaction norm but its position. For example, we can distinguish between externally imposed, unavoidable sources of juvenile mortality and those that can be compensated for by changes in age at maturity. Increases in the parameter λ, representing changes in externally imposed juvenile mortality, shift the optimal reaction norm to the right without changing its shape (Fig. 6.17).

Increases in adult mortality rates, a_0, also shift the optimal reaction norms to the right, but the position of the reaction norms is not nearly as sensitive to changes in adult mortality as to

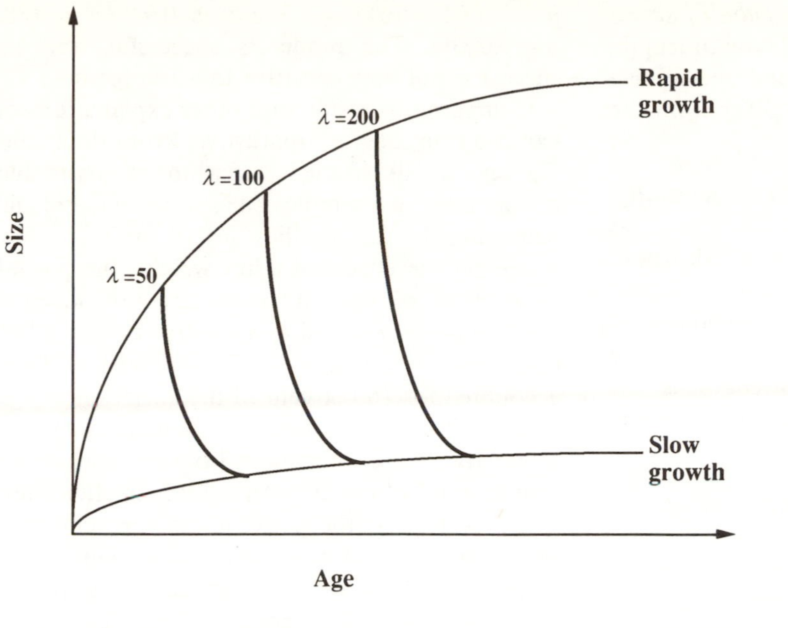

Figure 6.17 As *extrinsic* sources of juvenile mortality increase, the optimal reaction norm does not change shape but does shift to the right to take advantage of the compensating advantages in increased fecundity and decreased *intrinsic* juvenile mortality rate assumed to be associated with delayed maturity (after Stearns and Koella 1986).

changes in juvenile mortality. The position of the reaction norms is, however, very sensitive to changes in the size–fecundity relationship. If the slope of this relationship, F, increases, so that the organism gains fecundity more rapidly as it grows, then the optimal reaction norm shifts strongly to the left and maturation occurs earlier at a smaller size across the entire range of growth conditions. None of these shifts change the shape of the optimal reaction norm as much as they change its position.

Figure 6.17 could depict three reaction norms associated with three different levels of juvenile mortality representing the demographic conditions prevailing in three different places: here the model predicts geographic variation in reaction norms. Figure 6.17 could also represent a shift that occurred over time within a single population as the mortality caused by juvenile-selective predators increased: here the model describes how the average reaction norm shifts as the population compensates for increased predation pressure.

The diagram also clearly separates nature and nurture. The shape and position of the average reaction norm found in the population have been determined genetically by the history of natural selection. Shape and position will differ with both the internal constraints on the organisms, modelled by parameters like τ, σ, and γ, and with the externally imposed, unavoidable sources of adult and juvenile mortality, modelled by parameters like λ and a_0. Within a single population at a single point in time, the age and size at maturity achieved by a single average individual is predicted at the intersection of the reaction norm with that individual's growth curve. Variation caused by differences in growth rates among individuals within populations can be spoken of, in this simplified context, as environmentally determined.

Robustness of the predictions

The complexity of this model is justified by the many predictions that it makes; these place it at great risk. But complexity also raises doubts about the generality and robustness of the model. We therefore analysed four alternatives in each of which one basic assumption was changed.

1. Determinate growth

If the growth model is changed so that growth stops at maturation and fecundity remains constant throughout the reproductive period, there is no qualitative change in the predictions. All four types of reaction norms are found under the same conditions (cf. Fig. 6.14), and changes in the parameters produce the same kinds of shifts (cf. Fig. 6.17).

2. Fecundity increases with the cube of size
For exponential growth of fecundity with length, or even with weight, which is found among fish and salamanders, results did not differ from the basic model.

3. Juvenile mortality decreases exponentially with age
Here the juveniles are growing out of risk before they mature. The reaction norms become sigmoid even when juvenile mortality is independent of growth rate ($\sigma = 0$). This condition may be frequent in nature; the result strengthens the expectation that sigmoid reaction norms should be common.

4. Juvenile mortality decreases as size at maturity increases
In the basic model, instantaneous juvenile mortality rates depend on age at maturity rather than size at maturity. If we put in size dependency, an interesting qualitative effect occurs. In these optimal reaction norms, both age and size at maturity increase as growth rates decrease (e.g. the paraboloid reaction norm in Fig. 6.14). This suggests a condition that should be checked when such reaction norms, which we believe are rare, are found in nature. These are also the reaction norms predicted in Kozlowski's models (see below).

A quantitative test of the model

Correlations of predictions with observations of mean ages and sizes at maturity for populations and species of fish, salamanders, and lizards range from 0.6 to 0.9 (Stearns and Crandall 1981*a*; Roff 1984; Stearns and Crandall 1984). This model of reaction norms also predicts average maturation events as the point where the mean growth rate intersects the reaction norm. Stearns and Koella (1986) made such predictions for 19 populations and species of fish (Table 6.7). One parameter that could not be estimated from field data was γ, which determines the strength of the relationship between age at maturity and instantaneous juvenile mortality rate. We therefore made the predictions for two different values, 2.0 and 1.5, assuming that growth and mortality rates were not correlated ($\sigma, \tau = 0$). For $\gamma = 2$, the correlation of predicted with observed age at maturity was 0.93 ($r^2 = 0.86$, $p < 0.0001$); for $\gamma = 1.5$ it was 0.94 ($r^2 = 0.88$, $p < 0.0001$). The model is successful, and its success is not very sensitive to a change in γ.

Compare this result with other explanations of variation in age at maturity. Promislow and Harvey (1990) studied variation in many life history traits in a sample of 48 species of mammals belonging to 22 families in 10 orders. After removing the effects of adult weight, the partial correlation of age at maturity with juvenile mortality was $r = -0.42$ ($p = 0.07$); with adult mortality, it was $r = -0.54$ ($p = 0.02$). They accounted for 16 per cent of the variation in age at maturity in this sample of mammals in terms of variation in juvenile mortality rates and 25 per cent in terms of variation in adult mortality rates. When Roff used the peak in the $l_x m_x$ curve to predict age at maturity, he could explain about 61 per cent of the variation in age at maturity in a large sample of fish species. When Stearns and Koella used optimal r as the criterion for prediction, they could explain about 85 per cent of the variation in age at maturity in another sample of fish species. These figures mislead because they do not compare the success of different models on the same data set. They do indicate that one can gain some predictive power over simple correlations by modelling life histories, but they do not discriminate between the assumptions made and definitions of fitness used by Roff and Stearns.

Qualitative tests of the model: case studies of reaction norms

The model predicts the shape and position of reaction norms for age and size at maturity. In some case studies, the qualitative agreement of the model with the observations can be checked.

Fruit flies
Many experiments on *Drosophila melanogaster* have measured how fecundity and survival change with various environmental factors (e.g. Sang 1950; Robertson 1957; Shorrocks 1972). Flies growing at 27°C with abundant food at moderate densities start to reproduce at 11 days and at 1.0 mg. Crowded, poorly nourished flies start reproducing at 15 days or more and at 0.5 mg. Such shifts can be explained as adaptive reaction norms (Fig. 6.18).

Table 6.7 Predicted and observed age at maturity in years for 19 populations of fish. Either a linear or a cubic relation between fecundity and length was used, whichever fitted the data better. From Stearns and Koella (1986)

Species	Location	Reference	Model	Obs.	Predicted	
					$\gamma = 2$	$\gamma = 1.5$
Pike	River Stour, England	Mann (1976)	cubic	2.2	3.5	3.1
Roach	River Stour, England	Mann (1973)	cubic	4.5	6.5	5.4
Dace	River Stour, England	Mann (1974)	cubic	4.3	4.8	4.3
Dace	River Frome, England	Mann (1974)	cubic	3.9	5.0	4.2
Gudgeon	River Frome, England	Mann (1980)	cubic	2.1	3.0	2.4
Turbot	North Sea	Jones (1974)	cubic	4.0	6.0	6.0
Haddock	North Sea	Hislop (1984)	weight	3.4	4.8	3.9
Cod	North Sea	Hislop (1984)	weight	4.5	6.7	5.6
Upland Bully	New Zealand	Staples (1975)	linear	1.0	1.1	0.9
Haplochromis mloto	Lake Malawi	Tweddle and Turner (1977)	linear	1.4	1.4	1.0
H. intermedius	Lake Malawi	Tweddle and Turner (1977)	linear	1.9	1.7	1.2
Lethrinops parvidens	Lake Malawi	Tweddle and Turner (1977)	linear	3.1	1.6	1.5
Roach	Lake Volvi, Greece	Papageorgiou (1979)	cubic	3.8	4.3	4.0
Painted greenling	Monterey, USA	DeMartini and Anderson (1980)	cubic	3.0	4.0	3.0
Painted greenling	Seattle, USA	DeMartini and Anderson (1980)	cubic	3.2	4.7	3.7
Mosquito fish	Garden Island, Australia	Trendall (1981)	linear	0.11	0.09	0.09
Mosquito fish	Oakley Dam, Australia	Trendall (1981)	linear	0.21	0.15	0.10
Mosquito fish	Lake Leschenaultia, Aust.	Trendall (1981)	linear	0.17	0.13	0.12
Mosquito fish	Mill Pt. Rd., Australia	Trendall (1981)	linear	0.65	0.52	0.49

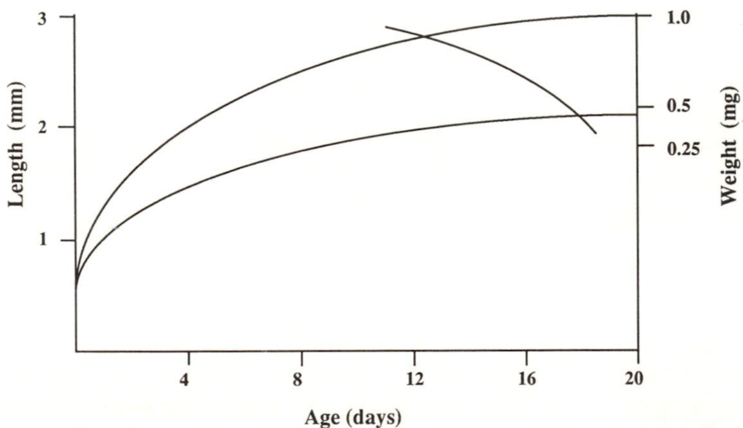

Figure 6.18 An optimal reaction norm for *Drosophila melanogaster* that is consistent with what is known about the size, fecundity, and larval mortality of the species. It agrees fairly closely with observation (after Stearns and Koella 1986).

Human females

Well-fed women mature 3–4 years earlier than poorly fed women. Frisch (1978) suggests that such phenotypic plasticity accounts for the difference in age at maturity in northern European females between the 19th and 20th centuries. Assuming that growth ceases at maturity, that fecundity increases slowly with body weight ($F = 0.0375$, $H = -1.25$), and that instantaneous juvenile mortality rates do decrease as age at maturity increases in a manner consistent with the shape of Fig. 6.5, one can predict optimal reaction norms for human females (Fig. 6.19).

Such case studies can be multiplied wherever data are available. For example, Stearns and Koella (1986) also predicted different reaction

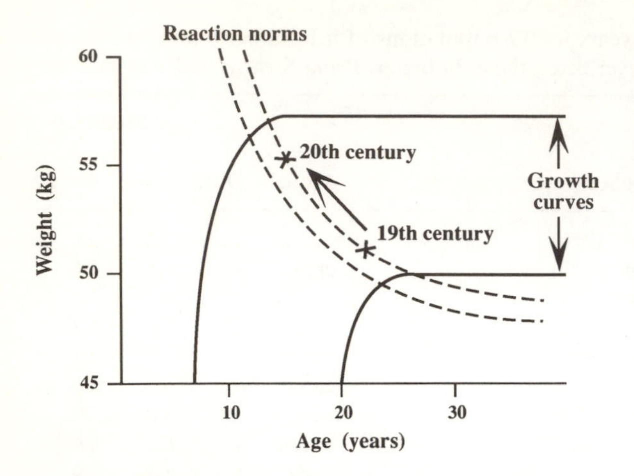

Figure 6.19 Two predicted reaction norms for age and size at maturity in human females. The upper norm describes the phenotypic reaction to better nutrition. The decrease in age at maturity of ca. 4 years and increase in weight at maturity of ca. 2 kg agree with the changes observed from the 19th to the 20th centuries. The lower norm describes the further shift towards younger ages at maturity expected to evolve if modern medicine keeps juvenile mortality rates below the historical average (after Stearns and Koella 1986).

norms for male and female red deer, and one could construct additional examples for various fish species. In all cases known to us, the predictions agree qualitatively with the observations, and the quantitative fit is usually not bad. However, additional case studies will not bring us much, for when one tests a model using data gathered for other purposes, some important parameter is always missing and values must be assumed. Only unusually complete data sets would contribute something that we do not already know. The limitations of the approach will be seen more clearly by checking assumptions than predictions. A central assumption is that the population consists of a clone with a single optimal reaction norm. Real populations usually consist of sexually reproducing, genetically unique individuals. The difference that this assumption makes to the descriptive power of the model can be seen in the next example.

Testing a central assumption: genetic variation in reaction norms

Gebhardt and Stearns (1988) measured developmental time (time to eclosion) and dry weight in 11 different crosses of *Drosophila mercatorum* raised on a series of yeast concentrations (1.5 per cent, 0.5 per cent, and 0.1 per cent) at 25°C. We crossed males from six isofemale lines from three field sites with females from two maternal lines that had been maintained parthenogenetically in the laboratory for a number of years. One cross failed. The results are summarized in Fig. 6.20.

Note how much genetic variation for this reaction norm is present. The maternal lines differ in their average effect, and there is considerable variation among the males. While the reaction norm envelopes show the predicted trend downward and to the right (Fig. 6.18), they describe a cloud of points around the predicted norm, which

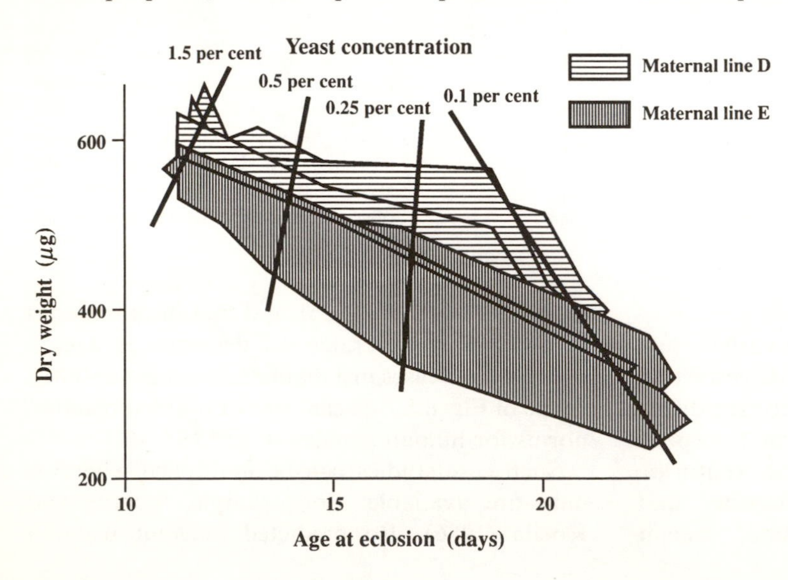

Figure 6.20 Reaction norms for age and dry weight at eclosion of female *Drosophila mercatorum* from 11 crosses constructed by mating males from six different isofemale lines collected in the field with females from two parthenogenetic maternal lines that had been in the lab for years. One cross failed. Each hatched polygon depicts the 95 per cent confidence envelope for the reaction norm from one cross. Not all crosses are depicted (after Gebhardt and Stearns 1988).

would be buried in the middle of the figure. The optimality model takes no account of important genetic variation. If the optimum norm for an average fly runs diagonally through the middle of the polygons, then individual flies deviated from that norm by as much as 100 μg dry weight and four days at eclosion. That deviation represents statistical uncertainty and the genetic and developmental constraints that keep individual flies from attaining the phenotypic optimum. Note also that the correlations between age and size at maturity change from positive to negative (cf. Chapter 3).

Summary of the first modelling approach

One can make successful predictions of age and size at maturity with r as the definition of fitness by assuming that maturity is delayed to gain fecundity or to reduce instantaneous juvenile mortality. The approach can predict reaction norms for the maturation event that are consistent with several case studies. These were the first reaction norms predicted from an underlying theory. The approach does not account for observed genetic variation in reaction norms, which is considerable.

PREDICTING OPTIMAL AGE AND SIZE AT MATURITY: MAXIMIZING EXPECTED NUMBER OF OFFSPRING

Kozlowski and his co-workers (Ziòlko and Kozlowski 1983; Kozlowski and Wiegert 1987; Kozlowski and Uchmanski 1987) and Roff (1981*b*, 1984, 1986) have developed a different approach to the problem. I concentrate here on Kozlowski's approach, to which Roff's has many parallels. Their assumptions differ from those made by Stearns, Crandall, and Koella, and some predictions also differ. Their key trade-off is between growth and reproduction. They assume that fecundity increases linearly with body size, that reproduction is continuous, that the population is not growing ($r = 0$), and that fitness is $R_O = \sum l_x m_x$. The reason for delayed maturity is an increase in *lifetime* reproductive success mediated by greater size at maturation.

Distinctive features of their work are its geometric interpretation (Fig. 6.21), its formulation as basic calculus, and its emphasis on growth.

Basic calculus

Fitness is measured as the volume $V = \sum l_x m_x$ of the solid in Fig. 6.21. This fitness measure only applies to stable populations. This adds a third condition, $r = 0$, to the two used previously—that the population be in stable age distribution and that $\partial r/\partial \alpha = 0$. The volume is given by

$$V = HS \tag{6.32}$$

where S, the area of the base, is a function of age at maturity (α), survival ($l(x)$), and lifespan (ω). The derivative of fitness with respect to age at maturity is

$$\frac{\mathrm{d}V}{\mathrm{d}\alpha} = \frac{\mathrm{d}H}{\mathrm{d}\alpha} S + H \frac{\mathrm{d}S}{\mathrm{d}\alpha}. \tag{6.33}$$

The height of the solid, the fecundity rate, is simply a function of size at maturity, $H(w(\alpha))$, so that

$$\frac{\mathrm{d}(H(w))}{\mathrm{d}\alpha} = \frac{\mathrm{d}(H(w))}{\mathrm{d}w} \frac{\mathrm{d}w}{\mathrm{d}\alpha}, \tag{6.34}$$

where the first term on the right-hand side is the rate of gain of fecundity with size and the second term is the rate of change of adult size with respect to age at maturity. If we assume that the area S is a function of body size, because mortality is size-dependent, and of age at maturity, because delaying maturity decreases the probability of surviving to maturation, then

$$\frac{\mathrm{d}S}{\mathrm{d}\alpha} = \frac{\partial S(w, \alpha)}{\partial w} f(w) + \frac{\partial S(w, \alpha)}{\partial \alpha} \tag{6.35}$$

where

$$f(w) = \frac{\mathrm{d}w}{\mathrm{d}\alpha}. \tag{6.36}$$

Now consider a small increase in age at maturity, $\Delta\alpha$. It will result in an increase in the adult survival rate because adult body size will be larger. There will be a gain to the area S, the life expectancy at maturity, due to better adult survival, and a loss from the area S due to later maturity (Fig. 6.22).

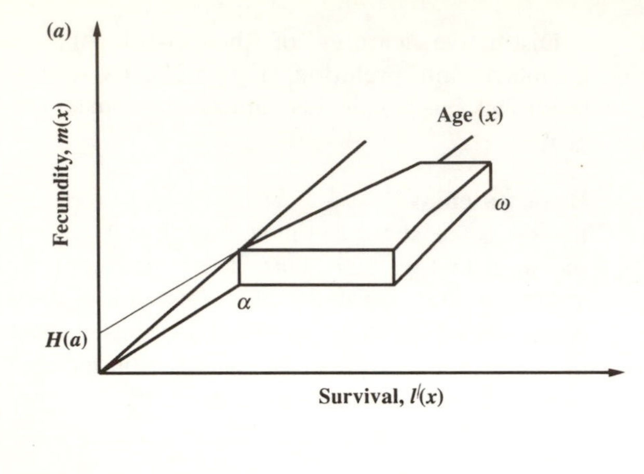

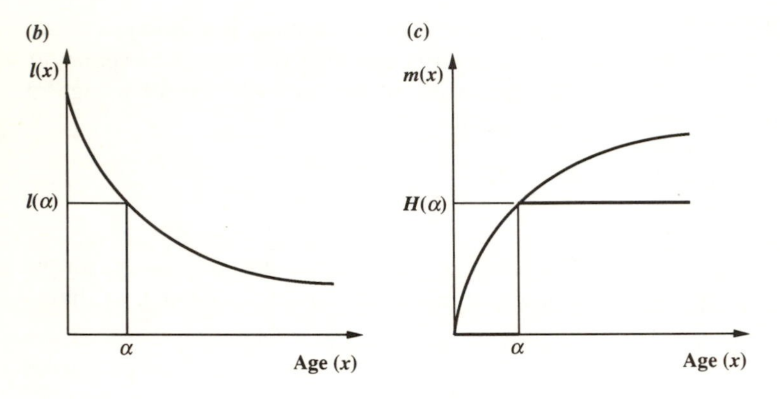

Figure 6.21 A geometric representation of fitness as number of offspring produced over the life of an annual species. (*a*) depicts the three-dimensional solid representing fitness, $l_x m_x$. (*b*) shows the projection of (*a*) on to the survival ($l(x)$) × age (x) plane. (*c*) gives the projection of (*a*) on to the fecundity ($m(x)$) × age (x) plane. $H(\alpha)$, the height of the solid, is the fecundity determined by size at maturation. It increases as maturity is delayed but remains constant thereafter. In this case, death occurs at the end of the season (at age ω) (after Kozlowski and Wiegert 1987).

The rate of change of the area showing losses is given by the second term of the right hand side of eqn (6.35), which describes the effect of age at maturity on the area under the survival curve. Because

$$S = \int_{\alpha}^{\omega} l(x)\,\mathrm{d}x \tag{6.37}$$

it follows that

$$\frac{\partial S(w, \alpha)}{\partial \alpha} = -l(\alpha), \tag{6.38}$$

where $l(\alpha)$ is the probability of surviving to maturity. If we now substitute eqns (6.34)–(6.38) into eqn (6.33) and set $\mathrm{d}V/\mathrm{d}\alpha = 0$, we get

$$G(w)f(w)E(w, \alpha) + \frac{\partial E(w, \alpha)}{\partial w} f(w) = 1, \tag{6.39}$$

where

$$G(w) = \frac{1}{H(w)} \frac{\mathrm{d}H(w)}{\mathrm{d}w} \tag{6.40}$$

and $E(w, \alpha)$ is life expectancy at maturity ($S/l(\alpha)$).

$G(w)$ is the rate of increase of reproductive rate with size. The second term in eqn (6.39), the partial derivative of life expectancy at maturation with respect to size, will be positive if mortality decreases with body size. At birth, the left-hand side of eqn (6.39) will be greater than one, and it will decrease as the organism grows. When it equals 1, it is optimal to stop growing and start reproducing. The factors delaying maturity are rapid growth (large $f(w)$), a rapid increase of reproductive rate with body size (large $G(w)$), long life expectancy at maturation (large E), and a life expectancy at maturation that increases with body size (positive $\partial E/\partial w$).

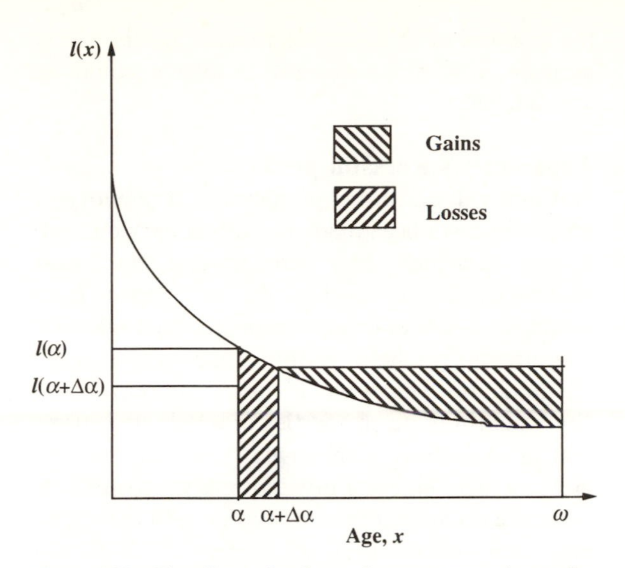

Figure 6.22 The effects of a change in age at maturity on the area S, the life expectancy at maturation (after Kozlowski and Wiegert 1987).

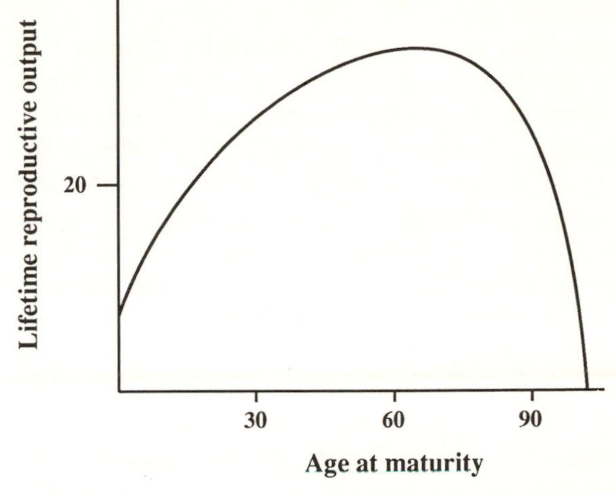

Figure 6.23 The value of V, lifetime reproductive output, plotted against α, age at maturity, for the example described in the text. At about 70 days, eqn (6.27) is satisfied and age at maturity is optimal (after Kozlowski and Wiegert 1987).

Fitness profile

To calculate a fitness profile from eqn (6.39), one needs concrete functions. If the initial body size w_0 is 1, the length of life ω is 100 days, fecundity is an exponential function of size ($H(w) = 0.1w^{0.75}$), and mortality decreases exponentially as size increases ($d(w) = 0.006w^{-0.25}$), then the fitness profile for age at maturity, expressed as average lifetime offspring production, has a maximum at about 70 days (Fig. 6.23).

Switching curves and reaction norms

Equation (6.39) can be solved for any combination of size and age at maturity (w, α); the result is a *switching curve*—to the left of the curve, the organism should grow; to the right of the curve, it should reproduce; on the curve it switches from growth to reproduction. The intersection of the switching curve with a set of growth curves describes an optimal reaction norm for maturation. When birth weight varies but growth rate does not (Fig. 6.24(a)), the switching curve—the reaction norm—is L-shaped, as in many of the cases analysed by Stearns and Koella (1986). When birth weight is fixed but growth rate varies, there is a different switching curve for each growth rate (Fig. 6.24(b)), and the reaction norm is nearly vertical, bending slightly to the right and concave upward. As growth rates increase, age at maturity

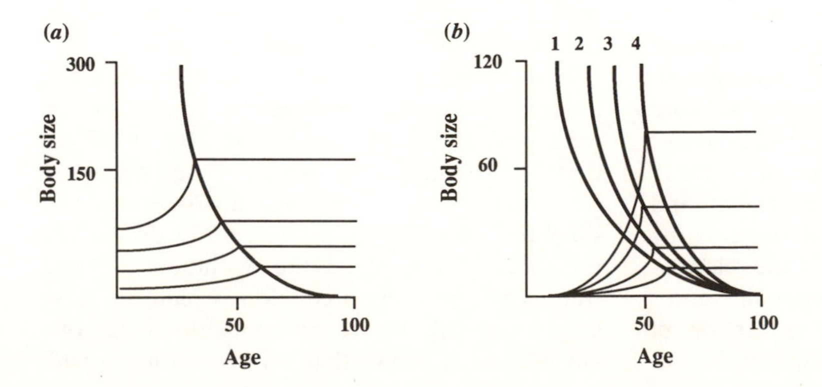

Figure 6.24 Dependence of optimal age and size at maturity on birth weight (a) and growth rate (b) in Kozlowski's model. The heavy lines are switching curves, the lighter ones are growth curves. Maturation is predicted to occur at the age and size where the two curves intersect (after Kozlowski and Wiegert 1987).

is delayed slightly and size at maturity increases dramatically.

In Stearns and Koella's modelling framework, such reaction norms were rather rare, occurring either when organisms grew out of risk (juvenile mortality decreasing with body size) or when there was a strong increase in juvenile mortality rate when growth rates decreased. In both cases, maturity was delayed to larger body sizes as growth rates *decreased*. Kozlowski and Wiegert predict a delay in maturity to larger body sizes as growth rates *increase*. The major difference in the assumptions of the two models is the definition of fitness—r for Stearns and Koella, $\sum l_x m_x$ for Kozlowski and Wiegert. If this difference in qualitative predictions stems only from the different definitions of fitness, then tests of these predictions should be especially illuminating.

DISCUSSION

Optimality models and genetic variation

These models buy the power to make phenotypic predictions by sacrificing genetics. The price is not too high because such predictions are useful and have not yet been made by models with more realistic genetics. The success of these predictions of age and size at maturity and of the quantitative predictions made with sex allocation theory justify the basic theory of evolutionary ecology (cf. Charnov 1982).

However, the predictions are far less detailed than the biological reality. A real population is a bundle of crossing reaction norms. With crossing reaction norms, the phenotypes associated with given genotypes change their rank order across environmental conditions. How far evolution can shape reaction norms towards the optimal ones predicted here given the ubiquity of crossing interactions is an important and unanswered question. The success of the qualitative case studies discussed above is reason for optimism.

Whether optimal reaction norms can evolve at all given the constraints of development and genetics is just one of several questions raised by these models. How should we model the relationship of genotype and phenotype? What differences do reaction norms make to the expression of genetic variation? Those are broad questions, but the analysis of this relatively small problem may indicate how more general problems could be approached.

Apparently successful predictions

Not many quantitative predictions of phenotypic properties have been made in evolutionary biology. Those described here are among the most successful. These models do not move from assumptions about environments through selection pressures to predictions about phenotypes. They are much more narrowly focused than that, based almost entirely on properties of whole organisms that are relatively disengaged from the environment, which enters only indirectly through its influence on mortality rates and growth rates. The predictions concern the adjustment of life history traits to one another, not to environmental conditions.

If the predictions had not been successful, then any number of things might have gone wrong—the list of costs and benefits might have been incomplete, the trade-off functions might have been improperly estimated, some constraint on the problem might not have been taken into account, the population might not have been in an evolutionary equilibrium because of recent environmental change or gene flow, and so forth. The match between prediction and observation was good, but that does not mean that ages and sizes at maturity are determined in nature for the reasons given here. There could be a circularity hidden in the modelling procedure, the success could stem from some feature of a function that has nothing to do with biology, and so forth. Positive results must also be treated with caution. Satou (1988), for example, has been particularly critical of a potential circularity stemming from the estimation of juvenile mortality rates.

If the results confirm the models, then the models adequately represent the constraints on organisms and the way selection acts. It follows that the assumptions made, including those concerning fitness, stable age distributions, and simplified genetics, do not seriously distort the processes important to the evolution of these traits. Otherwise the predictions would not have been successful. If true, this conclusion is important because it shows that we can get away with the massive simplification implicit in the assumptions.

CHAPTER SUMMARY

The benefits of earlier maturation are shorter generations and higher survival to maturity because the juvenile period is shorter. The benefits of delayed maturation are higher initial fecundity because of the greater body size permitted by a longer period of growth, lower instantaneous juvenile mortality rates, and higher overall lifetime fecundity.

Many organisms delay maturity, and variation in age at maturity can be detected among species, among populations within species, and among individuals within populations. There are strong phylogenetic effects on age and size at maturity. After the effects of size and phylogeny have been controlled, there is a strong pattern in both the birds and the mammals for organisms that delay maturity to have long lives and low fecundities. Bimaturism, where male and female ages and sizes at maturity differ, is common. Without social structure, females mature later and larger than males because they continue to gain appreciable fecundity after males have grown into the region of diminishing returns. Where males compete with each other to control access to females, males mature later and larger than females.

Reasonably successful predictions of age and size at maturity can be made if one predicts that organisms should mature at the peak of the $l_x m_x$ curve. Similar predictions of age and size at maturity can be made if one predicts that maturation should occur where lifetime reproductive success, measured either by r or by $\sum l_x m_x$, is optimized. Straightforward extensions of these models predict the shape and position of reaction norms for age and size at maturity. Only a few different shapes of reaction norms are expected. In most cases, maturity is delayed and occurs at smaller sizes as growth rates decrease.

Case studies indicate that the predictions of the models agree qualitatively with data on humans, fruit flies, red deer, and fish. The analysis of reaction norms permits an easily visualized separation of genetic and phenotypic effects.

PROBLEMS

1. Write a programme to solve eqn (6.31) numerically and use it to generate a reaction norm for age and size at maturity. Do the same for eqn (6.39) (see Kozlowski's papers for details). Compare assumptions and results.
2. Once successful predictions have been made from similar models for two or three different data sets, does it make more sense to publish further confirmations or to publish only those new cases in which the models are not confirmed? Would you invest a lot of effort, for example, in demonstrating that the models in this chapter also work for 40 frog species or 60 dragonfly species?
3. We assumed in this chapter that age and size at maturity are adjusted by natural selection to an intermediate optimum. This is equivalent to stating that the maturation event is under stabilizing selection. Under what circumstances might such an intermediate optimum *not* exist? Where, for example, might it always be better to mature earlier, or always to mature later? *Can* a trait like maturation be under directional selection?
4. Where are the lineage-specific effects in these models? Would it be possible to use them to make predictions for humans, flies, and fish if they were not somehow incorporated?
5. Repeat the analysis of optimal reaction norms, contrasting the results you get when you use r and R_0 as fitness definitions. Does changing fitness definitions make enough of a difference to the predictions that we could use the relative success of the predictions to judge the fitness definitions?

7

NUMBER AND SIZE OF OFFSPRING

The only check to a continued augmentation of fertility in each organism seems to be either the expenditure of more power and the greater risks run by the parents that produce a more numerous progeny, or the contingency of very numerous eggs and young being produced of smaller size, or less vigorous, or subsequently not so well nurtured. To strike a balance in any case between the disadvantages which follow from the production of a numerous progeny, and the advantages . . . is quite beyond our power of judgment

Charles Darwin, 1871

I believe that, in nidicolous species, the average clutch-size is ultimately determined by the average maximum number of young which the parents can successfully raise in the region and at the season in question

David Lack, 1947

The threshold at which it pays a parent to kill one of its offspring is higher than the threshold at which it pays one of the offspring to kill a sib

R. J. O'Connor, 1978

CHAPTER OVERVIEW

Two approaches dominate the analysis of the evolution of offspring number and size. The ornithologists have organized their understanding of offspring number around the *Lack clutch*, the clutch size that produces the most fledglings. In this tradition theory and experiment have aimed to understand what factors cause deviations, usually downward, from the Lack clutch. These include effects of clutch size on subsequent parental and offspring survival and reproductive performance and year-to-year variation in the optimal clutch size. An important theoretical tradition aims to optimize reproductive effort over the whole life history taking costs of reproduction into account. This approach is called either the *General Life History Problem* or the *Reproductive Effort Model*, and its predictions are well confirmed by experiments with poeciliid fish and with small, altricial, hole-nesting birds, but there are also substantial data on clutch sizes in sea birds, geese, ducks, and raptors and litter sizes in humans, rodents, deer, and other mammals. Recently the Lack tradition has been extended to parasitoid insects, where 'clutch' is taken to mean eggs laid per host.

The second approach is younger, not yet as thoroughly tested, and based on the recognition that parents and offspring may disagree over the clutch size and amount of parental care. Genes acting in the parent favour one level of investment, while genes acting in offspring may favour another. This approach asks two questions: who is in control, and, if a conflict over optimal clutch size is possible, who wins it—parents or offspring? If the parents have control, then the clutch size should be the Lack clutch amended by trade-offs. If selective abortion is possible, then parents should end up with the Lack clutch after overproducing enough to allow the optimal elimination of less fit progeny. However, offspring should prefer smaller clutches that reduce competition with sibs and the probability of being killed by a sib. If the parents are not completely in control, the interests of the offspring are in conflict with those of their parents, and the outcome of the evolutionary game is better analysed as an ESS (evolutionarily stable strategy) than as an optimum. In some cases the optimal clutch size for offspring is one.

Many analyses of clutch size fix offspring size and let reproductive effort vary, and this assumption often holds. However, offspring size sometimes varies as much as clutch size. Analyses of offspring size fix reproductive effort and let offspring and clutch size vary. As with clutch size, there are two approaches to offspring size. If the parents are in control, one seeks the combination of offspring size and number that maximizes parental fitness. If the offspring are in control, it pays to be a large offspring in a small clutch. Joint control may require a game theoretic solution. Recent manipulation experiments on offspring size are just starting to pay off and could grow into a tradition as substantial as that of clutch size manipulations.

INTRODUCTION

World records

The record for smallest clutch size—one—is shared by many organisms. The single offspring of a blue whale is as large at birth as an adult African elephant and grows to half its adult weight in the first six months of life feeding exclusively on fat-rich milk provided by a mother who is herself not able to feed, for the period from birth to weaning is spent in tropical waters poor in food and predators (Schaffer 1969). Although large, whale offspring are fairly small relative to the size of the mother. Bats have the largest offspring for their body size in the mammals, the record being held by *Pipistrellus pipistrellus*. It normally carries twins whose combined weight at birth is 50 per cent of the mother after parturition—and the mother must fly and feed while pregnant (Racey 1973, cited in Myers 1978). The kiwi, at 2 kg the size of a domestic chicken, lays the largest egg for its size of any bird. At 350–400 g, its egg is more than five times larger than the largest eggs from domestic chickens selected for large egg size (Calder 1979, 1984). The reproductive investment of the kiwi is, however, dwarfed by that of the caecilian *Dermophis mexicanus*, which can give birth to a clutch that weighs as much as 65 per cent of the post-parturition weight of the female (M. Wake, pers. comm.).

A clutch size of one is not surprising if the expected number of offspring per lifetime is large. But the small clutches of dung beetles—four or five—are astonishing (Hamilton, pers. comm.), for dung beetles are thought to reproduce only once. If so, their survival to maturity is as high as any arthropod's—about 0.5.

At the other end of the scale, many organisms have large numbers of very small offspring. The seeds of orchids number in the hundreds of millions to billions. Large bivalves and some large fish (sturgeon, large catfish, cod) release tens to hundreds of millions of eggs in each spawning. Many trees that live several hundred years produce well over 100 000 seeds per year. The species with the world record for worst juvenile survival is probably an orchid whose seed has a chance of about 10^{-9} of surviving to reproduce.

The Lack clutch

Life history theory views the number and size of the offspring produced in a single reproductive event as having been adjusted by evolution to optimize individual fitness, including evolved reactions to environmental conditions (adaptive plasticity). If nestling mortality rates increase with number of offspring produced and no other effects are important, the optimal clutch is the one that produces the most fledglings. We call this clutch the Lack clutch after Lack (1947), who suggested that altricial birds should lay the clutch that fledges the most offspring. Modern clutch size theory analyses the effects causing deviations from the Lack clutch. Trade-offs with other parental and offspring life history traits, temporal variation in optimal clutch size, parent–offspring conflict, and brood parasitism will all reduce the Lack clutch. Selective abortion and flexible brood reduction in response to fluctuating resource levels will increase the Lack clutch. Clutch size may also be constrained within lineages so that no variation is possible.

Lack noted that nestling survival decreases as clutch size increases. If N_f = number of offspring fledged, N_e = number of eggs laid, and $l_f(N_e)$ = survival to fledging as a function of number of eggs laid, then

$$N_f = N_e l_f \quad (7.1)$$

gives us the number of fledglings as a function of clutch size and survival rate, and, if we assume that survival is a linear function of clutch size, then

$$l_f = 1.0 - cN_e \quad (7.2)$$

gives us survival to fledging as a function of clutch size. Here c is the constant slope of the line relating fledgling survival to clutch size. Substituting eqn (7.2) in eqn (7.1) gives

$$N_f = N_e - cN_e^2. \quad (7.3)$$

The first derivative of fledglings with respect to clutch size is

$$\frac{dN_f}{dN_e} = 1 - 2cN_e, \quad (7.4)$$

which equals 0 when $N_e = 1/2c$. At this point, the negative second derivative $(-2c)$ indicates that the clutch is a local maximum. For example, if $c = 0.1$, then the most productive clutch is five, and it fledges 2.5 offspring (Fig. 7.1). More generally, if we represent the fitness gained by the parents through one clutch as the reproductive value of the clutch $V_c = N_e f(N_e)$, where $f(N_e)$ is the fitness produced by a clutch of N_e eggs, then

$$\frac{\partial V_c}{\partial N_e} = N_e f'(N_e) + f(N_e) \quad (7.5)$$

and

$$\hat{N}_e = \frac{-f(N_e)}{f'(N_e)} \quad (7.6)$$

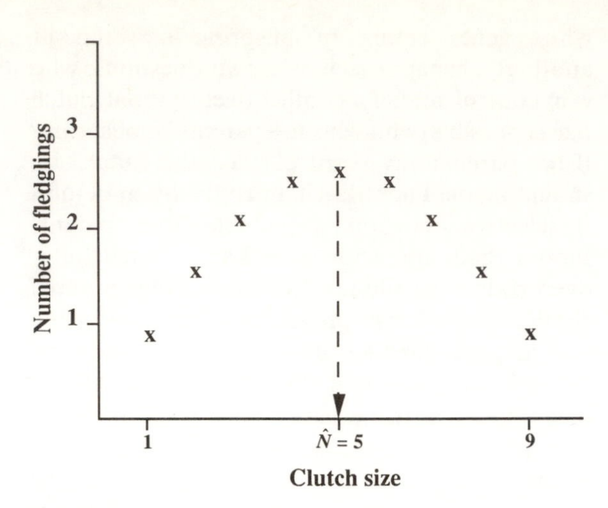

Figure 7.1 A model of Lack's Most Productive Clutch Hypothesis.

where $f'(N_e)$ is the first derivative of fitness with respect to clutch size (Parker and Courtney 1984). In Fig. 7.1, $f(N_e) = 1 - 0.1N_e$ and $f'(N_e) = -0.1$, giving an optimal clutch of 5.

In contrast to the ornithological tradition, where the Lack clutch means the clutch that fledges the most offspring, in general life history evolution the Lack clutch is the clutch that maximizes parental fitness when one considers effects of clutch size only on the fitness of offspring produced from that clutch and not on parental residual reproductive value (Charnov and Skinner 1984; Parker and Mock 1987).

Effects causing deviations from the Lack clutch

The effects causing deviations from the Lack clutch are implicit in the assumptions of the model: (1) no intergenerational trade-offs between clutch size and the reproductive traits of surviving offspring (age at maturity, clutch size, offspring survival); (2) no trade-off of clutch size with parental mortality; (3) no intragenerational trade-offs of clutch size with subsequent parental reproductive traits (time to next clutch, size of next clutch, size of subsequent offspring, survival of subsequent offspring); (4) no temporal variation in the optimal clutch size or perfect ability to track a changing environment; (5) no selection by the parents of the most fit offspring in this clutch; (6) no conflict between the parents and previous offspring over the investment in this clutch; (7) fixed offspring size or potential variation in offspring size too small to make any difference to clutch size; (8) clutch size not constrained to be some fixed number; and (9) no gene flow. After some background material, we relax each of these assumptions.

BACKGROUND: THE NATURAL HISTORY OF CLUTCH AND OFFSPRING SIZE

A qualification of the proximate–ultimate distinction

Lack distinguished between the proximate factors determining the number of eggs laid and the ultimate selection pressures that shaped the pattern of reproductive investment. Proximate environmental factors—things like temperature, day length, food availability—interact with physiological mechanisms that determine when a female starts to lay, whether she skips a day, and when she stops. The ultimate factor is natural selection optimizing clutch sizes through demography, genetics, and trade-offs. This distinction remains useful, but reactions to proximate factors are flexible adaptations that, like the reaction norms they resemble, vary genetically and respond to selection. The response to proximate factors has an ultimate component.

Lineage-specific effects

The approach one takes to explaining variation in number and size of offspring will be determined by developmental, physiological, and behavioural patterns fixed within lineages. These include resource storage, growth patterns, number of breeding attempts per lifetime, and developmental state at birth. How they affect the analysis of clutch size and reproductive investment is indicated by the following questions: are we dealing with an income breeder, a capital breeder, or something in between? Is growth in this species determinate or indeterminate? Is the species semelparous (monocarpic) or potentially iteroparous (polycarpic)? At what stage of development are the offspring born? Are they altricial or precocial? Is the number of offspring fixed or variable? Is the size of offspring fixed or variable?

Nidicolous (*altricial*) birds are born naked, blind, with poor thermoregulation, and are fed in the nest. *Nidifugous* (*precocial*) birds are born feathered, with open eyes, and some can fly on the first day of life. Similar patterns exist in the mammals, where ungulates, for example, are precocial, carnivores and primates are altricial, and marsupials are so altricial as to deserve a separate category. Case (1978) reviews the life history consequences of such patterns in birds and mammals. Similar patterns exist for parental care and viviparity.

In altricial organisms, the trade-off between number of young and juvenile mortality is more important than the trade-off between size of young and juvenile mortality, for parents can compensate for small offspring by feeding them more. In precocial organisms, the trade-off between size of young and juvenile mortality is more important because the offspring have to make it on their own.

Fish and amphibians vary greatly in stage of development at birth and egg size. In both groups the ovoviviparous forms eliminate larvae and produce fewer, larger, yolkier eggs, e.g. the coelacanth, *Latimeria*, which has eggs as large as oranges. In true viviparity special organs nourish the offspring as they develop inside the mother's body. The caecilians, members of the amphibian Order Apoda, include species (e.g. *Dermophis mexicanus*) in which the developing young are retained in the oviducts, exhaust their yolk supply at about 20 per cent of their length at birth, then feed and grow off of special secretions from the mother's oviductal epithelium (Wake 1982). Intrauterine embryonic cannibalism—adelphophagy—occurs in some viviparous sharks (e.g. *Eugomphodus taurus*, the sand tiger shark—(Springer 1948)) and results in a few, very large, offspring at birth (Wourms 1981).

Peripatuses (Phylum Onychophora) include species displaying the full range from egg-laying to full viviparity with a true placenta and a prolonged gestation of 12–15 months (Ghiselin 1985).

All tubenose birds (Order Procellariformes) lay one egg, in some species only every other year. These widely foraging seabirds vary in how far they travel to feed, in body size, and in age at maturity. The largest member of the order, the wandering albatross, takes up to 33 days and travels up to 15 000 km on a single foraging flight (Jouventin and Weimerskirch 1990). Here a clutch size of one is not surprising, for the single chick, which has a special starvation physiology, must wait up to a month for a meal. These birds can live 50–60 years and lay a total of 20–25 eggs per lifetime. Whether a clutch size of one is best for all species in a group of birds of some ecological diversity is not clear (Ricklefs 1990).

Nearly all lizards in the Family Gekkonidae have a clutch size of two, and all lizards in the Subfamily Anolinae have a clutch size of one. In both cases, females produce eggs fairly continuously when food and temperature permit, laying many eggs per year and giving them no parental care. These small clutches are part of a pattern of scattering many eggs widely in space and time; they are not puzzlingly small total reproductive investments.

Which trade-offs are important depends on development, physiology, and behaviour. Much work on reproductive investment has been done on easily manipulated seabirds and small hole-nesting birds. They are iteroparous income breeders with determinate growth, altricial young, extensive parental care, and more variation in number than in size of young. We cannot expect conclusions drawn from one special group to apply generally.

Physiological constraints

To make offspring, parents must acquire food or withdraw materials and energy from food stored earlier. Morphology and physiology limit the rate at which they can feed, the amount they can store, and how fast they can mobilize reserves. Whether the energy gathered goes into the parent before the offspring are born or into the offspring during a period of parental care, physiological constraints limit the number of young that can be produced. Apparently birds and mammals cannot work for extended periods at more than four times their basal metabolic rates without dying (Drent and Daan 1980), and the rate at which poikilotherms develop is a direct function of temperature.

Geographical trends

Geographic variation within species indicates local adaptation, either through genetically isolated demes or through reaction norms that adapt one population to a broad range of conditions. That clutch size increases with latitude has been recognized at least since the 1830s (Rensch 1938): the European wren has 3 eggs on Sicily, 5–6 in England, 6 in Germany, and 6–8 in Russia (Klomp 1970). Clutch sizes increase away from the equator in North and South America and to the east within North America and Eurasia, are smaller on islands than on neighbouring continents, and are smaller in forests than in savannas (Lack 1968; Klomp 1970). Litter size in mammals varies in similar ways (Cockburn *et al.* 1983; Conaway *et al.* 1974; Fleming and Rauscher 1978; Innes 1978).

The latitudinal trend in birds and mammals is reversed in some echinoderms and molluscs. The number of species per group that either brood their fewer, larger young, rather than releasing them into the plankton, or are viviparous increases with latitude (Thorson 1950; Rohder 1985; Strathmann 1977, 1985, 1990). Viviparity and smaller clutches are also commoner at higher latitudes and altitudes in snakes, lizards, and salamanders. For latitudinal variation in a copepod (Allan 1984) and altitudinal variation in the wood frog (Berven 1982, 1987) we have an idea of how much of the latitudinal variation is genetic and how much is due to phenotypic plasticity.

Geographic variation in reproductive investment is so well established that there is little sense in repeating such studies if the aim is just to document the existence of differences. It has been hard to use descriptions of such patterns to discover what causes the variation.

Variation within populations

Reproductive investment responds to population density, food supply, and temperature. Fruit flies lay more eggs when well fed, when kept at higher temperatures on adequate food, and at lower population densities. Birds of prey lay more eggs in years when prey population densities are high. Many fish reduce fecundity under food stress at high population densities and increase it when well fed and growing fast at low population densities.

Klomp (1970) lists 55 bird species in which clutch size declines right from the start of the *season*, and another 15 in which it first rises, then falls. No temperate birds increase their clutch size as the season progresses. The fecundity of organisms of the same age usually increases with *size*, especially in iteroparous organisms with indeterminate growth. *Age* also affects fecundity. The fecundity of organisms of the same size often increases just following maturity, then declines towards the end of life: e.g. in fruit flies and red deer (cf. Clutton-Brock 1988). In songbirds, age effects are harder to detect, and the effects of age can be confounded with the effects of selection. Where individuals with lower fecundity have higher mortality, average fecundity should increase with age simply because the weaker individuals have died off.

The considerable *genetic* variation for clutch-size, litter size, and fecundity is documented in Mousseau and Roff (1987). Heritabilities vary from 0–90 per cent, and values of the order of 20–40 per cent are common. Where genetic variation contributes to differences in clutch size among individuals within populations, clutch sizes of relatives are not statistically independent.

Clutch size thus varies among individuals within populations because they differ genetically, in age and size, and in physiological performance for environmental reasons, including food, temperature, and parasites. Any study of the adaptive significance of clutch size confronts these questions: to what degree can the mean clutch size in the population be explained as an adaptation to average conditions, to what degree can the flexible phenotypic responses of individuals to temperature, food, and parasites be explained as adaptive, how are both constrained, and what aspects of the patterns are not adaptive?

EFFECTS CAUSING DEVIATIONS FROM THE LACK CLUTCH

Evidence

When Klomp summarized the evidence available in 1970, he concluded that observed clutches were consistently smaller than the Lack clutch. Work done since 1970 has not changed this conclusion but has refined it considerably. Lindén and Møller (1989) and Dijkstra *et al.* (1990) reviewed 55 studies in which the consequences of manipulating the number of eggs or nestlings in various birds were studied (Table 7.1).

Table 7.2 summarizes Table 7.1. In 40 of 53 cases, the birds could raise more nestlings than they chose to raise themselves. The most frequent clutch was smaller than the most productive. This demands explanation.

Lack's basic assumption also often holds: in 28 of 44 studies nestling survival was poorer in enlarged broods, *and in no study was it better.* However, survival to fledging overestimates survival to breeding. Of 40 studies that reported nestling condition, 27 found that it was significantly worse in enlarged broods, 13 noted no difference, and *none found chicks in better condition in enlarged broods.* Chicks fledged from enlarged broods appear to suffer higher mortality before breeding than those that fledged from broods the size the parents chose to lay. Such effects on fledgling survival (recruitment) have been suggested in 8 of 15 cases and reliably documented in the collared flycatcher (Gustafsson and Sutherland 1988) and in the great tit (Pettifor *et al.* 1988; Smith *et al.* 1989). In both cases, probability of recruitment into the breeding population was lower in enlarged broods.

In the three studies in which one looked for effects of enlarged clutches on the future reproductive performance of offspring, they were found. In 5 of 14 cases, effects of parental survival were found; in 9 of 14 cases they were not. In 8 of 14 cases effects on the future reproductive performance of the parents were found; in 6 of 14 cases they were not.

In 82 per cent of the cases reported, costs of enlarged broods were found in offspring traits; in only 44 per cent of the cases were they found in parental traits. The difference is significant (*G*-test, $p < 0.025$). Tuomi (1990) shows that this difference is consistent with an extension of the theory of the Lack clutch: when increases in clutch size have larger effects on juvenile survival than on adult survival and both are plausible non-linear functions of clutch size, females can maximize their current reproductive success and their fitness without sacrificing their own survival.

In this table, in contrast to Lindén and Møller (1989), there is no significant difference in the frequency of adult survival trade-offs and adult fecundity trade-offs.

Table 7.1 The effects (+, 0, −) of increasing clutch or brood size in birds on N_f = number of fledglings, M_f = body mass of fledglings, S_0 = survival of nestlings until fledging, S_f = survival of fledglings until autumn or the next breeding season, B_0 = future breeding performance of offspring M_p = body mass of parents, S_p = survival of parents, B_p = future breeding performance of parents. From Lindén and Møller (1989) and Dijkstra *et al.* (1990)

	Offspring					Parents			
Order/species	N_f	M_f	S_0	S_f	B_0	M_p	S_p	B_p	Reference
Procellariformes									
Diomedea immutabilis	−		−						Rice and Kenyon (1962)
Oceanodroma castro	−		−						Harris (1969)
Oceanodroma leucorrhoa	0	−	−						Huntingdon (in Lack 1966)
Puffinus puffinus	−	−	−			0			Harris (1966)
Puffinus tenuirostris	0	−	−						Norman and Gottish (1969)
Pelecaniformes									
Sula sula	−	−							Nelson (1966)
Sula bassana	+		−						Nelson (1966)
Sula capensis	+	−				0			Jarvis (1974)
Anseriformes									
Branta canadensis	0	0	0	0	−	−		−	Lessells (1986)
Falconiformes									
Accipiter rufiventris		−	−						Simmons (1986)
Falco tinnunculus	+	−	−	0		−	−	0	Dijkstra *et al.* (1990)
Strigiformes									
Aegolius funereus	+	0	0			0	0	0	Korpimäki (1988)
Charadriiformes									
Stercorarius longicaudus	+		0						Andersson (1976)
Larus argentatus	+	0	0						Haymes and Morris (1977)
Larus glaucescens	+	−							Ward (1973)
Larus glaucescens	+		0						Vermeer (1963)
Larus glaucescens						−	−	0	Reid (1987)
Larus fuscus	+	0	0						Harris and Plumb (1965)
Rissa tridactyla	+	−	0						Coulson (in Lack 1966)
Creagus furcatus	+	0	0			0	0		Harris (1970)
Fratercula arctica	+		−						Corkhill (1973)
Fratercula arctica	−		−						Nettleship (1972)
Cepphus grylle	+	0	0						Asbirk (1979)
Alca torda	+	−	−						Lloyd (1977)
Alca torda	+	−	−						Plumb (1965)
Columbiformes									
Columba palumbus	+	−	−	0					Murton *et al.* (1974)
Apodiformes									
Apus apus	−	−							Perrins (1964)
Aerodramus spodiopygus	0	−	−						Tarburton (1987)
Passeriformes									
Delichon urbica	+	0	0						Bryant (1975)
Delichon urbica	+	−	−						Bryant and Westerterp (1983)

Table 7.1 (*continued*)

Order/species	Offspring					Parents			Reference
	N_f	M_f	S_0	S_f	B_0	M_p	S_p	B_p	
Passeriformes (*continued*)									
Iridoprocne bicolor	+	–	0	0		0	0		DeSteven (1980)
Troglodytes aedon	+	0	0			0		0	Finke *et al.* (1987)
Ficedula hypoleuca	0	–	–			–	–		Askenmo (1977, 1979)
Ficedula hypoleuca	+	–	0	0					von Hartman (1954)
Ficedula albicollis	+			–	–		0	–	Gustafsson and Sutherland (1988)
Tardus pilaris	+	0	0						Slagsvold (1982)
Parus major	+	–	–	–				–	Smith *et al.* (1987, 1989)
Parus major	+	–	–	–		0	0	–	Tinbergen (1987, unpublished)
Parus major	+			0			0		Boyce and Perrins (1987)
Parus major	+			–		0		0	Pettifor *et al.* (1988)
Parus major	+	–	–	–	–	–	–	–	Lindén (1987, 1988)
Parus major	+	–	–	–				–	Slagsvold (1984)
Parus montanus	+	–		–		0	0		Orell and Koivula (1988)
Parus caeruleus	+		0	0		–	–	0	Nur (1984*a*, *b*, 1988)
Plectrophenax nivalis	+	0	–			–			Hussell (1972)
Agelaius phoenicus	+	–	–						Cronmiller and Thompson (1980)
Pyrrhula pyrrhula	+		–	–					Newton (in Lack 1966)
Passer domesticus	+	–	–						Schifferli (1978)
Passer domesticus	+	–	0			0	0	–	Hegner and Wingfield (1987)
Quelea quelea	0	0	–						Ward (1965)
Sturnus vulgaris	+	–							Crossner (1977)
Sturnus cinerareus	+	–							Kuroda (1959)
Pica pica	–	0	–						Högstedt (1980)
Corvus corone	+	0	–						Loman (1980)
Corvus frugileus	+						0	–	Røskaft (1985)

Table 7.2 A summary of the clutch size studies listed in Table 7.1. Effects refer to increases in clutch or brood sizes. In many cases, clutch or brood sizes were also decreased

Trait	Number of studies		Effect of increase			% negative
	Reported	Not reported	+	–	0	
Offspring						
N_f = number fledged	53	2	40	7	6	
M_f = weight of fledglings	40	15	0	27	13	68
S_0 = survival in nest	44	11	0	28	16	64
S_f = survival to next season	15	40	0	8	7	53
B_0 = future reproduction	3	52	0	3	0	100
Parents						
M_p = weight of parents	17	38	0	7	10	41
S_p = survival to next season	14	41	0	5	9	36
B_p = future reproduction	14	41	0	8	6	57

Several factors make some of these results difficult to interpret (Lindén and Møller 1989). In some studies, individuals were assigned clutches very much larger or smaller than those they had chosen to lay (Reid 1987; Nur 1984*a*,*b*, 1988; Tinbergen 1987; Smith *et al.* 1987, 1989; Lindén 1988; Slagsvold 1984; Pettifor *et al.* 1988; Røskaft 1985). The option facing the bird is to lay one more or one fewer egg rather than many more or fewer; responses to such manipulations may be unnatural.

Whereas virtually all the studies done in nest boxes were carried out at unnaturally low densities of ectoparasites (chiefly mites and lice), addition of mites to barn swallow nests did decrease reproductive success (Møller 1990). Clutch size manipulations also decrease the risk of predation to adults and offspring, for human presence scares off predators and the nest boxes themselves are designed to reduce predation risk. Most studies using nest boxes were also carried out on populations that were either increasing (because nesting holes had been a limiting resource until the nest boxes were provided) or at abnormally high densities (because so many nest boxes had been available for so long). If the population was increasing, clutch size manipulations may have had little effect because food resources were not limiting. If it was at abnormally high density, then manipulations may have had large effects because food resources were limiting (Lindén and Møller 1989).

Finally, all brood size manipulations ignore costs incurred earlier in the life history—in particular the cost of producing eggs, which might well affect the results of brood manipulation experiments (L. Partridge, pers. comm.).

Effects of asymmetric offspring survival curves

Our simple model of the Lack clutch (Fig. 7.1) used a linear decline in offspring survival with increasing clutch size to produce a symmetrical recruitment curve. A more plausible relation between survival rate and clutch size for small hole-nesting birds in temperate areas would show low survival rates for very small clutches because of problems with thermoregulation, almost no difference in the higher survival rates across a range of intermediate clutch sizes, then sharply lower survival rates for large clutches (Fig. 7.2). The result is an asymmetrical recruitment curve in which the penalties for having a clutch smaller than the most productive are not as severe as those for having a larger clutch (Fig. 7.2(*b*)).

This detail is important. In Fig. 7.2(*b*), the optimal clutch of six fledges 5.4 offspring. A clutch of five fledges 4.9 offspring with a penalty of 0.5, but a clutch of seven fledges only 4.2 with a penalty of 1.2. The cost of having a clutch one egg too large is 2.4 times as large as the cost of having a clutch that is one egg too small. In such cases, any force tending to shift the optimal clutch size away from the one most productive in terms of fledgling survival will meet less resistance to the left than to the right of the optimum. Most perturbations of the Lack clutch move it to the left.

Intergenerational effects of enlarged clutches

Even when offspring of enlarged broods do not suffer higher mortality, they may pay other costs.

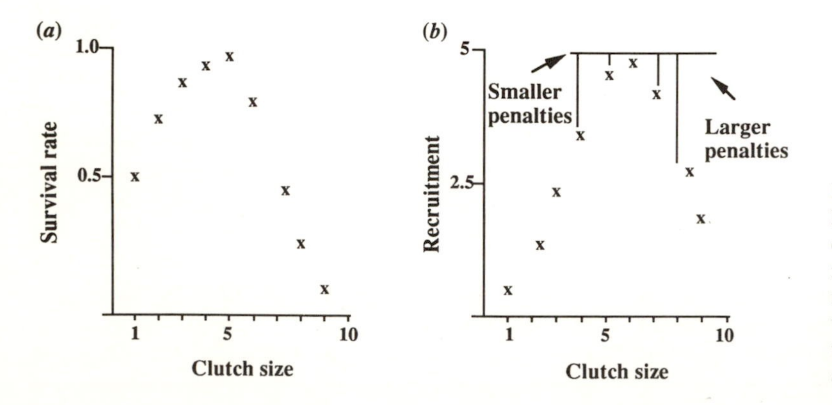

Figure 7.2 (*a*) A more realistic non-linear relation between survival rates and clutch size for small, altricial, hole-nesting birds in temperate areas. (*b*) The resulting asymmetrical recruitment curve with smaller penalties for clutches smaller than the most productive than for equivalently larger clutches.

Recall that Williams (1966*b*) partitioned Fisher's (1930) reproductive value into two parts, the present reproductive value of this clutch and the residual reproductive value of the parents (all later contributions to fitness):

$$V = V_c + V_r \qquad (7.7)$$

where

V_c = clutch size

× the reproductive value of one egg (V_0).

The reproductive value of one egg contains intergenerational terms, for it is the sum of the probability that the egg survives to reproduce once, times its first clutch size, plus the probability that it survives to reproduce twice, times its second clutch size, and so forth for all clutches that it might have, with each clutch discounted by $\lambda = e^r$ raised to a power equal to minus the age at which the clutch is produced (cf. Ch. 2):

$$V_0 = l_\alpha b_1 \lambda^{-1} + l_\alpha p_\alpha b_2 \lambda^{-2} + l_\alpha p_\alpha p_{\alpha+1} b_3 \lambda^{-3} + \cdots. \qquad (7.8)$$

The Lack clutch is predicted by assuming that variation in clutch size affects only the first term, l_α (and of that only survival to fledging), with no further effects.

However, intergenerational trade-offs have been demonstrated. In hymenopteran parasitoids, Klomp and Teerink (1967) and Waage and Ng (1984) found that daughters from larger clutches were smaller and had lower fecundities and shorter lives. In collared flycatchers (Gustafsson and Sutherland 1988) and great tits (Lindén 1990), offspring from enlarged broods were smaller, suffered higher mortality, and produced fewer fledglings from their first broods. Kestrels from enlarged broods were less likely to breed in their first season (Dijkstra *et al.* 1990).

Andersson (1978) worked out the consequences for optimal clutch size if individual offspring from enlarged broods delayed maturity, had smaller clutches, or themselves produced offspring of lower quality. He found, for example, that effects of clutch size on the probability that offspring breed in their first season could reduce the optimal clutch to only 40–62 per cent of the Lack clutch.

Table 7.3 Performance of children and mothers of singleton and twin births in rural Finland from 1769 to 1850. Sample sizes in parentheses (after Haukioja *et al.* 1989)

Parameter	Singleton	Twin	*P*
Total no. offspring born to mother	5.9 ± 2.2 (316)	6.4 ± 2.4 (168)	<0.05
Offspring survival to age 15 (%)	70.6 (299)	33.7 (344)	<0.0001
Survival of that mother's (other) singletons to age 15 (%)	72.8 (1166)	71.2 (410)	NS
Survival of mothers for the first year after delivery (%)	99.1 (293)	95.6 (159)	0.011

Intragenerational effects

Increased reproduction may reduce parental survival and subsequent reproductive performance (Williams 1966*b*; Gadgil and Bossert 1970; Schaffer 1974*a*; Charlesworth and Léon 1976; Charlesworth 1980). Such *reproductive effort* models link the evolution of clutch size to the evolution of life span. Charnov and Krebs (1974) showed that when parental survival is taken into account, the clutch size that optimizes lifetime fitness is smaller than the clutch size that fledges the most young. In capital breeders, reproduction also affects subsequent reproductive performance.

Both inter- and intragenerational trade-offs explain why humans normally have one child (Table 7.3: Haukioja *et al.* 1989; Anderson 1990*a*). Juvenile mortality rates are higher in twins than singletons, and mortality rates of the mothers of twins are also higher. Here a litter size of one produces the best survival rates for both parents and offspring and higher lifetime reproductive success than a litter of two. As recently as 140 years ago, there was strong natural selection against twinning in humans living in rural Finland.

A case study in optimal clutch size—the kestrel

Intragenerational trade-offs make the observed clutch size about equal to the optimal clutch size in kestrels (Daan *et al.* 1990). The kestrel study yielded data of unusually good quality, illustrated both intergenerational and intragenerational trade-offs, and applied the theory in unusual detail.

We assume that birds lay clutches that optimize individual fitness by responding flexibly to food supply and laying date and that reproductive value and fitness are equivalent. This is not strictly true. Fisher (1930) mentioned that the strength of selection on an age class should be proportional to its reproductive value. Schaffer (1974*a*) and Taylor *et al.* (1974) proved that, in a certain sense, maximizing the reproductive value of each age class also maximizes fitness, and a discussion between Caswell (1980), Yodzis (1981), and Schaffer (1981) clarified the sense in which that is true. The equivalence only holds when the reproductive value of each age class is maximized with respect to the reproductive effort expended in that age class, holding the reproductive efforts in the other age classes constant. Hamilton (1966) and Charlesworth (1980, p. 216) emphasized that reproductive value and fitness are not equivalent. For the evolution of senescence, assuming that they are equivalent will mislead. Caswell (1989*a*, pp. 176–7) provides a precise technical statement of the difference between maximizing r and maximizing reproductive value. However, assuming their equivalence has stimulated biologists to look for all the direct and indirect effects on individual fitness that arise through any trait listed in the expression for reproductive value. This has proved informative.

Daan, Dijkstra, and Tinbergen calculated the reproductive value of the clutch (V_c) and the residual reproductive value of the parents (V_r) for reduced and enlarged clutches. The total reproductive value ($V_t = V_c + V_r$) of the clutches actually laid should be larger than that of either reduced or enlarged clutches. The residual reproductive value of one parent, V_p, is

$$V_p = \frac{1}{l_t} \sum_{x=t}^{\infty} \lambda^{-(x-t)} b_x l_x. \tag{7.9}$$

These kestrels live near the North Sea in the Netherlands, where survival is good and clutch sizes are large. Survival rates per year from the third year on are fairly constant at about 0.7 ($p_3 = 0.698$, $p_4 = 0.721$, $p_5 = 0.697$, p_x $(x > 5) = 0.704$) and after the first year, clutch sizes are also fairly constant: b_x $(x > 1) = 5.33$. However, p_0, p_1, p_2, and b_1 vary with laying date, as does the probability that a nest will succeed. When these factors are accounted for (Daan *et al.* 1990), the consequences of increased or decreased reproduction are clear (Table 7.4). The clutch size actually laid was the one that yielded the highest reproductive value. This is a remarkable result because the decision on which effects to measure was made

Table 7.4 Computation of reproductive values in experimentally reduced, unchanged, and enlarged kestrel broods (Daan *et al.* 1990). Brood manipulations were done at 10 days

	Reduced	Control	Enlarged
Brood			
Number of broods	28	54	20
Mean clutch size	5.25	5.19	5.40
Mean brood size (at 10 days)	4.53	4.20	4.83
Mean change in brood size	−1.74	0.00	+2.51
Mean number fledged	2.60	3.95	5.84
V_c—reproductive value of clutch	2.52	4.20	5.59
Parents			
Number of males and females	49	85	35
Local parental survival, p_1	0.653	0.588	0.429
V_p—residual reproductive value	9.88	8.89	6.49
Total			
V—total reproductive value	12.40	13.09	12.08

years before the calculation of the optimal clutch and virtually all plausible costs of reproduction were taken into account.

The Reproductive Effort Model and the General Life History Problem

If the only effect of the life history theory developed during the 1970s had been to inspire the excellent work on reproductive trade-offs done by Pettifor *et al.* (1988), Gustaffson and Sutherland (1988), Dijkstra (1988), Haukioja *et al.* (1989), Lindén (1990), Daan *et al.* (1990), and others, it would have achieved a lot. However, those who developed the theory had something more general in mind: understanding how to optimize the allocation of resources to growth, survival, and reproduction from birth to death. Schaffer (1983) refers to this as 'the general life history problem', Charlesworth (1980, 1990*b*) to 'reproductive effort models'. Such analyses optimize age and size at maturity and the number of offspring produced at each age under the assumption of trade-offs between current reproduction and subsequent adult reproductive performance. Intergenerational trade-offs have not yet figured prominently in this tradition, but studies like those of Lindén (1990) and Dijkstra (1988) should stimulate their inclusion. The goal is to predict as much as possible of the equilibrium life history phenotype in a constant environment.

The technique used to solve such problems is optimal control theory. It focuses on the organism's state at a given age (weight, reproductive potential, condition), and assumes that this state is determined by the sequence of decisions made in each such state since birth. Associated with each age class is a control variable that keeps track of the decisions that could be made. It takes on values from 0 to 1 representing reproductive efforts ranging from 0 (no clutch laid, good adult survival and/or growth) to 1 (maximum possible reproductive effort, large clutch, no growth, adult dies before next breeding season). To calculate the optimal reproductive effort, one assumes a maximum possible age at which the final reproductive effort should be maximal because no adults that old will survive anyway. One then works back, calculating the optimal reproductive effort for the next to last age class, and so forth, until the entire life has been worked through in reverse from death to birth. At some point, the optimal reproductive effort switches from intermediate to zero, and that defines age at maturity. The need to assume a fixed maximum age (or size) makes this version of the technique inappropriate for the analysis of the coevolution of reproductive effort with lifespan and senescence. Bellman (1957) indicates how to handle the case with no maximum age, but this has not been followed up.

Schaffer (1974*a*), Taylor *et al.* (1984), Charlesworth and Léon (1976), Léon (1976), Charlesworth (1980), and Mangel and Clark (1988) all used optimal control theory to attack various aspects of the general life history problem. Schaffer (1983) gives a particularly clear exposition of the general method, and Kozlowski and Wiegert (1987) analyse an illuminating case (see below).

Before presenting that case, I summarize the main conclusions gleaned from the other studies. Recall the distinction between convex and concave profit and cost curves from Chapter 4 (Fig. 7.3). Gadgil and Bossert (1970) found that if the profit function is convex or the cost function is concave, an intermediate level of reproductive effort is optimal and repeated reproduction is expected (iteroparity, polycarpy: most vertebrates, most trees). Otherwise, a maximal reproductive effort is optimal, and one expects a single, suicidal reproductive event (semelparity: Pacific salmon, annual plants, agaves, dung beetles, bamboo, periodic cicadas). Taylor *et al.* (1974) and Schaffer (1974*a*) confirmed this result, and Schaffer extended it to more complex shapes of the profit and cost curves. For some cases Schaffer found multiple stable equilibria: either a semelparous or an iteroparous life history would evolve depending on initial conditions. The contrast of the semelparous Pacific salmon with the ecologically similar but iteroparous steelhead trout may be such a case.

It has proven very difficult to detect curvature in the cost curve (cf. Chapter 4). Most examples of profit curves are convex, as is consistent with a trade-off of clutch size with offspring survival (the Lack hypothesis). This sort of prediction can only be plausibly tested for its iteroparous part, since truly semelparous organisms show no variation in reproductive effort (maximal) or survival rate (zero) as adults.

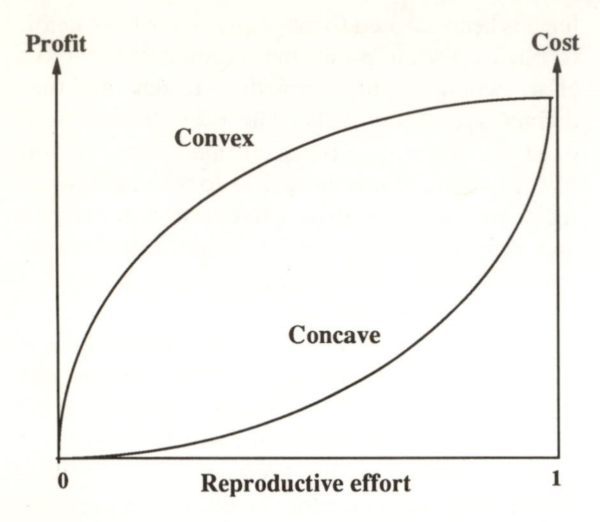

Figure 7.3 In reproductive effort models, the shape of the profit and cost curves makes a decisive difference to the level of reproductive effort that is optimal at a given age.

Gadgil and Bossert found that a uniform change in mortality rates in all age classes has no effect on the optimal age at maturity or on the optimal age distribution of reproductive effort. This was confirmed by Taylor *et al.* While this may be true in these models, where the investigator can in principle change one factor at a time, in the real world, a uniform increase in mortality rates in all age classes will change population density. This in turn will change the resources available per individual, and there will be both a plastic response and a genetic response that is not reflected in their models. In Kozlowski and Uchmanski's (1987) model, in which resources are represented, a uniform increase in mortality rates resulted in earlier maturity and a greater proportion of the growing season devoted to reproduction. Resource-based models are more realistic and get different results.

Gadgil and Bossert also found that if mortality rates are increased following a certain age, then the optimal reproductive effort will increase before that age and decrease after it. Michod (1979) proved a stronger version: if mortality rates increase in *one* age class, then the optimal reproductive effort increases before that age and decreases after it.

The most widely discussed of Gadgil and Bossert's findings was their confirmation through simulation of Williams' (1966*b*) claim that reproductive effort should always increase with age. Charlesworth and Léon (1976) found an analytical condition governing the increase or decrease of reproductive effort with age: it does not have to increase. If the population growth rate is high, if 'survival and growth do not fall off too fast with increasing effort, and if the maximal fecundity per unit size is low, . . . then reproductive effort must decrease with age' (p. 452). Red deer increase reproductive effort with age (Clutton-Brock *et al.* 1982).

A case study in optimal growth, reproductive effort, and survival

Kozlowski and his colleagues have modelled the optimal lifetime energy allocation to growth and reproduction relatively straightforwardly (Kozlowski and Wiegert 1987; Kozlowski and Uchmanski 1987; Kozlowski and Ziòlko 1988). Consider a perennial species with indeterminate growth, living in a seasonal environment with a stable population; define fitness as expected number of offspring per lifetime (R_0); assume adult mortality is independent of size; and assume that the switch from growth to reproduction is complete but reversible and occurs at most once per year (Kozlowski and Uchmanski 1987).

We begin with the simple case where the organisms live at most two years and can reproduce either only in the second year or in both. The average number of offspring produced by such organisms can be represented geometrically as the sum of two solids, each solid having a volume determined by integrating over the age interval of one reproductive season the probability of having survived to that age times the rate of production of offspring at that age (Fig. 7.4). The space between the two solids represents the winter when the organism can neither grow nor reproduce plus the spring of the second season when the organism can grow but has not yet started to reproduce. The problem is to maximize fitness as represented by the volume of the two solids. To do so, we must work backwards from the end of life and assume a final body size, w_2. Let V = lifetime reproductive success, V_1 = offspring production in first year, V_2 = offspring production in second year, t_1 = time to switch from growth

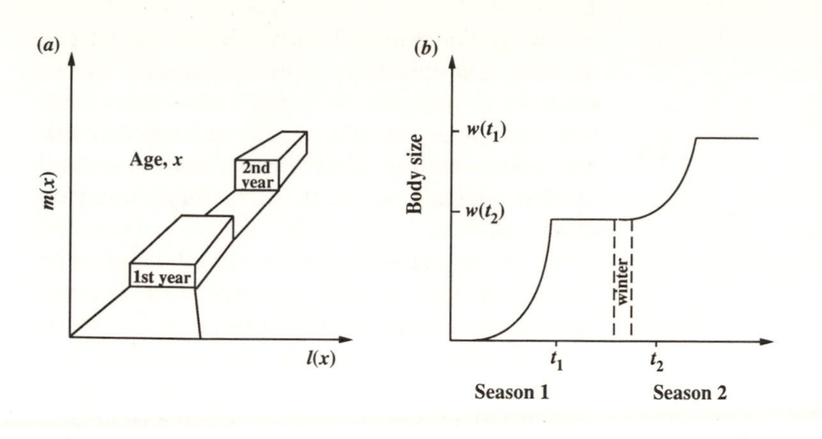

Figure 7.4 Representation of the lifetime energy allocation to reproduction and growth in a species with a maximum lifetime of two years and reproducing in both years. (*a*) The axes are age x, probability of surviving to age x ($l(x)$), and rate of energy allocation to reproduction—or rate of offspring—$m(x)$. Fitness is taken to be the sum of the volumes of the two solids representing reproductive success in the first and second years. (*b*) The associated dynamics of growth, where t_1 and t_2 represent the times of switching from growth to reproduction in the first and second year (after Kozlowski and Uchmanski 1987).

to reproduction in the first year, and t_2 = time to switch from growth to reproduction in the second year. Then

$$V = V_1 + V_2 \tag{7.10}$$

and V is maximized when the switching times are so chosen that the appropriate derivatives equal zero:

$$\frac{dV_1}{dt_1} + \frac{\partial V_2}{\partial t_1} = 0 \tag{7.11}$$

and

$$\frac{\partial V_2}{\partial t_2} = 0. \tag{7.12}$$

The volume of each solid is the product of its height—the offspring production rate, a function of body size $H(w)$—and the area of its base S—the area under the survival curve or the integral of $l(x)$ over the age interval of offspring production. Body size is itself a function of the switching time, and we let $f(w) = dw/dt$ represent growth rate at size w. That gives us

$$V_1 = H(w_1)S_1 \tag{7.13}$$

$$V_2 = H(w_2)S_2 \tag{7.14}$$

and

$$\frac{dV_1}{dt_1} = \frac{dH(w_1)}{dw_1} f(w_1)S_1 + H(w_1)\frac{dS_1}{dt_1}. \tag{7.15}$$

Because S_1 is the area under the survival curve, equal to the integral of $l(x)$,

$$\frac{dS_1}{dt_1} = -l(t_1) \tag{7.16}$$

and if we represent the derivative of reproductive output with respect to weight—the slope of the curve of fecundity against weight—as

$$H'(w) = \frac{dH(w)}{dw} \tag{7.17}$$

then we get for the first volume

$$\frac{dV_1}{dt_1} = H'(w_1)f(w_1)S_1 - l(t_1)H(w_1). \tag{7.18}$$

For the second volume,

$$\frac{\partial V_2}{\partial t_1} = \frac{\partial H(w_2)}{\partial t_1} S_2 + H(w_2)\frac{\partial S_2}{\partial t_1} \tag{7.19}$$

but because we assumed that adult mortality is independent of size, we know that

$$\frac{\partial S_2}{\partial t_1} = 0 \tag{7.20}$$

and if we assume that growth rate depends on size rather than age, then

$$\frac{\partial H(w_2)}{\partial t_1} = \frac{\partial H(w_2)}{\partial t_2}. \tag{7.21}$$

This means that the size attained at the switch from growth to reproduction in the second year does not change if the growth period is extended in the first year by the same amount by which it is shortened in the second year. Taking the last two equations into account gives us:

$$\frac{\partial V_2}{\partial t_1} = H'(w_2)f(w_2)S_2. \tag{7.22}$$

The partial derivative of the offspring production in the second year with respect to switching time in the second year is:

$$\frac{\partial V_2}{\partial t_2} = \frac{\partial H(w_2)}{\partial w_2} f(w_2)S_2 + H(w_2)\frac{dS_2}{dt_2} \tag{7.23}$$

or, by the same reasoning as above,

$$\frac{dV_2}{dt_2} = H'(w_2)f(w_2)S_2 - l(t_2)H(w_2). \tag{7.24}$$

Fitness is maximized when the sum of eqns (7.18) and (7.22) equals zero and when, at the same time, eqn (7.24) equals zero. This gives us the optimality conditions for switching from growth to reproduction in both years:

$$H'(w_1)f(w_1)S_1 + H'(w_2)f(w_2)S_2 = H(w_1)l(t_1) \tag{7.25a}$$

$$H'(w_2)f(w_2)S_2 = H(w_2)l(t_2). \tag{7.25b}$$

Dividing the first by $l(t_1)$ and the second by $l(t_2)$, we get

$$H'(w_1)f(w_1)E_1 + H'(w_2)f(w_2)pE_2 = H(w_1) \tag{7.26a}$$

$$H'(w_2)f(w_2)E_2 = H(w_2). \tag{7.26b}$$

Here p is the probability of surviving from the onset of reproduction in the first year to the onset of reproduction in the second year; E_1 and E_2 are the number of reproductive days expected at the onset of reproduction in each year (life expectancy minus winter days). $H'(w)$ is the rate of gain of fecundity with weight at the onset of reproduction, $H(w)$ is the rate of reproduction as a function of weight, and $f(w)$ is the somatic growth rate at body size w when all surplus energy is allocated to growth.

If we divide eqn (7.26*b*) by $H(w_2)$, we get eqn (6.27), the equation for optimal age and size at maturity in annuals. In other words, the second year of a two-year life history behaves like the first year of an annual life history, and in general the last year of an *n*-year life history should do so as well.

Equation (7.26*b*) must be solved before eqn (7.26*a*) can be solved; i.e., we must proceed backwards, assuming a final body size w_2. At the end of the second season, the expectation of productive life (E_2) is zero, as is the left side of eqn (7.26*b*), making it smaller than the right side. This means that it is optimal to reproduce. As we work our way back from the end of the second season, E_2 increases while $H(w_2)$ and $f(w_2)$ hold constant because all surplus energy is allocated to reproduction and the organism does not grow. It is likely—but not necessary—that before the beginning of the second growing season is reached, the left and right sides will become equal. That gives the optimal time t_2 to switch from growth to reproduction in the second year. Given t_2 we can solve the growth equation $f(w)$ backwards to the beginning of the second growing season, giving us the final body size in the first year, w_1, and allowing us to repeat the procedure for the first year, since w_2 and E_2 in eqn (7.26*a*) are now known. Once t_1 is known, the growth equation can be solved backward to get size at birth. If size at birth differs from that observed, adjust w_2 and repeat the procedure. If there is no t_1 that satisfies eqn (7.26*a*), it is optimal to devote the entire first year to growth if the left side is larger than the right side. If there is no t_2 that satisfies eqn (7.26*b*), then it is optimal to devote the entire second year to reproduction if the left side of eqn (7.26*b*) is smaller than the right side.

The procedure generalizes to a life span of any length m (Kozlowski and Uchmanski 1987), and the condition for optimal switching in year i is:

$$H'(w_i)f(w_i)E_i + p_{i,i+1}H(w_{i+1}) = H(w_i) \tag{7.27}$$

where $p_{i,i+1}$ is the probability of surviving from the onset of reproduction in year i to the onset of reproduction in year $i + 1$. For a perennial species that reproduces once per year, $p_{i,ij}$ is the probability of surviving from the end of the ith season to the end of the jth season, and

$$p_{ij} = q^{j-i}, \tag{7.28}$$

where q is the yearly survival rate.

Now consider a perennial species, reproducing at the end of each season after maturation, that can live for a maximum of 10 years. It has a surplus production rate for both growth and reproduction of $f(w) = H(w) = 0.05w^{0.75}$. Figure 7.5 shows the optimal fraction of each season devoted to reproduction (a) and the corresponding growth curves (b) for yearly survival rates $q = 0.41$, 0.65, and 0.72.

Note first that the method predicts both the optimal reproductive schedule (Fig. 7.5(a)) and the optimal growth curve (Fig. 7.5(b)). Like Schaffer (1974a), Taylor *et al.* (1974), Charlesworth and Léon (1976), and Michod (1979), it predicts that a decrease in the adult survival rate leads to earlier maturity, larger reproductive effort earlier in life, and smaller adult body size.

A decrease in the rate of increase of production with body size, expressed as $f(w) = H(w) = 0.05w^{0.68}$ (not depicted here), results in earlier maturation, a larger proportion of seasons devoted to reproduction, and much smaller body sizes—about one fifth as large as the body sizes attained when production is more efficient. This is because a decrease in the rate of production devalues juvenile growth and makes it less rewarding to grow rather than mature and reproduce.

Kozlowski and Uchmanski (1987) discuss applications to cone shells in Hawaii, scallops in Iceland, and arctic char in Labrador, where the model performed well.

An exemplary case study: the Trinidad guppies

More convincing (Charlesworth 1990b) is the confirmation of reproductive effort models by the field manipulation experiments done by Reznick and his colleagues on the guppy, *Poecilia reticulata* (Reznick 1982a, b, 1983, 1989, 1990; Reznick and Bryga 1987; Reznick and Endler 1982; Reznick *et al.* 1990).

Guppies are small (1–4 cm), viviparous, sexually dimorphic fish that live in shallow streams in

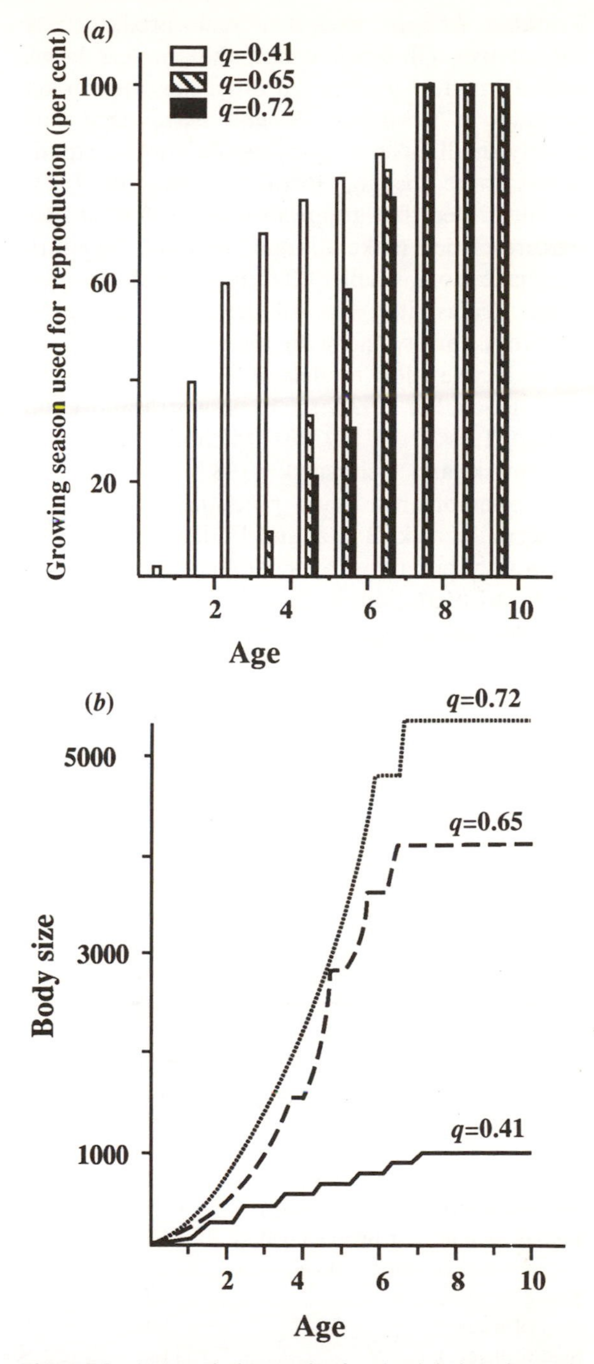

Figure 7.5 A numerical example of optimal timing of growth and reproduction in a species with a maximum life of 10 years that reproduces at the end of the season. q indicates the different levels of yearly survival rate used. (a) The optimal fraction of the season devoted to reproduction. The appearance of the first bar marks age at maturity. (b) Growth curves of individuals that follow the optimal reproduction schedule (after Kozlowski and Uchmanski 1987).

Trinidad. At some sites their main predator is a cichlid fish, *Crenicichla alta*, that prefers large, sexually mature guppies. At other sites, their main predator is a killifish, *Rivulus hartii*, that eats mostly small, juvenile guppies. Before the experiments were begun, Reznick (1982*a*, *b*, 1983) demonstrated that guppies from *Crenicichla* sites mature earlier, make a larger reproductive effort, and have more, smaller offspring than do guppies from *Rivulus* sites. The differences are heritable, and they correspond with the predictions of the reproductive effort models of Gadgil and Bossert (1970), Schaffer (1974*a*), Charlesworth and Léon (1976), Law *et al.* (1979*b*), Michod (1979), and Kozlowski and Uchmanski (1987).

To demonstrate that predation caused the pattern, Reznick, Bryga, and Endler manipulated natural populations by changing predation on adults to predation on juveniles. This decreased the age- and size-specific mortality rates for adults and increased them for juveniles. After 11 years, or 30–60 generations, they observed significant life history evolution in the predicted direction. In Table 7.5, the pair of columns on the right describes guppy life histories before manipulation at evolutionary equilibrium under relatively high adult mortality (*Crenicichla*) or relatively high juvenile mortality (*Rivulus*). Under higher adult mortality, both males and females mature earlier and smaller, the females have larger first litters (but no difference in second or third litters), smaller young, a shorter interbrood interval, and make a larger reproductive effort. The pair of columns on the left describes the results of the manipulation, which began in 1976. Guppies were taken from a *Crenicichla* site (the control) and introduced to a *Rivulus* site, releasing them from predation on adults and exposing them to predation on juveniles. Eleven years later, guppies reared in the laboratory through two generations (to eliminate environmental and maternal effects) had evolved most of the differences seen in populations under similar selection pressures for a much longer time. Only offspring size in the third litter, the reproductive interval, and reproductive effort had not yet responded as expected. Life history traits early in life—age and size at maturity, size of first litter, reproductive effort early in life, size of offspring in first two litters—had already changed and corresponded qualitatively and quantitatively to the differences found in the unmanipulated populations.

Table 7.5 Life history patterns in guppies after a field manipulation experiment that used natural predators to increase juvenile mortality rates and decrease adult mortality rates, contrasted with those observed under the same two natural predators without manipulation. From Reznick *et al.* (1990)

	Reznick *et al.* (1990)			Reznick (1982*a*)		
Life history trait	Control (*Crenicichla*)		Introduction (*Rivulus*)	*Crenicichla*		*Rivulus*
Male age at maturity (days)	48.5	$p < 0.01$	58.2	51.8	$p < 0.01$	58.8
Male size at maturity (mg-wet)	67.5	$p < 0.01$	76.1	87.7	$p < 0.01$	99.7
Female age at first birth (days)	85.7	$p < 0.05$	92.3	71.5	$p < 0.01$	81.9
Female size at first birth (mg-wet)	161.5	$p < 0.01$	185.6	218.0	$p < 0.01$	270.0
Size of litter 1	4.5	$p < 0.05$	3.3	5.2	$p < 0.01$	3.2
Size of litter 2	8.1	NS	7.5	10.9	NS	10.2
Size of litter 3	11.4	NS	11.5	16.1	NS	16.0
Offspring weight (mg-dry) litter 1	0.87	$p < 0.10$	0.95	0.84	$p < 0.01$	0.99
Offspring weight litter 2	0.90	$p < 0.05$	1.02	0.95	$p < 0.05$	1.05
Offspring weight litter 3	1.10	NS	1.17	1.03	$p < 0.01$	1.17
Interlitter interval (days)	24.5	NS	25.2	22.8	NS	25.0
Reproductive effort (%)	22.0	NS	18.5	25.1	$p < 0.05$	19.2

* The sum of four consecutive estimates of dry weight of brood divided by dry weight of maternal soma.

Recently Reznick (pers. comm.) has shown further that when *Crenicichla* was introduced to streams in which it had not previously occurred, the mortality rates of larger guppies increased, and this was followed by the evolutionary response in life history traits expected from the reproductive effort model.

In this case, changes in age and size specific mortality rates caused by a shift in predation pressure caused rapid evolutionary change in life history traits in the direction predicted here and in Chapter 6. The rapidity of the change is striking. If the unmanipulated populations were at evolutionary equilibrium, then the reattainment of equilibrium following a major perturbation took 30–60 generations, 11 years elapsed time, for most of the traits involved. The traits that changed most were those occurring early in life, confirming a basic tenet of demographic theory (Chapter 2) central to the evolution of senescence (Chapter 8): selection on reproductive performance earlier in life is stronger than selection on reproductive performance later in life.

A parasite-induced shift in reproductive effort
The case study of guppies in Trinidad documents a genetic response to known selection pressures. The following example documents a plastic response that has itself surely evolved and is now part of the physiology of each individual. Parasites usually shorten the expected lifespan of their hosts, and many parasites castrate their hosts. Infection is a dependable cue that the host's birth and death rates will change for the worse. On detecting infection, the host should mature and start reproducing, if still juvenile, and increase reproductive effort, if adult (Minchella 1985). *Biomphalaria* is a snail found in East Africa as the intermediate host of the trematode parasite *Schistosoma*, an important human pest. Snails infected by trematodes suffer parasitic castration, then death. Infected snails respond by increasing reproductive output (Fig. 7.6). Exposed but uninfected snails still show the response. In these survivors there is a reduction in fecundity later in life relative to unexposed controls (Minchella and Loverde 1981). This elegant, minimal manipulation demonstrates that infection is not necessary. The snails need only be exposed to water in which

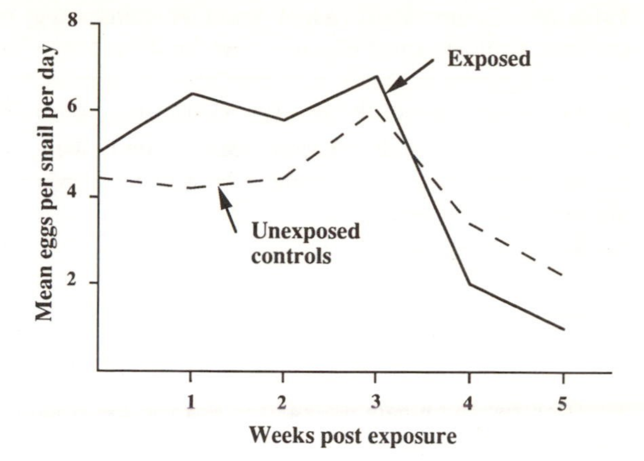

Figure 7.6 A snail's reaction to being exposed to, but not infected by, a trematode parasite that would castrate and kill it if it had infected it (after Minchella and Loverde 1981).

parasites have been held, and they react with precisely the adaptive plasticity predicted by reproductive effort models.

Skelly and Werner (1990) have recently performed a similar experiment on the reaction of toad tadpoles to the presence of predacious odonate larvae and have found a significant reduction in size at metamorphosis in the presence of the predator mediated by a behavioural response—lower activity levels. Lüning (pers. comm.) has evidence that *Daphnia* responds similarly to its predator *Chaoborus*.

Reproductive investment in plants
Willson (1983) reviews the substantial body of work on reproductive investment in plants relevant to the ideas discussed here.

Temporal variation in the optimal reproductive investment

Gillespie (1977) showed that when there is variation in the optimal number of offspring from generation to generation, then fitness is measured by the geometric means of reproductive success taken over generations.

If the mean reproductive success evolves independently of the variance, then the fittest type is the one with the lowest variance for a given mean (Table 7.6). This argument does not rely on any costs of reproduction. Ekbohm *et al.* (1980) noted that when the mean is not independent of

Table 7.6 Comparison of a high-risk, high-gain strategy with a low-risk, low-gain clutch-size strategy. The low-risk, low-gain strategy wins (as can be seen by comparing either their products or their geometric means)

(a) The clutch sizes of the two types compared for a series of good and bad years

Year:	Good	Bad	Good	Bad	Good	Bad	Good	Bad	Good	Bad	Good	Bad
(i) High-risk, high-gain	5	2	5	2	5	2	5	2	5	2	5	2
(ii) Low-risk, low-gain	4	3	4	3	4	3	4	3	4	3	4	3

(b) Calculation of mean values of clutch size over time

	Sum	Product	Arithmetic mean	Geometric mean	Variance
(i) High-risk, high-gain	42	$(5 \times 2 = 10)^6$	3.5	3.16	1.57
(ii) Low-risk, low-gain	42	$(3 \times 4 = 12)^6$	3.5	3.46	0.52

the variance, a non-zero variance in offspring number can be maintained. Neither paper analysed the case in which the reproductive effort is determined by a reaction norm optimized to deal with the temporal variation under the prevailing environmental constraints on growth and survival.

Murphy (1968) and Schaffer (1974*b*) also analysed the effects of temporal fluctuations in mortality rates on reproductive investment. (Because of their consequences for the evolution of the lifespan, they are dealt with in more detail in the next chapter.) If there is variation among reproductive events in the probability that offspring will survive to breed, and if reproductive effort and adult survival trade off, then it pays to reduce reproductive effort in order to live longer and reproduce more times, sampling a larger number of reproductive conditions and increasing the number of offspring born into good conditions. This pattern has been called *bet-hedging* after the practice of gamblers at race tracks who bet their money on several horses in the same race to increase their average gains (see Phillipi and Seger 1989, for a recent perspective on bet-hedging).

Here we concentrate on Gillespie's idea, which assumes no cost of reproduction in terms of adult mortality. If individuals laying larger clutches also experience greater variance in reproductive success, then they may have lower fitness than individuals laying smaller clutches. Boyce and Perrins (1987) applied this idea to data from the long-term study on great tits in Wytham Wood at Oxford. They first summarized 22 years of data demonstrating that the clutch sizes actually laid were considerably smaller than the clutch sizes that produced the most offspring (Fig. 7.7).

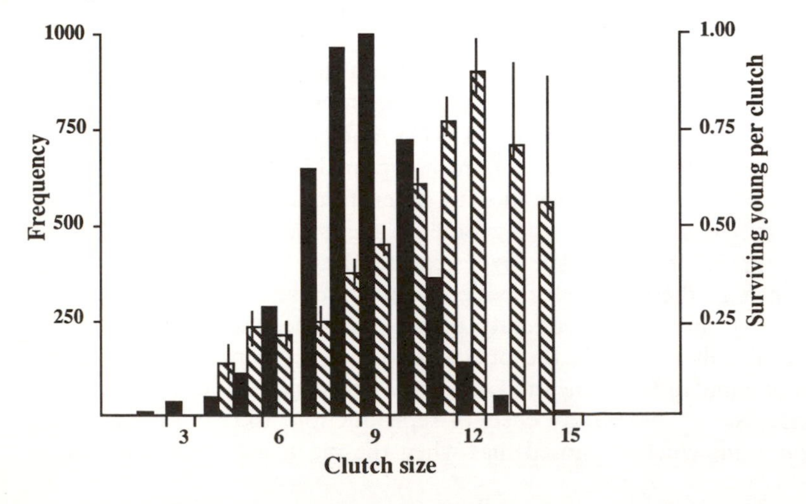

Figure 7.7 The frequencies of clutch sizes (solid bars) for 4489 great tit clutches in Wytham Wood, Oxford, from 1960 to 1982. The hatched bars represent the number of young per clutch surviving at least to the next season. Vertical lines are standard errors (after Boyce and Perrins 1987).

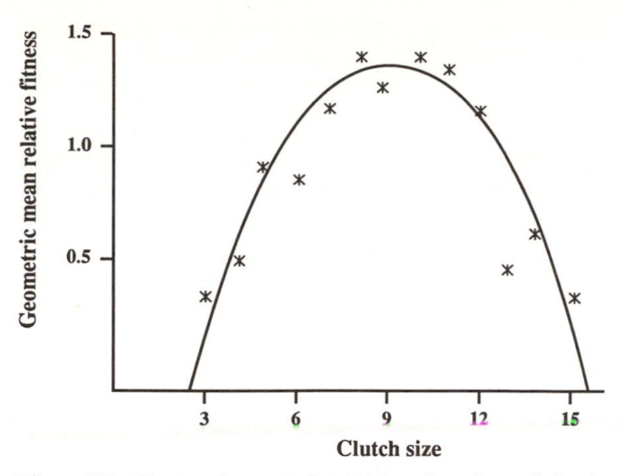

Figure 7.8 Geometric mean fitness as a function of clutch size for great tits in Wytham Wood, Oxford, from 1960 to 1982. Clutches of eight had highest fitness. A least-squares fit of a quadratic line yields an optimum clutch of about nine eggs. The observed mean clutch size was 8.53 (after Boyce and Perrins 1987).

They found no evidence in this case that the deviation of the average clutch size from the Lack clutch size could be explained either by costs in adult survival or subsequent reproductive performance or by asymmetrical recruitment curves of the sort depicted in Fig. 7.2. Therefore they concentrated on the possibility that birds with intermediate clutches had higher geometric mean fitness. Assuming age-specific survival rates and accounting for the fact that a given female does not lay a clutch of the same size every year, they calculated the geometric mean relative fitness of each clutch size across the 22 year sampling period (Fig. 7.8). The optimal clutch of 9.01 was not significantly different from the observed mean clutch of 8.53 ± 1.79; the modal observed clutch was 9 (Fig. 7.7).

Bad years affect larger clutches much more than smaller clutches. This reduces the mean and increases the variance in fitness for individuals laying larger clutches more than for those laying smaller clutches. It nicely accounts for the downward deviation of the observed mean clutch from the Lack clutch.

The effects of brood parasitism

Power *et al.* (1989) argue that egg parasitism also reduces the optimal clutch below the Lack clutch. This is an extension of the Lack hypothesis (Fig. 7.1) with brood parasitism having the effect of making the negative slope of the line relating own-nestling survival and total clutch size steeper—parasitism increases own-offspring mortality to fledging for a given clutch size. They illustrate the effect with a study of European starlings (*Sturnus vulgaris*) in which the most common clutch was five, the most productive was six, and a clutch of seven was overcrowded. Brood parasitism was common, with one-third of the clutches receiving foreign eggs. Their data do not rule out other hypotheses, but the effect of egg parasitism may have been real, and many clutch size studies have not taken it into consideration.

Summary of the ornithological tradition

In a stable environment where the only cost of producing more young is a decrease in their survival to maturity, an intermediate clutch size, the Lack clutch, is optimal. Observed clutches are often smaller than the Lack clutch. The reasons for the downward deviations include intergenerational trade-offs of clutch size with offspring survival and reproductive performance, intragenerational trade-offs of clutch size with parental survival and subsequent reproductive performance, and temporal variation in the optimal clutch size. In the latter case, the geometric mean clutch size is the appropriate measure of fitness, and there is selection to reduce the variance in clutch size by reducing the mean size below the Lack clutch.

Search-time trade-offs

Insects whose offspring develop in pupae, seeds, or fruit encounter problems of reproductive investment unlike those faced by birds or mammals. The female searches for sites to deposit her eggs. These sites vary in the number of larvae that can be raised to eclosion depending on the size and quality of the resource, the number of other organisms that are competing for it, and the fighting and predation that occurs within it.

If a female only laid once in her life, then she should lay the Lack clutch as modified by the appropriate trade-offs. Many of the same trade-offs exist in insects that have been found in birds: the survivorship of larvae declines with the number of eggs laid at the site, the size of the adults produced also declines with clutch size, and fecundity increases with adult size, so that there

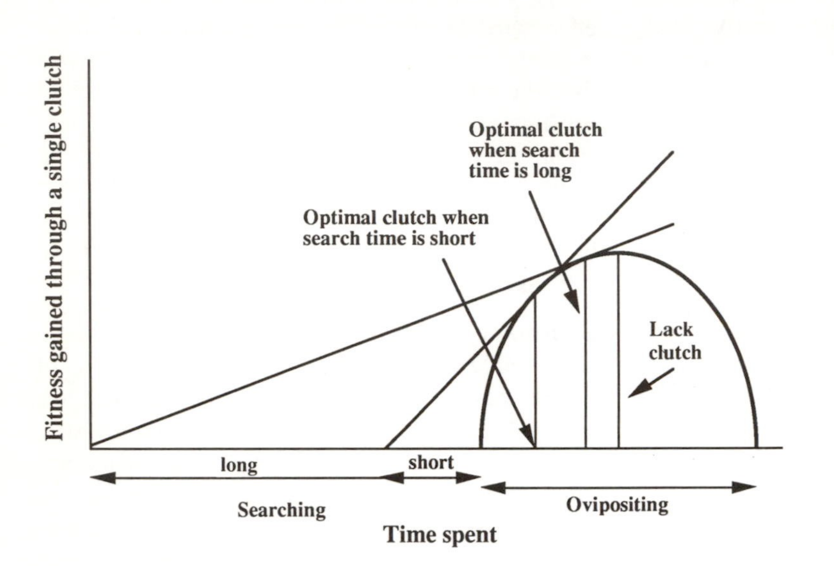

Figure 7.9 The impact of search time on optimal clutch size. The curve at the right describes the fitness gained through a single clutch, where clutch size is determined by the time spent ovipositing. Fitness is assumed to depend on clutch size in the fashion given for a Lack clutch. To mazimize the fitness gained per unit time, we need an expression of the form (offspring produced)/(time spent). Fix a point on the time axis and consider the slope of the line drawn from that point to somewhere on the clutch size curve. The slope of the line is (offspring produced)/(time spent), exactly what we are looking for. The line that has the steepest slope—the one tangent to the clutch size curve—maximizes net production rate. The optimal time spent ovipositing is longer—clutch sizes are larger—when search times are longer, simply because the point fixed on the time axis for long search times is farther away from the clutch size curve than it is for short search times (after Charnov and Skinner 1984).

are strong intergenerational trade-offs between size of clutch and offspring survival and future reproductive performance (Klomp and Teerink 1967; Charnov and Skinner 1984).

However, the female does not lay once in her life, but often several times per day, and she may have to search longer for better sites. To take a simple example (Charnov and Skinner 1984), if it takes her 65 minutes to find and handle a host in which she can lay 30 eggs, but 40 minutes to find one in which she can lay 20 eggs, then she lays 0.5 eggs per minute in the second host but only 0.46 eggs per minute in the first. By analogy to foraging theory, Charnov and Skinner predicted that clutch size should vary in direct proportion to search time: short searches—smaller clutches, long searches—larger clutches (Fig. 7.9). On an average host, the rate of fitness gain is rapid at first but drops off monotonically. She should leave the host before she lays the Lack clutch when the marginal rate of fitness gain drops to the point where she could gain fitness more rapidly on the next host, given her search time. For a certain search time, the optimal time to spend on the host is found by drawing a tangent to the clutch size curve from the search time (cf. Fig. 7.1).

If she can afford to spend unlimited time finding a site to lay eggs, then the optimal clutch is the Lack clutch, but when she is time-limited, the clutch size laid should be smaller than the Lack clutch (Charnov and Skinner 1984; Skinner 1985; Parker and Courtney 1984; Waage and Godfray 1985). It should also be smaller than the Lack clutch when she is egg-limited (Godfray 1987*a*, *b*). For example (Charnov and Skinner 1988), in many insects the female matures eggs as she searches for an oviposition site. The number of eggs that are mature and ready to lay when a site is found is a function of the developmental rate of the eggs and the time spent between sites. If sites are rare and search times are long, then the female will arrive at the site with many mature eggs. If they equal or exceed the Lack clutch, she should lay the Lack clutch and depart. If hosts are abundant and handling time is negligible, then clutch sizes of one are optimal (Skinner 1985), for the fitness gained from the first egg laid is always higher than fitness gained from the second under the Lack assumptions. When search times are very short, the female can afford to search after each egg and still get the higher rate of fitness gain from first eggs than she would if she stayed on the site and laid a second egg. For intermediate search times, egg limitation introduces a time limit. Above it, the female should lay all her currently mature eggs and leave. Below it, she should lay all her eggs but remain to mature and lay more eggs until either the threshold time or the Lack clutch is reached.

Effects that increase the Lack clutch: (1) Flexible brood reduction

When optimal brood size varies unpredictably among breeding attempts, excess zygotes should be produced so that the number of independent offspring can be flexibly adjusted downward to the Lack clutch for that attempt (Lack 1954; Howe 1978; Lloyd 1980, 1987; Anderson 1990*b*). Examples include fruit abortion in many plants and flexible brood reduction in owls and eagles. Temme and Charnov (1987) and Kozlowski and Stearns (1989) investigated the conditions under which zygote overproduction is optimal (Fig. 7.10).

A substantial overproduction of zygotes should occur when the unit cost of producing an aborted embryo is low (but well within the bounds of biological realism) and the variability of environmental conditions important for breeding success is high. When the cost of an aborted embryo tends towards zero, the optimal zygote number tends to infinity because there is some chance that a year will come along in which an extremely large number of offspring can be raised.

Effects that increase the Lack clutch: (2) Selective abortion

If zygotes differ in fitness in ways the parents can identify, then parents should overproduce zygotes, select those with the highest fitness expectations and kill or abandon those with lower fitness to concentrate investment in the offspring with the best prospects (Lloyd 1980; Stearns 1987). Examples include consecutive spontaneous abortions in humans and rats when parents share identical alleles at the major histocompatibility locus (Beer and Quebbeman 1982), and sex ratio adjustment in polygynous mammals, where females in good condition produce excess male and females in poor condition produce excess female offspring (Clutton-Brock and Iason 1986; Austad and Sunquist 1986). The lower the cost of the aborted offspring, the higher the expected overproduction of zygotes (Fig. 7.11).

For two reasons, selective abortion best explains overproduction of zygotes when fitness differences can be identified while the offspring are still very young. When the offspring are still young, costs are low to the parent, and as the offspring age,

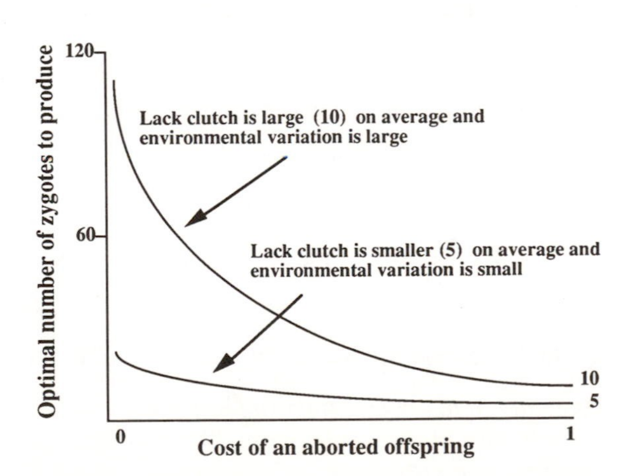

Figure 7.10 When the parents are in control, all offspring have equal fitness, and offspring are discarded at random, then clutch-to-clutch variation in the optimal clutch size leads to an optimal level of zygote overproduction that depends on both the average and the variance in the Lack clutch (after Kozlowski and Stearns 1989).

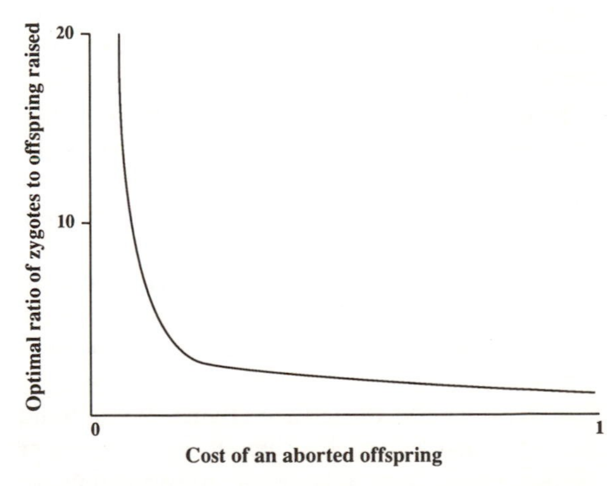

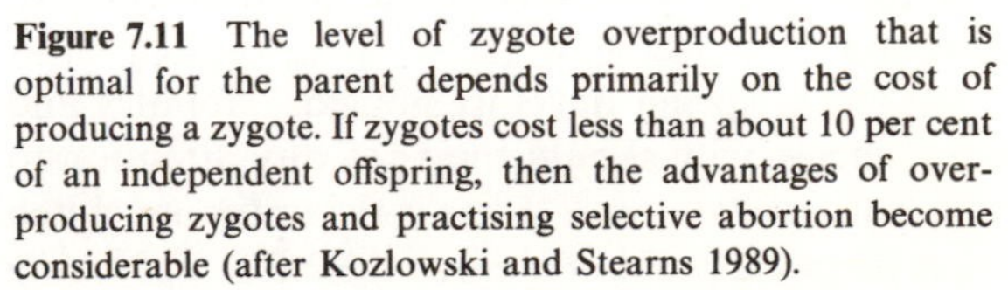
Figure 7.11 The level of zygote overproduction that is optimal for the parent depends primarily on the cost of producing a zygote. If zygotes cost less than about 10 per cent of an independent offspring, then the advantages of overproducing zygotes and practising selective abortion become considerable (after Kozlowski and Stearns 1989).

they get harder to control. It is not in the interests of the offspring to be killed by a fitness-maximizing parent unless their indirect fitness gains through siblings are more than twice as high as their own would be if they had not been aborted. Selective abortion works most clearly when the parents win any conflict of interest.

Parent-offspring conflict over clutch size

Flexible brood reduction and selective abortion are general hypotheses for overproduction of zygotes; they say nothing about mechanisms. Both predict the Lack clutch defined by the parents' interests after whatever reduction takes place. This clutch size should be

$$\hat{N}_e = \frac{-f(N_e)}{f'(N_e)}, \tag{7.6}$$

where $f'(N_e)$ is the first derivative of offspring fitness with respect to clutch size. The equation incorporates all effects of clutch size on offspring fitness but no costs of reproduction on parental survival and reproduction. This optimal parental clutch could be achieved by siblicide if there are no costs of zygote overproduction and if performance of offspring in the siblicidal arena is a good predictor of lifetime fitness. Suppose the parents do not control their offspring, as is the case for parasitoid wasps that depart after laying their eggs in a host. It does not contradict the parents' interests to let their offspring carry out the selective abortion for them.

Siblicide is common in raptors. Gargett (1978) documents the particularly gruesome case of the black eagle in Rhodesia, and Edwards and Collopy (1983) review several case studies. Sibling cannibalism is also widespread in snails (Baur 1991). Hatchling snails (*Arianta arbustorum*) do not distinguish sibs from non-sibs (Baur 1987), and hatchlings fed conspecific eggs grew faster and survived to maturity almost twice as well as hatchlings reared on lettuce (Bauer 1990). Sand shark siblings eat each other *in utero* (Springer 1948). There definitely is a phenomenon to explain.

If the offspring develop siblicidal abilities and tendencies, why should they pay any attention to the parents' interests? What is the optimum clutch size from the point of view of the offspring, taking into account their relation to the sibs they may kill? If it differs from the parents', how is the conflict resolved? These questions were addressed by O'Connor (1978), Lazarus and Inglis (1986), Parker and Mock (1987), and Godfray (1987*a*). I have relied mostly on the latter two.

Parker and Mock find the clutch optimal for offspring in a model of sib competition among nestling birds or parasitoid larvae. If some sibs are older and larger than others, then the older ones could eliminate the younger ones. How many should they eliminate? Suppose there are two sorts of sibs—stronger and weaker—and that the stronger ones are invulnerable. Let n_m = the number of offspring produced by the mother before any siblicide occurs, n_e = the number of siblings eliminated by the stronger sibs, and $n = n_m - n_e$ = the number of offspring remaining in the clutch after the elimination has been completed. Number the stronger sibs 1, 2, 3, . . . in order of strength. Now we ask, how many weaker siblings should each of the stronger siblings eliminate? Label the number of weaker siblings eliminated by each of the stronger siblings $n_{e1}, n_{e2}, n_{e3}, \ldots$. In a clutch of full sibs, we take the first offspring to have a rare mutant allele determining its value for n_{e1}. Each of its sibs have probability 0.5 of sharing this allele, and the other stronger offspring will eliminate $n_{e2}, n_{e3}, \ldots$ weaker sibs. The fitness of the first offspring is then

$$\begin{aligned} w_1 &= f(n_m - n_{e1} - n_{e2} - n_{e3} - \cdots) \\ &\quad + 0.5(n_m - n_{e1} - n_{e2} - n_{e3} - \cdots - 1) \\ &\quad \times f(n_m - n_{e1} - n_{e2} - n_{e3} - \cdots) \\ &= \text{its own fitness} \\ &\quad + \text{the fitness of that allele in the full sibs.} \end{aligned} \tag{7.29}$$

To find the best n_{e1} we set $\partial w_1/\partial n_{e1} = 0$ and get

$$n_{e1} + n_{e2} + n_{e3} + \cdots = f(n)/f'(n) + n_m + 1. \tag{7.30}$$

How many should the second strong sib eliminate? The same equations apply, as they do for all the rest of the older sibs. This means that there will be no conflict among the stronger sibs over the *total* number of weaker sibs that should be eliminated. Noting that the optimal clutch for the strong sibs is $n = n_m - n_e$, we can rearrange the last equation to get

$$n + 1 = \frac{-f(n)}{f'(n)}. \tag{7.31}$$

Thus the strong full sibs prefer a clutch, after elimination, that is one offspring smaller than that preferred by the parents. This result ignores the cost of damaging resistance by weaker sibs and the benefit of not just killing them but eating them.

Using a similar procedure, Parker and Mock then calculate the optimal clutch when every sib has a different father. This gives the maximum level of siblicide when the offspring are as unrelated as possible. The result is

$$n + 3 = \frac{-f(n)}{f'(n)}. \tag{7.32}$$

Thus strong half-sibs prefer a clutch, after elimination, that is three offspring smaller than that preferred by the parents. These effects are small in large clutches, large in small clutches, and depend on the way offspring fitness drops off as clutch size increases, $f(n)$. Parker and Mock concluded that conflict is most intense when there is a rapid initial drop in survivorship with increased clutch size and with high levels of multiple paternity.

Should the conflict be resolved in favour of the parents or the offspring? Parker and Mock claim that if the offspring are programmed to kill all sibs in excess of their optimal clutch size, it can never pay the parents to produce a clutch that is larger than the sib optimum. They reason that offspring in excess of that number cost the parents something in terms of future survival and reproduction. However (cf. Appendix 2), such costs are apparently not present in many species. Moreover, siblicide may have the effect of selective abortion, increasing the average lifetime performance of survivors, changing the $f(n)$ curve in the parents' interest. The results of such effects have not yet been worked out.

Godfray (1987*a*) considered the fate of a rare mutant allele that causes its bearer to kill all its sibs, a case suggested by parasitoid wasps. For negative exponential $f(n)$, he found that such an allele would spread when the parents' preferred clutch was less than four, driving the evolution of clutch size down to one. It spread more easily for half sibs than for full sibs. Such fighting genes produce locally absorbing states, for the conditions for spread of a tolerant allele in a population of fighters were more stringent than the conditions for the spread of a fighting allele in a population of tolerants. For the tolerant gene to spread in Godfray's model, an individual must have higher fitness as one of a pair than when alone; the spread of a tolerant gene is more likely if fighting is costly.

Godfray and Parker (1991) predict that the optimal parental clutch size should be reduced by sib competition. They note that all-male broods are smaller than all-female broods in parasitic wasps, where the coefficient of relatedness of the sisters is higher. Their interpretation is that the less closely related brothers compete with each other more intensely than the more closely related sisters, and therefore it does not pay the mother to produce as large a clutch of all-male offspring, because sib competition implies a law of strongly diminishing returns. Sex allocation theory offers an alternative. If male reproductive success increases more rapidly with size than does female reproductive success, then the mother of an all-male brood should make sure the brood is smaller than a female brood so that each emerging male will be larger than an emerging female.

These analyses agree that parent–offspring conflict over clutch size exists and is more intense with multiple paternity and small clutches. Siblicide is clearly sometimes in the parents' interests—when broods are reduced to adjust to unpredictable changes in resource levels, when selective abortion concentrates investment in the progeny with the highest fitness expectations—and sometimes in the offspring's interests—when the clutch size is small, when offspring fitness drops off very rapidly with increasing clutch size or when offspring are half sibs, both of which devalue the component of inclusive fitness realized through sibs. Simply observing siblicide does not demonstrate that the offspring have won. One must test both assumptions and predictions to have some hope of identifying who won the game. When the parameters lie in the region where siblicide is in the interests of both parties, parents and offspring agree that it is a good thing.

Casper (1990) tested the hypothesis that the offspring have won the parent–offspring conflict over brood size in the semi-desert perennial plant *Crypantha flava*, in which most fruits mature a single seed but a few mature more than one. She showed that a seed benefits initially by being the only one in the fruit: emergence was greater for single-seeded fruits, although two-seeded fruits were more likely to yield at least one seedling. However, the number of offspring that survived to maturity did not differ between one-seeded and two-seeded fruits, for mortality was higher for individuals growing in pairs than for single individuals. Thus it was not clear that parents and offspring were in conflict over brood size. Casper points out that in organisms with modular construction, even if the mother loses the conflict over number of seeds per fruit, she can compensate by producing more fruit to achieve the brood size that is in her interests.

Local adaptation prevented by gene flow

When the habitat is patchy, movement of organisms among patches is frequent, and optimal reaction norms have not yet evolved, then one should not expect locally optimal clutch sizes. Such effects will be particularly strong in small patches that are quite different from neighbouring large patches. The small patches are then simply swamped by an influx of inappropriate genes, and the reverse flow of genes into the large patches is not strong enough to permit an optimal reaction norm to evolve. Stearns and Sage (1980) documented a case in which a small freshwater population of mosquito fish was maladapted to the freshwater environment; there was strong electrophoretic evidence for gene flow from the large, neighbouring, brackish habitat. Dhondt *et al.* (1990) demonstrate in great tits and blue tits that in poorer quality habitats the most productive clutch sizes were smaller than the most frequent ones, as expected if gene flow is strong enough to prevent local adaptation.

SIZE OF OFFSPRING

Background

Analyses of clutch size alone are only possible because clutch size is usually much more variable than offspring size. Then reproductive effort is almost directly proportional to clutch size. For example, in response to the density effects of intraspecific competition, seed number in wheat varied 81-fold but mean weight per seed only 1.03 fold (Puckridge and Donald 1967). Mustard seeds retain a virtually constant weight of 4–5 mg while plant weight varies 20-fold, but sunflower seeds increase in weight as plant size increases, their size being limited by the Fibonacci spiral of the flowering head (Milthorpe and Moorby 1979). Birds vary clutch size much more than egg size (Lack 1954, 1968).

In an elegant analysis of evolutionary divergence in the egg size of closely related sea urchins on both sides of the isthmus of Panama, Lessios (1990) shows that while there has been a trend toward smaller egg sizes in the more productive Pacific, even after 3.5 million years membership in a particular genus remains a stronger effect on egg size than living in a particular ocean.

However, size and number of offspring do trade off with each other at the level of mammal orders (Fig. 5.13, cf. Read and Harvey 1989), and the same pattern exists among 35 species in three groups of Hawaiian fruit flies (Fig. 7.12).

Offspring number and size also trade off across 42 species of teleost fish belonging to 26 families, where the correlation of relative clutch size with relative egg volume (after removing the effects of body size) is $r = -0.88$ (Elgar 1990). Salthe (1969) reviews egg size variation in amphibians, and Lack (1968) reviews egg size variation in birds. The size–number trade-off can also sometimes be found within populations and individuals (see Ware (1975) for fish, Harper (1977) and Milthorpe and Moorby (1979) for plants). It exists within populations of grasshoppers in which individuals lay fewer, larger eggs in the lab than in the field (Fig. 4.3).

Just the opposite is found within well-fed individual *Daphnia magna*, where females have more *and* larger offspring as they age and grow (Fig. 7.13). When *Daphnia* are poorly fed, offspring size increases as clutch size decreases (Green 1954; Ebert, pers. comm.). This raises two problems that have not been solved: what should be the reaction norm for number and size of young, simultaneously, and why should size of offspring increase with age of mother in some species but not others?

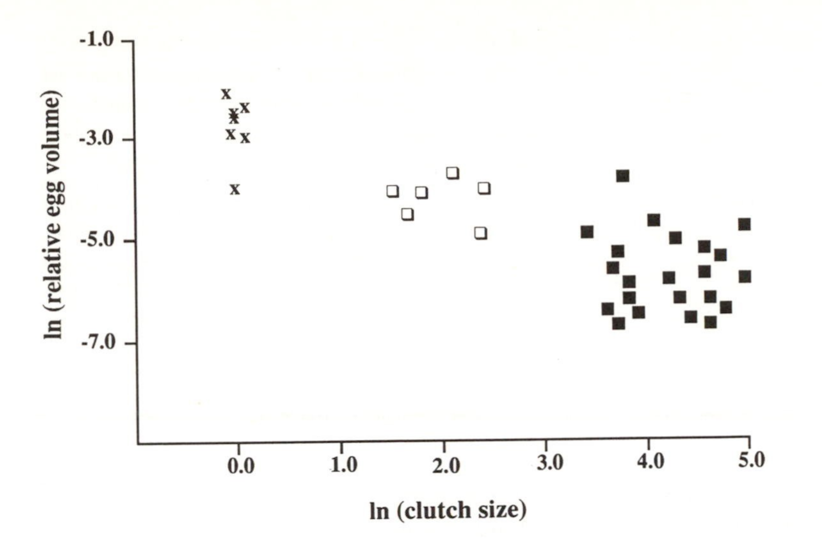

Figure 7.12 Size vs. number of young for 36 species of Hawaiian *Drosophila*. Group 1 (×) contains species with few ovarioles, alternating ovariole development, with larvae specialized on a restricted food source, flower pollen. Group 2 (□) contains species with intermediate ovariole number, with larvae that use several types of abundant but nutritionally poor food (bacteria within decaying leaves). Group 3 (■) contains species with many ovarioles, a large range of eggs per ovariole, utilizing unpredictable but rich larval food (yeast within rotting bark, stems, and fruits) (after Montague *et al.* 1981).

Small differences in size at birth can become big differences later in life for several reasons. Growth is often multiplicative—the growth rate of many organisms depends on their size, and sometimes the bigger an organism is the more efficiently it can acquire resources. For example, sib competition reinforces small size differences, the large sibs sequestering disproportionately more resources until size differences become large. In arthropods constrained by their exoskeleton to grow by moulting, it is particularly clear that small differences in size at birth can translate into big differences in life histories. Consider two newborn sibs of slightly different sizes but with the same growth conditions and the same rule for maturation—initiate maturation as soon as a threshold size has been exceeded (Fig. 7.14).

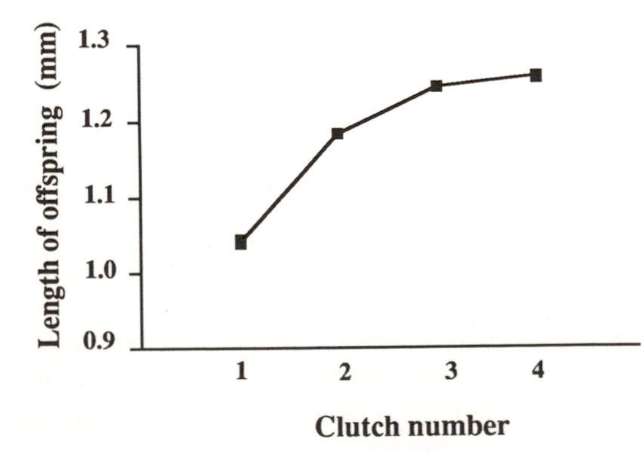

Figure 7.13 Offspring size increases with clutch number in *Daphnia magna*. The mean *volume* of offspring from fourth clutches is about 1.7 times that of offspring from first clutches (Ebert 1991).

The larger of the two exceeds the threshold in the second juvenile instar, initiates maturation, and pays the price of slower growth as resources are diverted to reproduction. At the end of the second juvenile instar the smaller of the two is still smaller than the threshold. It goes through an extra instar, maturing later and at a larger size than its sib that was larger at birth. The difference in size at maturity is translated into a difference in clutch size and offspring size that lasts throughout life. A little variation in size at birth, coupled with the constraint of having to moult and a maturation rule based on a size threshold, generates variation in lifetime reproductive patterns several fold larger than the variation in size at birth (Ebert 1991). The same pattern is generated by seasonality in continuously growing animals that have a size threshold for maturity.

The central ideas on the evolution of offspring size

Chapter 4 of Clutton-Brock (1991) reviews the evolution of offspring size.

Size–number trade-offs

The first of the three main ideas on the evolution of offspring size concentrates on the size–number trade-off (Smith and Fretwell 1974; Brockelman 1975; Wilbur 1977; Parker and Begon 1986; Lloyd

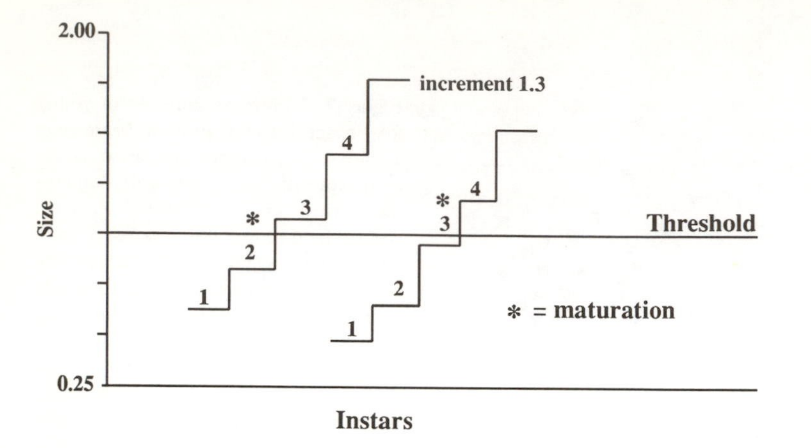

Figure 7.14 Two newborn offspring are slightly different in length at birth—0.5 vs. 0.6 mm. Both grow at the same rate, increase 1.3 times in length at each moult, and both have the same maturation rule—mature in the instar in which a threshold of 1.0 mm has been exceeded. The larger achieves this length in its second moult, but after its second moult the smaller is still slightly smaller than 1.0 mm. It matures after its third moult at a larger size than the first. The growth curves of the two offspring are offset for clarity (after Ebert, in prep.).

1987; Winkler and Wallin 1987; Lloyd 1988; McGinley and Charnov 1988): *at the evolutionarily stable equilibrium point, the proportional increase in fitness resulting from the production of slightly larger offspring just equals the proportional decrease in fitness resulting from the fewer offspring that result from the increased expenditure on each offspring* (Lloyd 1987).

This resembles the stability condition for the Lack clutch, where the marginal increase in parental fitness resulting from the production of more offspring equals the marginal reduction in fitness resulting from their lower individual survival rates.

Lloyd (1987) generalizes this condition to all size–number trade-offs. He does so by comparing the fitness of two phenotypes and asking, under what condition does one phenotype have no advantage over the other because both have chosen the best allocation? Suppose an individual can produce a continuously varying size and number of products by subdividing a fixed expenditure in different ways. Each product contributes independently to the fitness of the individual, and its success s is a function $f(q)$ of the amount of resources q invested in it, $s = k_s f_s(q)$, where k is a proportionality constant incorporating all factors affecting success that have nothing to do with the investment q. The number of products is also a function $f_n(q)$ of the investment in each product, $n = k_n f_n(q)$, and we get the total fitness w_i of an individual i by multiplying the number of products times their success:

$$w_i = k_n k_s f_n(q_i) f_s(q_i). \tag{7.33}$$

Now if we compare the fitness of two individuals, 1 and 2, we can write

$$w_2 - w_1 = k_n k_s \{ f_n(q_2) f_s(q_2) - f_n(q_1) f_s(q_1) \}. \tag{7.34}$$

The fitness advantage of the second individual changes with its investment in number of products and in the individual quality of each. The way it does so is given by

$$\frac{\partial(w_2 - w_1)}{\partial q_2} = k_n k_s \{ f_n(q_2) f'_s(q_2) + f'_n(q_2) f_s(q_2) \}. \tag{7.35}$$

The stable investment occurs when

$$\frac{\partial(w_2 - w_1)}{\partial q_2} = 0 \tag{7.36}$$

and this implies that

$$\frac{f'_s(\hat{q})}{f_s(\hat{q})} = -\frac{f'_n(\hat{q})}{f_n(\hat{q})} = \frac{1}{\text{the Lack clutch}}. \tag{7.37}$$

This is the mathematical restatement of the conclusion in italics above. Note its resemblance to eqn (7.6). It is a general statement of the evolutionary equilibrium expected for all size–number trade-offs, some of which are listed in Lloyd (1987).

In organisms with small clutches, the continuous approximation fails (Ricklefs 1977) and a discrete model is needed. Suppose (Lloyd 1987), as above, that parental fitness equals the product of offspring number and offspring success ($w_i = n_i s_i$). If an increase in number of offspring up to n is advantageous but $n + 1$ offspring reduces parental fitness, then $w_{n+1} < w_n$, by definition, and $(n+1)s_{n+1} < ns_n$. If we set $s_{n+1} = s_n + \Delta s$, then $(n+1)(s_n + \Delta s) < ns_n$ and

$$\frac{1}{n+1} < -\frac{\Delta s}{s_n}. \tag{7.38}$$

Selection will increase clutch size until the marginal gain in parental fitness from the addition of one more individual is less than the marginal decrease in the success of each offspring caused by the drop in parental investment per offspring.

Parker and Begon (1986) applied these ideas to a range of ecological situations, getting a marginal value theorem for offspring size that is formally almost identical to the one Charnov and Skinner (1984) derived for clutch size in a foraging insect. They found that increasing intensities of sib competition favoured smaller clutches of larger eggs, that increasing intensities of non-sib competition favoured smaller clutches of larger eggs, and that in seasonal environments, slower rates of development and shorter season lengths favour larger eggs and smaller clutches. Their paper views the adjustment of offspring size and number to given environmental conditions as a flexible response achieved through phenotypic plasticity, a view supported by the data on field grasshoppers (*Chorthippus*) and water fleas (*Daphnia*).

McGinley and Charnov (1988) relaxed the assumption of a single resource pool. When there are several resources, the optimal allocations depend on the sizes of the resource pools. They predict that if seed size reflects mostly carbon allocation, then seed size should increase as the ratio of carbon to nitrogen available for investment in seeds increases, and seed size and absolute amount of nitrogen per seed should be negatively correlated.

The safe harbour hypothesis

The second major idea on the evolution of size of offspring was suggested by Williams (1966*a*), formalized by Shine (1978), and amended by Sargent *et al.* (1987): *selection should increase the proportion of time spent in the safest developmental stages.* The survival of offspring from fertilization to age at maturity depends on the instantaneous mortality rates of two developmental stages, eggs and juveniles, and the length of time spent in each stage. Shine assumed that larger eggs take longer to hatch but that the increased time spent in the egg stage is subtracted from time spent as a juvenile. He also assumed that the instantaneous rate of juvenile survival is not related to egg size: offspring hatching from larger eggs had better overall juvenile survival only because they spent less time as juveniles. He predicted that if the instantaneous mortality rate of eggs is lower than that of juveniles, parents should increase egg size and decrease the amount of time spent in the juvenile stage. Where instantaneous mortality rates were higher for eggs than juveniles, the parents should reduce egg size.

This model clarified the issues, but it does not fit the data very well (Clutton-Brock 1991). Sargent *et al.* (1987) repeated the analysis while incorporating the assumption that larger juveniles have lower instantaneous mortality rates. This allows the longer development of larger eggs to be compensated by the higher survival rates of larger juveniles, and all predictions become dependent on the quantitative relationships between size, developmental rates, mortality rates, and time spent in each stage. Under appropriate conditions, Sargent *et al.*'s model predicts increases in egg size even when the instantaneous mortality rate of eggs is higher than that of juveniles.

Variation in offspring size

The third major idea on the evolution of offspring size concerns not means but variances. Kaplan and Cooper (1984) see intraclutch variation in offspring size as an adaptation to an unpredictable offspring environment brought about by selection acting on the environmental sensitivity of the developmental processes that determine the size of offspring produced, a trait that selection can adjust to an optimal level through mechanisms that give offspring size distributions the appearance of having been produced by a stochastic process. McGinley *et al.* (1987) show that the conditions needed to select for intraclutch variation in offspring size are not simply heterogeneous environments. To select for variation in offspring size, spatial heterogeneity must be combined with strong density-dependent effects on offspring fitness and parental control of the dispersal of offspring to appropriate habitats. Temporal variation can select for variation in offspring size if one uses a geometric mean measure of fitness. However, McGinley *et al.* remain impressed by the broad range of environmental conditions that do *not* select for variation in offspring size.

Intraclutch variation in offspring size that consists of a normal distribution about a single mean does not force one to resort to adaptive explanations; non-adaptive alternatives are plausible and may be difficult to exclude. However, where there is a stable, widespread, and persistent pattern of bimodal or multimodal distributions with clearly separated means, as is the case in a number of plants (cf. Venable and Burquez 1989), and where members of each size class are capable of normal growth and maturation in some natural environment, then adaptation may explain the variation. Kaplan and Cooper (1984) and McGinley *et al.* (1987) suggest different processes that could maintain such variation; convincing empirical demonstrations remain lacking.

An exemplary case study—manipulations of the egg size of lizards

Reproductive effort models find their culmination in field manipulation experiments. Similar experiments on offspring size appeared to be beyond the reach of technique until Sinervo (1990), extending previous work on sea urchin eggs (Sinervo and McEdward 1988), showed that he could remove yolk from lizard eggs that survived, hatched, and produced smaller offspring than unmanipulated eggs. This development should have just as far-reaching significance for the study of offspring size as did the discovery that one could manipulate clutch size.

Sinervo worked with a widespread iguanid lizard, *Sceloporus occidentalis*, that has larger clutches and smaller eggs in Washington than in California (12 vs. 7 eggs per clutch, 0.4 g vs. 0.65 g). Larger unmanipulated eggs needed a longer incubation time and produced larger hatchlings that grew more slowly but had a higher sprint speed, important for avoiding predators. Laboratory comparisons revealed genetic differences between populations in growth rate. When Sinervo reduced egg and hatchling size by withdrawing up to 50 per cent of the yolk, the manipulated eggs had shorter incubation times, produced smaller hatchlings, and these smaller hatchlings had slower sprint speeds. The slower sprint speed of Washington hatchlings could be largely explained by their smaller eggs, as showed by reducing California eggs to the size of Washington eggs (Sinervo and Huey 1990).

Egg size manipulations produce phenocopies, mimicking the effects of genetic change. The success of these manipulations has inspired long-term experiments, currently in progress, on the survival and lifetime reproductive success of hatchlings of different sizes. They might also inspire manipulations of the fat reserves of vertebrate capital breeders using liposuction, thus directly manipulating the storage component of residual reproductive value. This would follow the lead of Hahn and Tinkle (1965), who removed fat bodies from early oestrus female lizards and thereby inhibited follicular development.

CHAPTER SUMMARY

One approach to clutch and offspring size emphasizes trade-offs, the other conflicts of interest. Both start from the notion of the Lack clutch—the clutch size that optimizes parental fitness taking into account effects of clutch size on offspring survival and reproductive performance but ignoring such effects on parental survival and reproductive performance. The trade-off approach to clutch size is organized around additional effects, whether in offspring or parents, that lead to an optimal clutch smaller than the Lack clutch.

The most complete framework for analysing such trade-offs is the Reproductive Effort Model or General Life History Model. Several analyses agree on the following conclusions: decreasing returns from reproductive investment (a convex profit function of reproductive effort) or mortality that accelerates with reproductive investment (a concave cost function) both lead to intermediate levels of reproductive effort. Otherwise a single, suicidal reproductive event is expected. If mortality rates increase in one adult age class, optimal reproductive effort increases before that age and decreases after it, a prediction well confirmed by field manipulation of mortality rates in fish.

Flexible brood reduction in the face of an unpredictably fluctuating resource supply and selective abortion both lead to zygote overproduction. After adjustment or selection, the number of offspring eliminated should be just enough to produce the Lack clutch as adjusted by reproductive effort considerations.

When parent–offspring conflict over clutch size is won by the progeny, the optimal clutch size for the progeny is always smaller than the optimal clutch size for the parents. If siblicide evolves, there will be especially strong selection in small clutches for large newborn with well-developed siblicidal abilities, and a clutch size of one can become an absorbing state. Conflicts among sibs are stronger with multiple paternity and a rapid initial drop in juvenile survival with increasing clutch size.

The trade-off between number and size of young should be at evolutionary equilibrium when the clutch size has been increased and the size of offspring has been decreased to the point where the marginal gain in parental fitness from the addition of one more offspring is less than the marginal decrease in the success of each offspring that results from lower investment per offspring.

We have learned most about clutch and offspring size from field experiments that have followed the consequences of manipulating mortality rates and of increasing or decreasing clutch or offspring size. The evidence available strongly suggests that organisms flexibly adjust their reproduction to optimize individual fitness.

RECOMMENDED READING

Dijkstra, D. (1990). Reproductive tactics in the kestrel, *Falco tinnunculus*. Ph.D. thesis. (University of Groningen).

Godfray, H. C. J. (1987). The evolution of clutch size in invertebrates. *Oxford Surv. Evol. Biol.* **4**, 117–54.

Lindén, M. (1990). Reproductive investment and its fitness consequences in the great tit, *Parus major*. Ph.D. thesis. (University of Uppsala).

Parker, G. A. and Mock, D. (1987). Parent–offspring conflict over clutch size. *Evol. Ecol.* **1**, 161–74.

PROBLEMS

1. Given clutch size of parents c_p, clutch size of offspring c_0, l_α = survival of offspring to maturity, predict the Lack clutch of iteroparous birds:

   (a) in which survival of offspring to maturity (l_α) can be described by the relation $l_\alpha = 1 - k_1 c$ (see Fig. 7.1);

   (b) in addition where the proportion of offspring breeding at α is p and at $\alpha + 1$ is $(1 - p)$, where p is also a linearly decreasing function of clutch size, $p = 1 - k_2 c$ (see Andersson 1978);

   (c) in addition where the clutch size of the offspring decreases linearly as the clutch size of the parents increases beyond a given threshold t, $c_0 = t - k_2 c_p$.
2. Using optimal control theory as described by Schaffer (1983) and the benefits and costs reported in Daan *et al.* (1990), see whether kestrels maximize their fitness r—not just their reproductive value—by laying the clutch they actually lay.
3. The basic insight of evolutionary ecology is that the phenotype is designed by natural selection to solve ecological problems. What is an ecological problem? Does the organism define its ecological problems by being the kind of organism it is?
4. Some have claimed that clutch size evolves to balance the mortality rate. Others have claimed that mortality rate evolves to balance clutch size. Do you agree with the first, the claim, or neither? Why?
5. Polyembryony is a pattern of development found in parasitoids in which a single egg laid into a host splits up into tens to hundreds of genetically identical siblings. In some cases, a subset of these siblings have a first larval instar that is specialized for fighting and killing other larvae but that cannot develop further. These fighting larvae appear to attack other larvae but not their sibs. Discuss in terms of clutch size, kin selection, and parent–offspring conflicts of interest.

8

REPRODUCTIVE LIFESPAN AND AGEING

... it is not clear whether in animals and plants universally it is a single or diverse cause that makes some to be long-lived, others short-lived. Plants too have in some cases a long life, while in others it lasts for but a year

Aristotle, *De Longitudine et Brevitate Vitae*

... the force of natural selection weakens with increasing age—even in a theoretically immortal population, provided only that it is exposed to real hazards of mortality. If a genetical disaster ... happens late enough in individual life, its consequences may be completely unimportant

Peter Medawar, 1952

I shall assume initially ... that senescence is an unfavourable character, and that its development is opposed by selection. To account for its prevalence, ... it is necessary to postulate another force that favours its development in such a way that the observed variations in senescence reflect variations in the balance between these two forces. I believe that this other force is an indirect effect of selection, and results from the selection of genes that have different effects on fitness at different ages.

George Williams, 1957

CHAPTER OVERVIEW

This chapter discusses the evolution of the reproductive lifespan as a balance between selection to increase the number of reproductive events per lifetime and effects that increase the intrinsic sources of mortality with age. The first lengthen life, the second shorten it. The increase in the intrinsic component of mortality rates with age is *senescence* or *ageing*. Age-specific selection pressures can adjust the length of life to an intermediate optimum determined by the interaction of selection with trade-offs intrinsic to the organism *whether there is any ageing or not*. Selection pressures that lengthen life decrease the value of offspring and increase the value of adults. These include lower adult mortality rates, higher juvenile mortality rates, and increased variation in juvenile mortality rates from one reproductive event to the next. The discussion of selection pressures that lengthen the reproductive lifespan starts with the distinction between *semelparity* (monocarpy in plants, one reproductive event per life) and *iteroparity* (polycarpy in plants, more than one reproductive event per life). Here the best evidence is descriptive and comparative information on fish and plants.

Superimposed on the adaptive pattern determined by optimal allocation of resources to maintenance, growth, and reproduction are the maladaptive effects of ageing. The effects of ageing *increase mortality rates* and *decrease fecundity rates* in late life beyond the levels predicted from optimal allocation. Organisms that reproduce sexually, have a clear separation of germ line from soma, and return from a multi- to a single-cell stage at some point in their life cycle are subject to ageing. For organisms that reproduce by simple division, the force of selection does not decrease with the age of the individual, and the reproductive lifespan is potentially unlimited. For the rest, evolutionary theory states clearly that ageing is unavoidable. The best evidence comes from genetic experiments on fruit flies and nematodes.

INTRODUCTION

Which organisms age?

In Weismannian organisms, the force of selection decreases with age because selection acts directly on the germ line but indirectly on the soma. As the organism continues to reproduce, more and more copies of its germ line are in its offspring rather than in itself. In organisms reproducing by symmetrical division, mothers and daughters cannot be distinguished, one cannot decide which half is older and which is younger, and the genes have equal chances of producing offspring in both halves. Such organisms are potentially immortal. In higher plants, which age but have no clear ontogenetic distinction between germ line and soma, the critical feature that permits ageing is the return to a single-celled zygote, a trait that appears to have evolved to defend the interests of the germ line against those of the soma. It is more widespread than the ontogenetic distinction originally used by Weismann (Buss 1987; Maynard Smith, pers. comm.).

The balance of forces acting on lifespans

The evolutionary viewpoint on lifespan and ageing is summarized in Fig. 8.1. It sees the forces that extend the lifespan as external factors that change the relative value of offspring and adults. If the investment in the adult soma tends to maintain its value as the organism ages because adult mortality is low and constant relative to juvenile mortality, then selection for further survival and reproduction is strong. If adult soma loses value as it ages because adult mortality is high and variable relative to juvenile mortality,

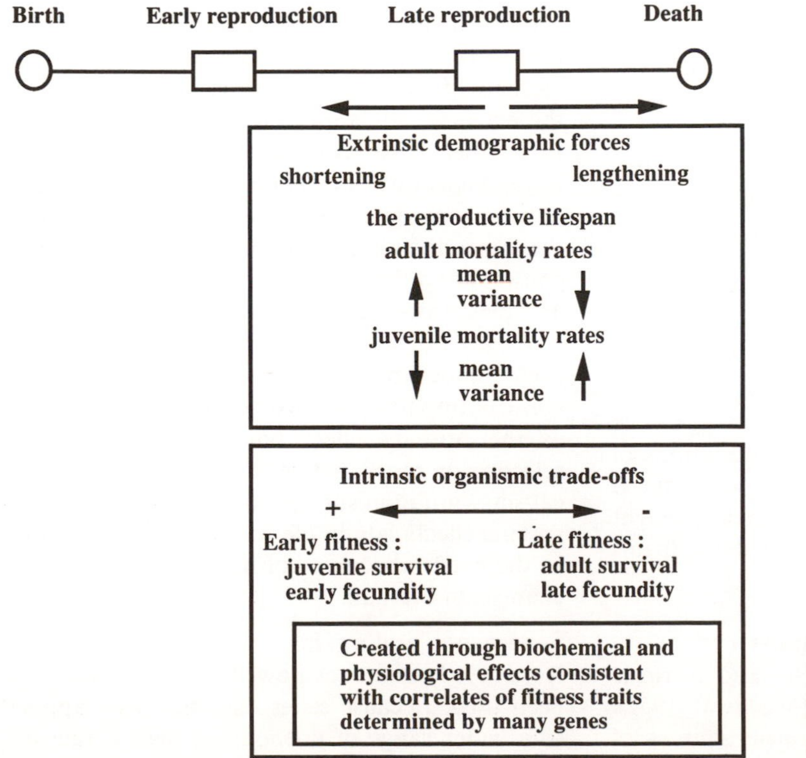

Figure 8.1 A diagrammatic representation of evolutionary thought on lifespan and ageing. The figure portrays the balance of forces acting on the reproductive lifespan of a model organism.

then selection for further survival and reproduction is weak, and the effects we call ageing can accumulate. Note that increases in either the mean or the variance in adult mortality rates devalue adults relative to juveniles, and that decreases in either the mean or the variance of juvenile mortality rates increase the value of juveniles relative to adults. The division of the lifespan can be any age, not just the maturation event: any consistent increase in the mean or variance of mortality rates after a given post-maturational age devalues all subsequent age classes relative to all earlier age classes. Note, however, that ageing is not possible before maturity: it can only start after the first reproductive event.

The lifespan is shortened by the byproducts of selection on traits more important to fitness. Traits are bound together by trade-offs, of which the most important are the physiological links between early and later survival and reproduction. Early survival and reproduction make larger contributions to fitness than later survival and reproduction, and improvements in fitness components early in life are often coupled to deterioration in fitness components later in life. These physiological links frame the evolution of genetic effects on ageing.

One effect is *antagonistic pleiotropy*, where the same genes that have positive effects on fitness components early in life have negative effects on survival late in life. Another is called *mutation accumulation*, where detrimental mutations that only act late in life will not be as efficiently eliminated by selection as those that act only early in life and can accumulate in the population. Antagonistic pleiotropy (Williams 1957) and mutation accumulation (Medawar 1952) are not mutually exclusive. Both are made possible by the decline in the force of selection with age. The physiological and biochemical manifestations of ageing have evolved as the result of reduced investment in maintenance and repair brought about by the two effects mentioned (Williams 1966*a*, Kirkwood 1981).

Intrinsic versus extrinsic mortality

The distinction between extrinsic and intrinsic sources of mortality is crucial (Medawar 1952). For the theory of ageing, extrinsic sources of mortality are those that are not sensitive to changes in reproductive decisions. Intrinsic sources of mortality can be influenced by changes in allocations among reproduction, maintenance, and defensive structures and behaviours. This distinction allows us to define ageing as an increase in the intrinsic component of mortality with age. The problem is to separate the intrinsic from the extrinsic effects, easier in theory than in practice. Promislow (1991) discusses one way to measure increases in age-specific mortality rates from life tables.

Lifespan versus ageing

It is possible to imagine an evolutionary scenario in which, for whatever reason, there were no internal trade-offs between lifespan and other fitness traits, no intrinsic sources of mortality. Nevertheless, variation in lifespan could still evolve because of variation in extrinsic mortality. Consider two populations that both start with the same intermediate lifespan. Extrinsic mortality rates change for both, but in different ways. For the first, adult mortality rates rise drastically, the value of the adult soma declines to very low levels, and the lifespan decreases simply because the extrinsic adult mortality rate has increased. For the second, adult mortality rates drop. The lifespan of the second increases simply because the extrinsic adult mortality rate has decreased. In neither case can one detect any internal, physiological or biochemical signs of senescence, nor can one detect any rise in mortality among the reproducing adults. Adult mortality holds constant at different levels in the two populations, and lifespans differ for purely extrinsic reasons.

We are not likely to observe this case, for once such demographic differences had evolved, genes contributing to ageing would rapidly be fixed, not as the primary effect but as a byproduct of selection on reproductive performance. Mutations causing antagonistic pleiotropy or with age-specific effects late in life will accumulate as soon as the external sources of mortality have caused changes in lifespan.

Senescence and ageing

Rose (1991) discusses how these two terms have been used. Because senescence has been applied to a wider range of phenomena than is relevant to the evolutionary theory of lifespan—including physiological and biochemical changes in structures like leaves and teeth—I use ageing to describe all internal changes that increase intrinsic mortality with age.

Why not immortality?

Williams (1957) emphasized that ageing is a phenomenon of the soma. Where there is no distinction between germ line and soma, as in all prokaryotes, many protozoa and algae, and in all organisms that reproduce asexually by simple and equal division, there should be no ageing. This is one of the strongest predictions of evolutionary theory (Rose 1991). *The cost of specialization into germ line and soma is death.* Some clones of grass are estimated to be 15 000 years old (Noodén 1988). Some clones of blue–green algae and bacteria are probably so old that the age of the clones exceeds the lifespan of most Weismannian species. They have no separate germ line, and they can live a very long time. Bell (1984*c*) tested this predicted difference between Weismannian and non-Weismannian organisms, but the results were inconclusive. It would be worth making more extensive experimental comparisons of ageing effects in the two types of organisms.

In a special sense the germ line itself ages (Bell 1988). Without sexual reproduction, mutations accumulate in clonal lineages. Experiments on clones of ciliate protozoans maintained for up to 11 000 generations—done early this century and reinterpreted by Bell—support the view that recombination functions in part as an endogenous repair mechanism using information copied from one strand of DNA to the other. Sexual outcrossing allows exogenous repair by the elimination of deleterious mutations present in some offspring but not others. Thus the evolution of sex can be viewed as the evolution of the mechanisms preventing the ageing of the germ line. For further discussion, see Bell (1988) and Michod and Levin (1987). Here we concentrate on the evolution of lifespan in Weismannian individuals where ageing occurs in the soma and the germ line is so frequently rejuvenated by bouts of sexual reproduction that no ageing effects in the germ line itself can be detected.

A PHYLOGENETIC PERSPECTIVE ON LIFESPAN

Animals

Comfort (1979) lists the maximum recorded lifespans for many invertebrates. A sample is given in Table 8.1.

The distribution of lifespans on Comfort's list (Fig. 8.2) of 274 species from 11 phyla, most of them molluscs, is broad. The longest lived invertebrates are sea anemones, lobsters, and bivalves. The shortest lived are rotifers, insects, and some crustaceans. Parasites of long-lived vertebrates tend also to be long lived. Species have maximum possible lifespans ranging across four orders of magnitude (days to centuries). This needs explanation.

Heller (1990) analysed correlates of variation in lifespan in the molluscs with a sample of 547 species in which lifespans ranged from two months to 220 years, one of the broadest in any animal phylum. The longest lived molluscs are marine bivalves: *Arctica islandica* can live 220 years, and *Crenomytilus grayanus* can live 150 years. Molluscs with shells live much longer than molluscs that either lack shells or have an external shell that is semi-transparent. Molluscs living in environments with high solar radiation and high temperatures and in environments predictable enough for reproduction to occur at least once per year also had short lifespans.

Beverton and Holt (1959) gave estimated maximum lifespans for many fish and noted that it underestimates the ages achieved in captivity. A sample emphasizes the natural variation in lifespan within the group (Table 8.2).

The lifespans of other vertebrates also vary considerably (Table 8.3).

Eisenberg (1981) lists the maximum captive lifespans for mammals in the world's zoos. The record is held by the African elephant at 57 years, with the domestic horse and the spiny echidna (!) tied for second at 50 years. This frequency distribution (Fig. 8.3) poses the same question as did that for the invertebrates: why the tremendous variation in lifespan within and among lineages and species?

Mean and maximal lifespan in birds and mammals vary with body size and taxonomic

Table 8.1 Selected maximum recorded lifespans in invertebrates. From Comfort (1979)

	Scientific name	Common name	Maximum lifespan (years)
Coelenterates	*Actinia mesembryanthemum*	Sea anemone	65–70
	Cereus pedunculatus	Sea anemone	85–90
Flatworms	*Taeniorrhynchus saginatus*	Tapeworm	35+
	Planaria maculata	Planaria	6–7
Rotifers	*Asplanchna sieboldii*	Rotifer	0.03–0.06
	Keratella culeata	Rotifer	0.08
Annelids	*Lumbricus terrestris*	Earthworm	5–6
	Allelobophora longa	Earthworm	5–10
Crustaceans	*Homarus* spp.	Lobster	50
	Balanus balanoides	Barnacle	5+
Termites	*Neotermes castaneus*	Termite	25+
Butterflies	*Maniola jurtina*	Butterfly	0.2–0.75
Beetles	*Blaps gigas*	Giant beetle	10+
	Carabus auratus	Scarab beetle	1–3
Bees and ants	*Apis mellifera*	Honeybee queen	5+
	Formica fusca	Ant queen	15+
Echinoderms	*Echinus esculentis*	Edible sea urchin	8+
	Asteria rubens	Sea star	5–6
Molluscs	*Chiton tuberculatus*	Chiton	12
	Haliotis rufescens	Red abalone	13+
	Limax flavus	Slug	3
	Cepaea nemoralis	Striped garden snail	5–6
	Mytilus edulis	Edible mussel	8–10
	Pecten maximus	Scallop	22
	Anodonta piscinalis	River mussel	10–15
	Loligo pealii	Squid	3–4
	Octopus vulgaris	Octopus	3–4

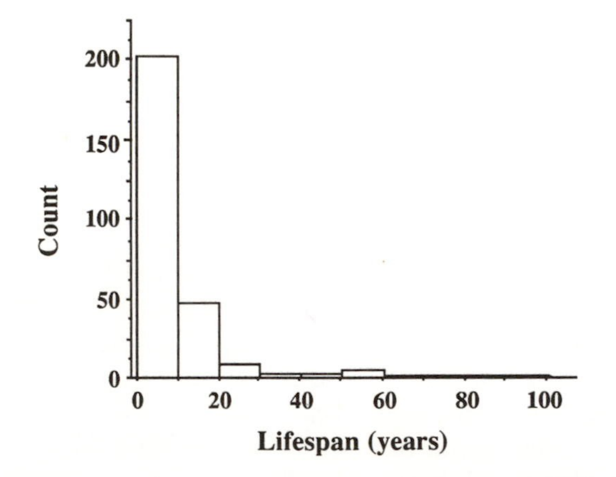

Figure 8.2 The frequency distribution of maximum recorded lifespans of 274 species of invertebrates (after tables in Comfort 1979).

group (Prothero and Jürgens 1987). Lifespan increases with body size and is higher in primates and birds than in most mammals, but highest of all, for a given body size, in bats (Fig. 8.4). This holds even for tropical bats that do not experience periods of torpidity. The ability to fly appears to pay off in better survival.

The distribution of lifespans in metazoans ranges from about a week in rotifers to about a century in some turtles, primates, bivalves, and large fish. Most metazoans are short-lived, reaching their maximum lifespan at less than 30 years, but long-lived Weismannian metazoans with lifespans approaching or exceeding a century have evolved repeatedly and independently in the bivalves, fish, reptiles, and mammals.

Table 8.2 A sample of maximum recorded lifespans of fish, from Beverton and Holt (1959)

	Scientific name	Common name	Maximum lifespan (years)
Herring family	*Clupea harengus*	Atlantic herring	22
	Sardinops caerulea	California sardine	13
Cod family	*Gadus callarias*	Cod	85
	Melanogrammus aeflefinus	Haddock	26
Flatfish	*Hippoglossus vulgaris*	Halibut	30
	Solea vulgaris	Sole	8+
Salmon family	*Coregonus clupeaformis*	Whitefish	27
	Onchorhynchus nerka	Sockeye salmon	6
	Salvelinus alpinus	Char	24+
Sturgeon	*Acipenser fulvescens*	Lake sturgeon	82
	Acipenser medirostris	White sturgeon	30
Sea horses	*Hippocampus hudsonius*	Sea horse	1
Tuna family	*Neothunnus macropterus*	Yellowfin tuna	5
	Thunnus thunnus	Bluefin tuna	13

Table 8.3 Maximum recorded lifespans for selected mammals, birds, reptiles, and amphibians. From Kirkwood (1985) after Altman and Dittmer (1972), Flower (1938), and Comfort (1979)

	Scientific name	Common name	Maximum lifespan (years)
Primates	*Macaca mulatta*	Rhesus monkey	29
	Pan troglodytes	Chimpanzee	44
	Homo sapiens	Man	115
Carnivores	*Felis catus*	Domestic cat	28
	Canis familiaris	Domestic dog	20
	Ursus arctos	Brown bear	36
Ungulates	*Ovis aries*	Sheep	20
	Sus scrofa	Swine	27
	Equus caballus	Horse	46
Proboscideans	*Elephas maximus*	Indian elephant	70
Rodents	*Mus musculus*	House mouse	3
	Sciurus carolinensis	Grey squirrel	15
	Hystrix brachyura	Porcupine	27
Bats	*Desmodus rotundus*	Vampire bat	13
Birds	*Streptopelia risoria*	Domestic dove	30
	Aquila chrysaetos	Golden eagle	46
	Bubo bubo	Eagle owl	68
Reptiles	*Eunectes murinus*	Anaconda	29
	Alligator sinensis	Chinese alligator	52
	Testudo elephantopus	Galapagos tortoise	100+
Amphibians	*Xenopus laevis*	African clawed toad	15
	Bufo bufo	Common toad	36

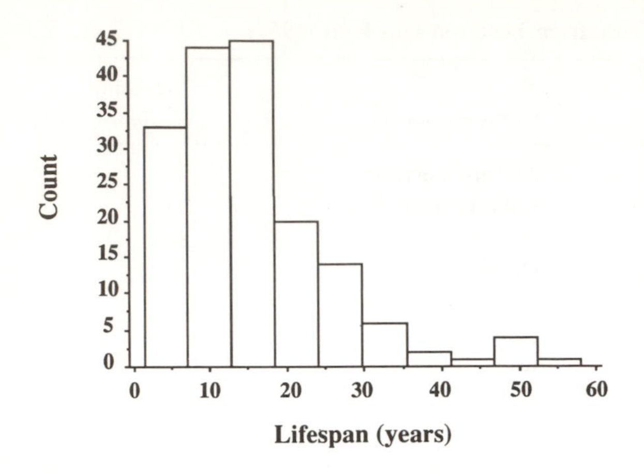

Figure 8.3 The frequency distribution of maximum recorded lifespans of 170 mammal species kept in the world's zoological parks (after tables in Eisenberg 1981).

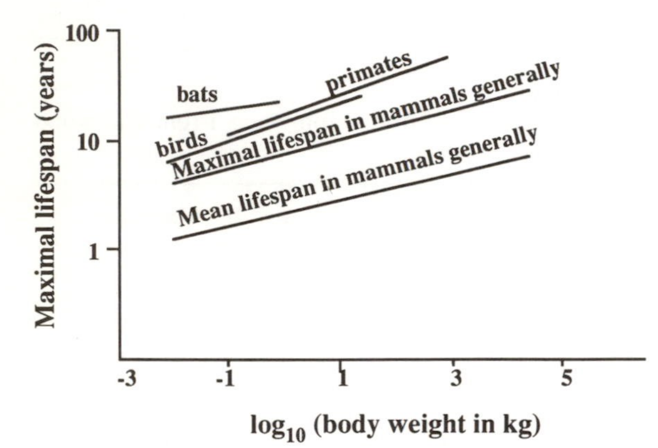

Figure 8.4 A log–log plot of lifespan vs. body weight for mammals and birds. Note that birds and primates have similar lifespans for their sizes, but that bats have the longest lifespans of all (after Prothero and Jürgens 1987).

Plants

Noordén (1988) lists the maximum recorded lifespans of a number of plants.

Reproductive lifespan and semelparity

A large maximum recorded lifespan does not necessarily indicate a long reproductive lifespan, for there are long-lived semelparous organisms. The 13- and 17-year cicadas spend all but a few weeks of their lives as grubs living below ground, eating the roots of trees. The semelparous *Lobelias* on Mount Kenya grow slowly for 40–60 years from germination to reproduction (Young 1990), then die within a few months of flowering. Some yuccas and agaves in the American southwest and the silverswords and greenswords of the Hawaiian mountains have a similar life cycle. All have iteroparous relatives. Variation among taxa in maximum reproductive lifespan can be as great as variation in maximum total lifespan.

Some semelparous organisms have evolved a hormonally controlled trade-off between resources allocated to reproduction and to repair. Pacific salmon that have been castrated live as much as 18 years longer than those that have not (Robertson 1961). Soya beans from which all fruits and flowers have been removed live long after their reproducing sibs are dead (Leopold 1961). Octopus from which the optic gland has been removed continue to feed and survive reproduction much longer than sham operated controls (Wodinsky 1977). The remarkable reproductive effort of male marsupial mice, *Antechina stuartii*, is associated with a strong rise in plasma androgen concentration and a breakdown in disease resistance (Bradley *et al.* 1980). Uncastrated males die shortly after mating before they are a year old. Castrated males have quite different behaviour and survive for a second or even a third season. Hormonal control of semelparity is clearly implicated in these cases. In the mite *Adactylidium*, the offspring hatch inside the body of the mother and eat their way out. These organisms are not programmed for death, but they are programmed for extreme reproductive efforts with death as a byproduct.

SELECTION FOR LONGER REPRODUCTIVE LIFESPAN 1: MEAN MORTALITY RATES

The evolution of the reproductive lifespan can be analysed into two parts: the extrinsic demographic forces that adjust the relative contributions of young and old organisms to fitness—the relative 'value' of offspring and adults—and the intrinsic trade-offs that couple traits expressed early in life with those expressed late in life (Fig. 8.1). The extrinsic demographic forces affect reproductive lifespan through effects either on mean mortality rates or on variation in mortality rates. This section deals with theory and data on the effects of changes in means, while the next section deals with the effects of changes in the variance of mortality rates. Both deal with the phenotypic and demographic aspects of lifespan, with the extrinsic selection pressures.

Theory

Cole's Paradox

Recent work on the evolution of reproductive lifespan has been dominated by explication of a minor point in Cole's (1954) classic paper that clarified the strong dependence of population growth rate on age at maturity. That point is called Cole's Paradox. Cole built a simple model to compare an annual with a perennial and found that 'for an annual species, the absolute gain in intrinsic population growth which could be achieved by changing to the perennial reproductive habit would be exactly equivalent to adding one individual to the average litter size' (p. 119), This raises the question of why there are any perennials at all.

Cole's result is no paradox because it depends on his assumptions that there is no juvenile mortality in either the annual or the perennial and that the only adult mortality is in the annuals, all of which die after reproducing—the perennials live forever. Gadgil and Bossert (1970) introduced juvenile mortality, Charnov and Schaffer (1973) introduced adult mortality in the perennial, and Bell (1976, 1980) and Young (1981) also relaxed assumptions about age at maturity and fecundity. Bell (1976) and Young (1990) both give tables comparing the results of various models. Young's table gives the same general impression as Bell's concerning the increasing complexity of the models. He manages to retain more of the simplicity of Charnov and Schaffer's (1973) approach and includes the effects of changing the age of onset of senescence in the iteroparous type and of changing the interval between reproductive episodes.

Let us examine Cole's analysis and Charnov and Schaffer's continuation. Cole modelled the semelparous type as an annual organism with no juvenile mortality. Let B_a represent its total number of offspring. Then

$$N_{t+1} = e^r N_t = B_a N_t \tag{8.1}$$

and

$$\ln(B_a) = r. \tag{8.2}$$

Compare it with an iteroparous type with number of offspring B_p and no adult mortality. Then

$$N_{t+1} = B_p N_t + N_t = (B_p + 1)N_t \tag{8.3}$$

and

$$\ln(B_p + 1) = r. \tag{8.4}$$

For the two types to have equal fitness,

$$B_a = B_p + 1. \tag{8.5}$$

Hence Cole's conclusion. In hindsight, the case was artificial and one wonders why it took two decades to clear up the confusion. Simple models are often a good idea, but this is a lesson in the limits of simplification. Leaving out mortality rates when doing life history theory is a bad idea.

Charnov and Schaffer (1973) put in juvenile survival rate P_a and adult survival rate (per year) P_p, and compared an annual semelparous type with a perennial iteroparous type also maturing at one year. For the semalparous type,

$$N_{t+1} = P_a B_a N_t \tag{8.6}$$

while for the iteroparous type

$$N_{t+1} = P_a B_p N_t + P_p N_t = (P_a B_p + P_p)N_t. \tag{8.7}$$

If the two are to have the same fitness, then

$$B_a = B_p + \frac{P_p}{P_a}. \tag{8.8}$$

In Cole's model, $P_a = P_p = 1$, so the two analyses are consistent. Charnov and Schaffer's result suggests that semelparity is favoured by high adult mortality and low juvenile mortality because that reduces the amount by which the semelparous reproductive output must be increased to match the fitness of the iteroparous alternative. They also analysed delayed reproduction.

If one concentrates just on the explication of Cole's result, the historical development is discouraging, for its message is that things are not as simple as one had hoped, many effects must be taken into account, and the equations comparing the advantages of semelparity and iteroparity get very complex when one allows variation in fecundity, age at maturity, and adult and juvenile mortality rates. If, however, one concentrates on two classes of effects, those tending to shorten and those tending to lengthen reproductive lifespan, the models, in all of their detail and diversity, agree on one main point: any change in fecundity, age at maturity, or age-specific mortality that reduces the value of juveniles and increases the value of adults will cause a shift away from semelparity and toward iteroparity (Young 1990).

Because hindsight has special clarity, two of these effects are intuitively clear: increasing the extrinsic adult mortality rate or decreasing the extrinsic juvenile mortality rate favours a shift towards semelparity by increasing the value of juveniles relative to adults (Gadgil and Bossert 1970; Charnov and Schaffer 1973; Bell 1976).

Two effects are not so clear, for their impact on the relative value of adults and juveniles is indirect. Earlier maturation or higher fecundity favour a shift towards semelparity, a reduction in reproductive lifespan. Delayed maturity and lower fecundity favour a shift in the direction of iteroparity, a lengthening of the reproductive lifespan. These changes in fecundity and age at maturity affect the strength of selection on reproductive lifespan because of situation-specific selection pressures. If fecundity is already high and maturity already comes early, then fitness is quite sensitive to further changes in age at maturity and fecundity early in life, but changes in the life history after the first reproductive event have little impact. Changes in maturation and fecundity revalue later reproductive events. Delayed maturation and low fecundity increase the importance of later reproduction and thus make it easier to lengthen further the reproductive lifespan.

None of these effects is qualitative. Whether semelparity or iteroparity evolves depends on the quantitative interplay of all the traits contributing to fitness.

Models for semelparity based on reproductive effort

Semelparity implies the maximum possible reproductive effort with death as a byproduct. It is natural to think of iteroparity evolving through a change in conditions such that a less than maximal reproductive effort is optimal on the first attempt and the reproducing organism survives to reproduce a second time, then a third time, and so forth. This line of thought runs through the reproductive effort models discussed in Chapter 7, and certain aspects of those models appear again here. I use a graphical approach suggested by Charnov (Stearns 1976) and based on Gadgil and Bossert (1970), Emlen (1973), and Schaffer (1974*a*).

If we isolate a single age group j from the Euler–Lotka equation, we get

$$1 = \sum_{\alpha}^{j-1} e^{-rx} l_x b_x + l_j e^{-rj}(b_j + p_j V_{j+1}). \tag{8.9}$$

Assuming that the only cost of reproduction is an effect on survival to the next age class—an effect of b_j on p_j—we can ask how rapidly future value changes with current reproductive effort. Assuming independence among age classes, we can write all terms that do not contain b_j or p_j as constants A, B, and C, getting

$$1 = A + B(b_j + p_j V_{j+1}) \tag{8.10}$$

and then

$$b_j = C - p_j V_{j+1}. \tag{8.11}$$

The partial derivative of reproductive effort with respect to adult survival is then

$$\frac{\partial b_j}{\partial p_j} = -V_{j+1}. \tag{8.12}$$

Under the assumptions made about the cost of reproduction, this equation implies that increases in current reproductive effort cause decreases in future reproductive value in such a way that r remains constant. It thus defines lines of equal fitness implicit in the demography of a stable age distribution that form a family of isoclines across which r increases upwards and to the right on a plot of reproductive effort versus survival. We take that as an extrinsic condition and ask how it interacts with an intrinsic trade-off between reproductive effort and adult survival. That trade-off could take several forms; a simple concave one is depicted in Fig. 8.5. Where the trade-off curve intersects with the solution to the demographic equation, both intrinsic and extrinsic

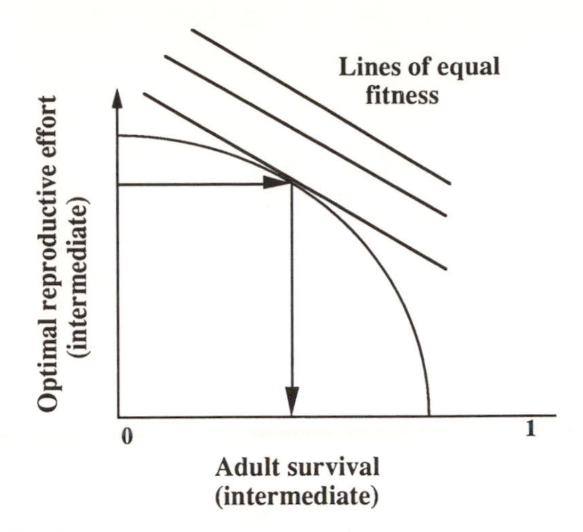

Figure 8.5 A graphical method for determining the level of reproductive effort in a given age class that maximizes fitness given that the only cost of reproduction in the age class is a concave effect on survival to the next age class (after Stearns 1976).

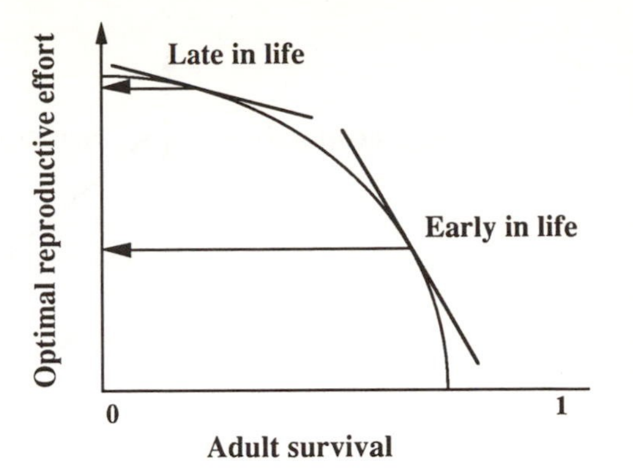

Figure 8.6 The simplest assumptions about the shape of the fitness function interact with a concave trade-off curve to predict increasing reproductive effort with age.

conditions are fulfilled and we have the combination of reproductive effort and survival for which fitness is maximal.

By drawing various trade-off curves, many possible optimal reproductive investments can be generated. Such an exercise is purely formal, but it does point out the importance of the shape of the trade-off curve and its interaction with the slope of the fitness function in determining the level of reproductive investment and the associated intrinsic portion of adult survival. Three points are important.

First, the slope of the fitness function, $-V_{j+1}$, roughly corresponds to residual reproductive value and should be steeply negative early in life, increasing to a relatively flat line with slope near zero for the last reproductive attempt. Thus without any change in the intrinsic trade-offs with age, the tangent of the fitness function to a concave trade-off curve should move steadily to the left as the organism ages (Fig. 8.6; cf. Pianka and Parker 1975). This prediction of increasing reproductive effort with age holds for concave trade-offs between reproductive effort and survival, not for all imaginable trade-off curves. The last reproductive decision of the iteroparous type is seen as conditioned by the same constraints that a semelparous type encounters on its first reproductive decision.

Second, convex trade-off curves favour semelparity and concave trade-off curves favour iteroparity, for it takes a perfectly flat fitness function to get semelparity from a concave trade-off curve (Fig. 8.7). When adult mortality is extremely high, fitness functions are quite flat because residual reproductive value is nearly zero, and then the trade-off curve makes no difference to the prediction: semelparity is always expected. One thinks of the many lepidoptera species that have no mouth parts and must die as soon as the energy stored in the larval phase is exhausted. Their semelparity is an absorbing state with no exit. Convex trade-off curves lead to predictions of either no reproduction at all (juvenile phase, intersection at lower right) or a single catastrophic breeding attempt (semelparity, intersection at upper left) for many fitness functions.

Third, a trade-off curve might be a property of a clade—all species in a genus of fish, for example. If the curve were convex–concave (Fig. 8.8), then two points of evolutionary equilibrium would be expected for fitness functions of the same slope, one semelparous and the other iteroparous. Schaffer (1974*a*) has suggested that this might explain the coexistence of semelparous Pacific salmon and iteroparous steelhead trout, species that otherwise have similar ecologies, both types spawning in the same freshwater streams and migrating to sea before returning to spawn. Which

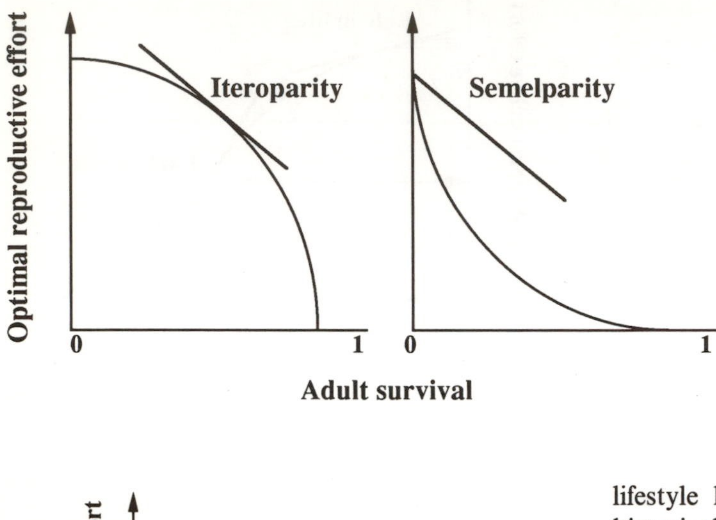

Figure 8.7 Concave trade-off curves interact with realistic fitness functions to predict iteroparity. Convex trade-off curves predict semelparity.

Figure 8.8 Complex trade-off curves interact with simple fitness functions to predict multiple stable evolutionary equilibria for reproductive lifespan (after Stearns 1976).

lifestyle happened to evolve would depend on historical accident.

Data

All models agree that in a comparison of closely related semelparous and iteroparous forms with otherwise similar life histories (age and size at maturity, physiological condition), the semelparous forms should have higher fecundity. This appears to be the case (Table 8.4).

Case study 1: Mount Kenya lobelias

Because semelparity should evolve more easily in early than in later maturing species, the long-lived semelparous rosette plants of tropical alpine and subalpine regions and of deserts are of particular

Table 8.4 Reproductive allocations of short-lived semelparous and iteroparous plants. The comparisons for *Oryza* and *Ipomopsis* are intraspecific; the rest are interspecific. From Young (1990)

Species	Semelparous/iteroparous	Reference
Oryza perennis	2.9	Sano *et al.* (1980)
Oryza perennis	5.3	Sano and Morashima (1982)
Ipomopsis aggregata	1.5–2.3	Marshall *et al.* (1985)
Gentiana spp.	2.2–3.5	Spira and Pollak (1986)
Helianthus spp.	1.7–4.0/10.0	Gaines *et al.* (1974)
Hypochoeris spp.	2.4–3.7	Fone (1989)
Lupinus spp.	2.2–3.2	Pitelka (1977)
Sesbiana spp.	2.1–2.3	Marshall *et al.* (1985)
Temperate herbs	2.8–2.9 (means)	Struik (1965)
Old field herbs	1.7 (mean)	Abrahamson (1979)

Table 8.5 Ecological, demographic, and reproductive traits of a semelparous and an iteroparous species of *Lobelia* on Mount Kenya. From Young (1990)

Trait	*Lobelia telekii*	*Lobelia keniensis*
Life history	Semelparous (monocarpic)	Iteroparous (polycarpic)
Habitat	Dry rocky slopes	Moist valley bottoms
Growth form	Unbranched	Branched
Reproductive output	Large inflorescences, more seeds	Smaller inflorescences, fewer seeds
Variation in size of inflorescence	Highly variable, increases with soil moisture	Relatively invariable, independent of soil moisture
Demography	Virtually no adult survival	Populations in drier sites have lower adult survival and less frequent reproduction
Variation in number of seeds per pod	Strongly positively correlated with inflorescence size	Independent of inflorescence size, positively correlated with number of rosettes

interest. They include species in the genera *Puya*, *Draba*, and *Espletia* in the Andes, *Argyroxiphium* (silverswords and greenswords) and *Lobelia* in Hawaii, *Echium* in the Canary Islands, *Harsiopanax* in New Guinea, *Frasera* and *Hymenoxys* in North America, *Yucca* and *Agave* in New World deserts, and *Tillandsia*, which is a neotropical epiphyte (Young 1990).

Young (1981, 1984, 1985, 1990) has done an exemplary long-term demographic study on the lobelias of Mount Kenya, where the semelparous *Lobelia telekii* lives near its close relative, the iteroparous *L. keniensis* (Table 8.5). Iteroparous *L. keniensis* flower only every 7–14 years. This requires an adaptation of the models used to compare the relative fitness of semelparity and iteroparity (Young 1981). He assumes that age at maturity is the same for the two types but may be greater than one year, that long-term population sizes are stable, and that the iteroparous species experiences no senescence. These assumptions appear to hold for the Mount Kenya lobelias; the predictions are relatively robust to deviations from the assumptions. The study, which has been under way since 1978, has not yet lasted long enough to produce reliable estimates of age at maturity.

If z is the mean number of years between reproduction for the iteroparous type, then p^z is the mean survivorship between reproductive episodes, and we get

$$\frac{B_s}{B_p} = \frac{1}{1 - p^z} \qquad (8.13)$$

as the condition demarcating the semelparous from the iteroparous life history as a function of mean yearly adult survivorship, years between flowering episodes, and relative fecundity. This is a straightforward generalization of the model developed by Charnov and Schaffer (1973). Near the geographical boundary between the species, the fecundity ratio B_s/B_p is 3–4 when estimated from seeds *pro inflorescence*.

Young has demographic data on three populations of the iteroparous species. The two better-studied populations, where he followed about 100 individuals for eight years and has confidence intervals for the parameters, are located, as they should be, on the iteroparous side of the demarcation line (Fig. 8.9). The outlying population, where he followed 30 individuals for 3.5 years, appears to lie close to the semelparous side of the line about at the limits of uncertainty in the estimates of the critical fecundity ratio. It is in drier habitat, within the ecological range of the semelparous species.

Case study 2: the colonial ascidian Botryllus schlosseri

Imagine a semelparous/iteroparous polymorphism with a demonstrated genetic basis. One exists within a single population of ascidians (Grosberg 1988). *Botryllus* is a clonal ascidian that does not reproduce by colony fission and behaves like an individual in that both the asexual production of zooids and the sexual production of gametes is synchronized among all zooids in a colony, which

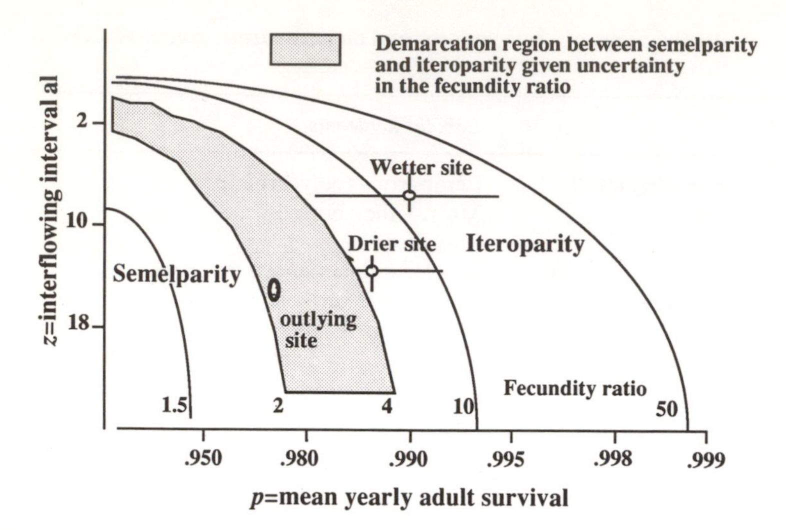

Figure 8.9 Three iteroparous populations of Mount Kenya lobelias (after Young 1990).

remain physiologically connected by a vascular system. Thus although *Botryllus* reproduces asexually and has no clear separation of soma and germ line, it may still meet the assumptions of the evolutionary theory of ageing. Certainly the semelparous colonies undergo a process indistinguishable from ageing after they reproduce.

After a tadpole larva settles and transforms into a zooid, the colony goes through 5–10 asexual cycles of about six days each in which zooids multiply by budding. Then all zooids reproduce sexually, the ova being fertilized by sperm brought in with the feeding current. Testes ripen after ova, and self-fertilization is probably rare. The fertilized eggs develop in the parental zooids until the zooids in the colony regress in synchrony, whereupon the sibling tadpole larvae swim out of the inhalent siphons as a temporally discrete clutch.

Grosberg reared the semelparous and iteroparous morphs through several generations in the laboratory and analysed the effects of temperature, food level, and maternal and paternal genotypes on life history variation. At all temperatures and food levels, the semelparous morphs matured earlier at a smaller size, had only one clutch, grew faster, and produced more embryos per zooid per clutch than did the iteroparous morphs (Table 8.6). Food level and life history morph had significant effects on all traits. Temperature effects were only significant on size at first reproduction. Food level × morph interactions had significant effects on all traits, and there were some interaction effects on reproductive effort.

Maternal and paternal genotypes made similar contributions to the life histories of the offspring, and during the common-garden experiment (Table 8.6) both morphs were exposed to the full range of natural variation in food and temperature, but none switched from one life history morph to the other. These two observations argue that the morphs are mainly genetically determined and are not polyphenisms that can be elicited from a single genotype by a developmental switch.

In the field, the proportion of semelparous morphs that survive to sexual maturity is high in early June but declines through the summer, changing from 25 per cent to about 5 per cent. The iteroparous morphs have poor survival to maturity throughout June and July but good survival in August and September, changing from about 5 per cent to about 25 per cent in the opposite direction. The power of temporal variation in selection pressures to maintain a genetic polymorphism depends on the number of generations over which the difference persists and the difference in the selection pressures among the regimes (e.g. Haldane and Jayakar 1963; Felsenstein 1976). If the differences in the fitnesses of the two genotypes between regimes are large enough (more than 10–20 per cent), then temporal variation could maintain a polymorphism, but with less intense selection the approach to equilibrium could take thousands of generations.

Table 8.6 The means of five life history traits for semelparous and iteroparous morphs raised in the laboratory (a) at three different temperatures and (b) at three different food levels. From Grosberg (1988). The life history traits correspond to (1) age at first reproduction, (2) size at first reproduction, (3) clutches per lifetime, (4) growth rate, and (5) reproductive effort

(a) Temperature effects: means across the three feeding levels

Morph	Temperature (°C)	Number of asexual cycles before maturity (1)	Number of zooids at maturity (2)	Number of clutches (3)	Buds per zooid per asexual cycle (4)	Embryos per zooid per clutch (5)
Semelparous	15	7.57	224.65	1	3.12	10.06
	20	7.56	227.68	1	3.11	9.88
	25	7.60	224.12	1	3.10	10.04
Iteroparous	15	12.81	672.53	4.66	1.86	2.85
	20	12.74	658.80	4.56	1.86	2.91
	25	12.85	646.68	4.62	1.84	2.95

(b) Food effects: means across the three temperatures

	Food Level	(traits as above)				
Semelparous	1 ×	10.26	237.88	1	1.99	10.02
	2 ×	6.26	217.61	1	3.64	9.99
	3 ×	6.20	220.96	1	3.70	9.97
Iteroparous	1 ×	15.45	654.22	4.12	1.62	2.11
	2 ×	11.37	659.69	4.84	1.99	3.24
	3 ×	11.58	664.17	4.89	1.97	3.37

If only the juvenile survival changed from spring to fall, the semelparous morphs would still have an advantage over the iteroparous morphs in the fall. Since they actually decrease in frequency, other fitness components must also change dramatically in the field, but we lack information on the life histories of the semelparous type in the fall and the iteroparous type in the spring because they are rare in those seasons. Because the study was conducted for only two seasons, not much can be said about whether the system is in equilibrium or not.

SELECTION FOR LONGER REPRODUCTIVE LIFESPAN 2: VARIATION IN MORTALITY RATES

Introduction

Variation from one reproductive attempt to the next in offspring survival rates produces a selection pressure for longer reproductive lifespan. There are two ways to think about this selection pressure. The first is to see it as a sampling problem faced by the parent. The environment varies in space and time with respect to offspring chances, and the more variable the environment, the more samples (clutches) must be taken by the parent to achieve a given probability that a certain mean number of offspring survive to reproduce (Stearns and Crandall 1981*b*). The second is to see the problem as a trade-off between the mean and the variance in the mean number of successful offspring produced per lifetime. If the variance can be reduced by reducing the mean, then a higher geometric mean fitness can be achieved by a lower mean per generation fitness if the reduction in the mean brings with it enough of a reduction in the variance (Gillespie 1977; see the discussion of bet-hedging in Chapter 7).

The second view is more fundamental, for the

sampling version is simply one mechanism for reducing the variance in mean number of surviving offspring per lifetime. The first may be more evocative, suggesting concrete models that are easy to visualize. Whatever the view, the process is called bet-hedging after the practice of professional gamblers, who have long known that the best way to produce a steady long-term profit is to spread their money over several bets. If one bets many times per year, this reduces the variance in mean income per year with the cost of reducing the maximum amount won on any single bet.

A brief history

The first analyses of life history evolution in unpredictable environments dealt with selection not to lengthen the reproductive lifespan but to 'lengthen the start of life'. The problem was diapause strategies. The seeds of desert annual plants, the eggs of annual fish, the ephippia of cladocera, and other diapausing forms all face a similar problem. When it rains, they cannot tell how much it has rained nor if the water will suffice to complete the life cycle. The decision to germinate or hatch is a bet with high stakes.

Cohen (1966, 1967) showed that variation in diapause among the offspring of a single parent plant, generated by variation in the response to environmental cues that break diapause, is adaptive in desert plants. By increasing the variance in this trait, the intergeneration variance in fitness is reduced. Since these are annual plants, there is no question of reducing the mean reproductive investment, which should always be maximal. An analogous argument applies to the annual fish studied by Wourms (1972), who found interactions between developmental mechanisms and genetic variation that ensure that all parental genotypes produce offspring that exhibit a range of diapause durations.

Murphy (1968) first saw that selection for a longer reproductive lifespan was generated by year to year variation in juvenile survival. He produced models that confirmed the effect, as did Schaffer (1974*b*) for a reproductive effort model with symmetrical variation in juvenile survival between good and bad years. I called the effect 'bet-hedging' (Stearns 1976) but did not see that the key feature was an increase in geometric mean fitness brought about by a reduction in intergeneration variance in fitness. This was made clear by Gillespie (1977). Increases in the variance of any component of reproductive success—e.g. clutch size—should produce effects similar to those produced by increases in the variance of juvenile mortality.

Schaffer's model was criticized by Hastings and Caswell (1979), that criticism was countered by Bulmer (1985), and Bulmer's commentary was amended by Orzack and Tuljapurkar (1989), who found that Schaffer's results do not readily generalize. Their approach is discussed below. Goodman (1984) found that a longer reproductive lifespan is favoured in a variable environment because it reduces the difference between the stochastic growth rate and the growth rate associated with the average Leslie matrix, the latter being larger. With Goodman (1984) and Orzack and Tuljapurkar (1989), the analysis of the problem reached a level of sophistication that cuts both ways. On the one hand, the mathematics has a power and generality that shows previous results to be special cases of limited application. On the other, the data that one would need to test the predictions of the more sophisticated approach is far beyond anything currently available or likely to be gathered in the near future, and the mathematics used drastically restricts the audience. This problem, well known in theoretical physics, is likely to become more common in life history evolution as more powerful theoreticians are attracted to the field.

Some theory

The following three theoretical approaches bring different insights. That taken by Stearns and Crandall (1981*b*) was a simulation study of trees or marine invertebrates living in ecological systems dominated by patch dynamics. A concrete view of life history evolution in sessile organisms that encounter both spatial and temporal variability, it sacrifices generality for realism. Bulmer's (1985) approach is direct and simple. He avoids complications by asking for the conditions under which rare alleles for semelparity or iteroparity will invade a population of the other type and sacrifices detail for generality. That taken by Orzack and Tuljapurkar (1989) is distinguished by its application of a fitness measure appropriate in non-equilibrium populations, by its generality and power, and by the complexity of their conclusions. All three approaches agree that temporal variation in mortality rates can select for longer reproductive lifespan, and that the effect is not always strong enough to effect the transition from semelparity to iteroparity.

For a good overview of bet-hedging in evolutionary biology, see the review by Seger and Brockmann (1987). They point out that it is a mistake to apply the term to genetic polymorphisms, which are maintained by many other forces. Frank and Slatkin (1990) analyse the impact of heterogeneous environments on gene-frequency change and phenotypic evolution and use it to compare the concepts of bet-hedging, risk aversion, and variance discounting.

Life history evolution of sessile organisms in a world dominated by patch dynamics

Stearns and Crandall (1981*b*) modelled sessile organisms on a two-dimensional plane where patches suitable for offspring open at random in space and time. The density of offspring dispersing from the parental organism on to the plane falls off exponentially with distance from the parent. Such dispersal interacts with the random pattern of suitable patches that open up in space in each generation to produce a curve relating the number of patches colonized per generation to the reproductive effort of the parent (Fig. 8.10). This curve increases monotonically at a decreasing rate. Note that a maximum of 1.44 patches per generation are successfully colonized, i.e. the maximum possible number of offspring that survive to reproduce under the conditions taken as a reference point is only 1.44. The process is homogeneous in time, i.e. the mean probability of a patch opening up is constant in time for a given set of conditions, but the number of patches opening up within the dispersal region of any one individual is quite variable from year to year.

The parental organisms have a cost of reproduction—their mortality increases as their reproductive effort increases (Fig. 8.11), and the cost is low at low and intermediate reproductive efforts. Otherwise they could not persist. Optimal reproductive effort was calculated for two fitness measures, r and the probability that the phenotype will be represented in the next generation, for values of parameters near the following, which are intended to approximate a forest tree: lifespan = 200 years, implying that patches open with frequency = 0.005 per year, patch area = 625 m^2, age at maturity = 50 years, juvenile mortality = 0.05 (given that a patch is found), adult mortality = 0.0033, seed output per year = 250, maximum seed output = 1000, and slope of the negative exponential offspring distribution $k = 0.015$, in

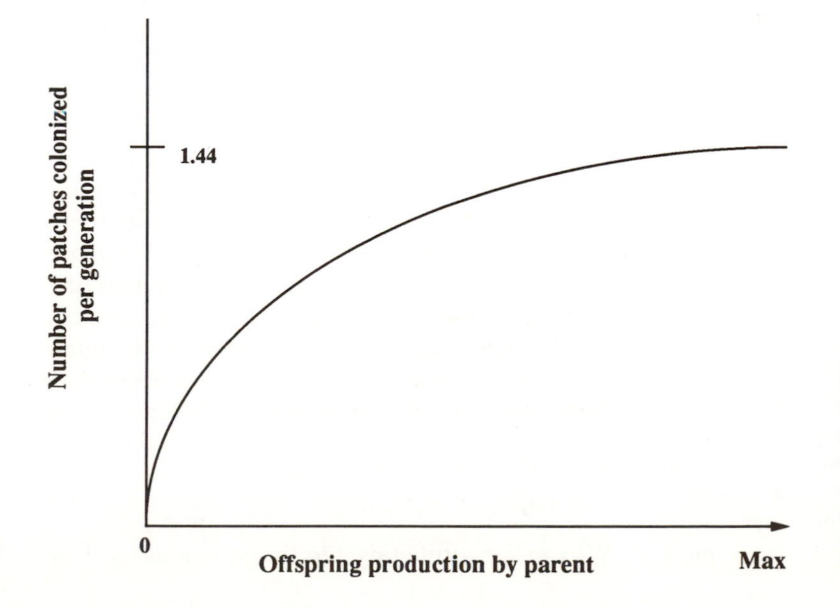

Figure 8.10 Number of patches colonized per generation as a function of reproductive effort (number of offspring) for an organism living in a planar environment where patches open up at random in space and time and the number of dispersed offspring falls off exponentially with distance from parent (after Stearns and Crandall 1981*b*).

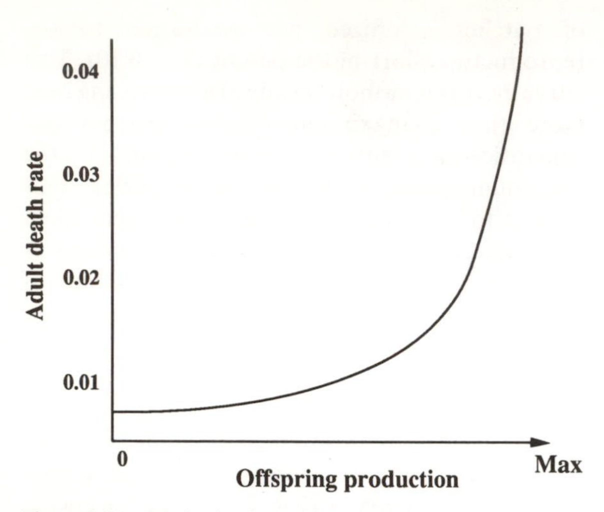

Figure 8.11 Adult death rate increases with reproductive effort, slowly at first, then rapidly, finally diverging to infinity at some level of maximum effort that kills any organism that attempts it (after Stearns and Crandall 1981*b*).

the middle of the range of values reported by Hubbell (1979).

Five results were of interest. First, the reproductive output that maximized r was always much greater than that which minimized the probability of extinction.

Second, for all values of the parameters that we used and for both fitness measures, iteroparity was optimal, but when we minimized the probability of extinction rather than maximizing r, we predicted lower reproductive effort and greater longevity. When we maximized r, the optimal reproductive effort was 40–80 per cent of the maximum and 0.03 per cent to 37 per cent of the organisms that matured lived another 200 years. When we minimized probability of extinction, the optimal reproductive effort was 1–10 per cent of the maximum and 2–82 per cent of the organisms that matured lived another 200 years.

Third, over much of the range of parameters explored, neither fitness measure was very sensitive to changes in reproductive effort: reproductive effort was nearly neutral over a large region of parameter space with a flat fitness profile (cf. Chapter 3). Thus either fitness measure could be optimized without much loss in the other.

Fourth, both fitness measures were most sensitive to changes in juvenile mortality and to the parameter that determined the pattern of offspring dispersal, k. Offspring dispersal became more important as juvenile mortality dropped.

Fifth, as the patch frequency dropped, the reproductive effort that maximized r fell, the organisms lived longer, and even fewer gaps opened up. This created positive feedback that selected organisms of great longevity, and may explain the life histories of trees and sessile marine invertebrates like clams, corals, and sponges. We did not get this result when we minimized probability of extinction, where a decrease in patch frequency led to increased seed output and shorter life. Much depends on what one thinks is a reasonable definition of fitness: the consensus favours r.

Selection for iteroparity in a variable environment

Bulmer (1985) asked whether environmental variation selects for iteroparity in two types of models. In the first, he started from Charnov and Schaffer's (1973) model,

$$N_{t+1} = P_a B_a N_t \tag{8.14}$$

for the annual and

$$N_{t+1} = (P_a B_p + P_p) N_t \tag{8.15}$$

for the perennial and used Chesson and Warner's (1981) lottery model to represent competition between annuals and perennials. He aimed to examine the effects of density dependence and environmental variation in a model simple enough to treat analytically. Suppose that the adult survival rate varies randomly from year to year, being $P_{p,t}$ in year t, representing the probability of surviving from year t to year $t + 1$, that there is space for exactly K organisms, and that the juveniles compete to fill the $K - P_{p,t}N_{p,t}$ spaces that become available for new organisms through death of old ones in year t. Thus the density-dependent juvenile survival rate in year $t + 1$ is

$$P_a = \frac{(K - P_{p,t}N_{p,t})}{(B_a N_{a,t} + B_p N_{p,t})}. \tag{8.16}$$

We can substitute this into the equations used by

Charnov and Schaffer to get

$$N_{a,t+1} = \frac{(K - P_{p,t}N_{p,t})B_a N_{a,t}}{(B_a N_{a,t} + B_p N_{p,t})} \tag{8.17}$$

for the annual and

$$N_{p,t+1} = \frac{(K - P_{p,t}N_{p,t})B_p N_{p,t}}{(B_a N_{a,t} + B_p N_{p,t})} + P_{p,t}N_{p,t} \tag{8.18}$$

for the perennial.

Bulmer got quick insight into the behaviour of this model by using invadability analysis (Turelli 1978), where a necessary and sufficient condition for a type to increase when rare is that its average logarithmic rate of increase be >0. When annuals are rare, their equation can be approximated as

$$N_{a,t+1} = (1 - P_{p,t})\frac{B_a}{B_p}N_{a,t} \tag{8.19}$$

and the expected value of their logarithmic rate of increase is

$$\Delta_a = E\left\{\log\left((1 - P_{p,t})\frac{B_a}{B_p}\right)\right\}. \tag{8.20}$$

When $\Delta_a > 0$, annuals increase when rare. When perennials are rare, their equation can be approximated as

$$N_{p,t+1} = \left(\frac{B_p}{B_a} + P_{p,t}\right)N_{p,t} \tag{8.21}$$

and the expected value of their logarithmic rate of increase is

$$\Delta_p = E\left\{\log\left(\frac{B_p}{B_a} + P_{p,t}\right)\right\}. \tag{8.22}$$

When $\Delta_p > 0$, perennials increase when rare.

There are four possible outcomes: (1) annuals and perennials coexist in a stable equilibrium distribution, like Grosberg's (1988) ascidians ($\Delta_a > 0, \Delta_p > 0$); (2) annuals persist and perennials die out ($\Delta_a > 0, \Delta_p < 0$); (3) annuals die out and perennials persist ($\Delta_a < 0$, $\Delta_p > 0$); (4) either annuals or perennials persist, but which one wins is a matter of choice ($\Delta_a < 0, \Delta_p < 0$). In the last case the common type usually wins. Chesson (1982) proved that the invadability criterion classifies these cases correctly for lottery models.

However, we have more information. Because Δ_a is an increasing function of $R = B_a/B_p$ and Δ_p is a decreasing function of R, the equations $\Delta_a = 0$ and $\Delta_p = 0$ have unique roots that we can designate R_a and R_p. If the variable adult mortality function has expected value P_p and variance σ^2, then from Jensen's inequality

$$R_p \leqslant \frac{1}{1 - P_p} \leqslant R_a \tag{8.23}$$

with the equalities holding only when $\sigma^2 = 0$. When $R > R_a$, annuals win; when $R < R_p$, perennials win. When R is intermediate, who wins is a matter of chance. Coexistence does not occur.

Variance in adult mortality works as follows. When perennials are rare, the adult mortality rate varies but the juvenile mortality rate does not. Being perennial is risky and the annuals win. When perennials are common, variation in the adult mortality rate generates corresponding variation in the juvenile mortality rate. This makes being an annual risky, for the perennials can spread their seeds over several years and increase their geometric mean fitness. 'Thus, there are competing risks, variability in the adult survival rate favouring annuals directly but favouring perennials indirectly through generating variability in the juvenile survival rate. The balance turns out to favour perennials, though this could not have been predicted without doing the calculations. Thus, in the lottery model, bet-hedging in a variable environment favours the more common form' (Bulmer 1985, p. 68).

In a second model, Bulmer studied the population dynamics of annuals and perennials. He found that they usually coexisted, but when the birth rate of the annuals exceeded that of perennials by a substantial margin, the numbers of annuals fluctuated much more violently than did the number of perennials, suggesting that under such conditions annuals should go extinct more readily than perennials.

The evolution of iteroparity with non-equilibrium demography

Orzack and Tuljapurkar have been exploring non-equilibrium models of demography for some years. They use them to examine the robustness of the assumptions made in equilibrium models and to define a measure of fitness applicable to the non-equilibrium situation where Cohen (1977) has shown that the geometric mean does not predict the stochastic growth rate. Their methods are beyond the scope of this book, but some of their results are important. They find, unlike Schaffer (1974*b*), that variance in the juvenile survival rate does not necessarily select for a longer reproductive lifespan, nor does variance in the adult survival rate necessarily select for a shorter reproductive life span. Sometimes they do, sometimes they don't. They do find that variable environments have two general effects on life histories. First, they often—but not always—select for a longer reproductive lifespan, and they almost always reduce the fitness of a strictly semelparous type. Second, like Bulmer (1985), they suggest that in variable environments semelparous types have more chaotic population dynamics than iteroparous types and go extinct more rapidly.

Summary of the theory

The experience with models for the evolution of iteroparity parallels that gathered for models of age and size at maturity and reproductive effort. The initial simplicity of the results attracted many to the field and is all that is remembered by some of those working in neighbouring areas. Later we learned that there are no general predictions, for when the early intuitive generalizations were made mathematically rigorous, everything depended on the quantitative details. One can still say that many types of environmental variation select for a longer reproductive lifespan, and that semelparity is a risky option in an uncertain world, but one should not be surprised if a semelparous organism is occasionally found surviving and doing well in a variable environment, for in a stochastic world that should happen from time to time. This may be how we can understand the otherwise paradoxical existence of a semelparous tropical canopy tree that matures at more than 50 years in the wet forests of Panama (Foster 1977).

Evidence

Murphy (1968) used data from herring-like fish to buttress his original argument for a role of variation in juvenile mortality rates in generating selection for longer reproductive lifespan. Mann and Mills (1979) extended that data set, finding a small positive interspecific correlation between reproductive lifespan and year-to-year variation in spawning success (Fig. 8.12).

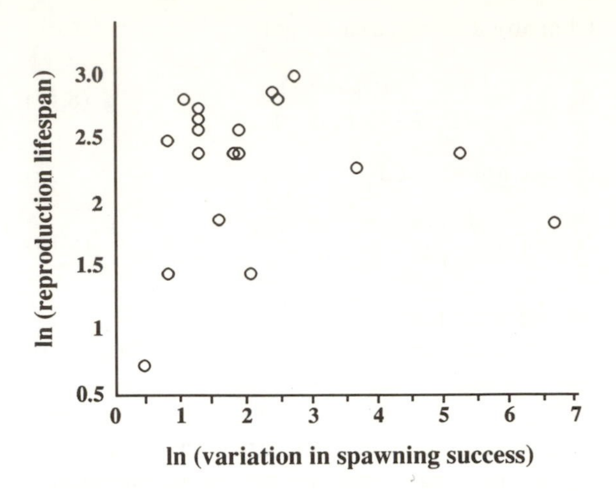

Figure 8.12 There is a weak correlation ($r^2 = 0.08$) between variation in spawning success and reproductive lifespan in 18 fish species (after Mann and Mills 1979).

While the correlation is weak, there appear to be few fish with high variation in spawning success and a reproductive lifespan of less than nine years (the lower right hand part of the plot is empty). On the other hand, one sees a range of reproductive lifespans for similar variation in reproductive success—from 10 to 14 years for four species that all are reported to have 10 × inter-year variation in reproductive success. Either the data on variation in reproductive success are rather inaccurate, which is not unlikely, or the fitness profile of reproductive lifespan is pretty flat across a broad range of variation, as is suggested by theory. Both may be true.

The pitfalls of correlation analysis were made clear by Roff (1981*a*). One can view reproductive lifespan as a trait directly under selection, as a correlated response to selection on other traits, or as part of the response of the entire life history. Roff chose to view it as a correlated response to selection on age at maturity. He showed that in 10 flatfish species there was no correlation between reproductive lifespan and the ratio of maximum to minimum reproductive success (Fig. 8.13). In this case there were fish that had short reproductive lifespans and relatively large

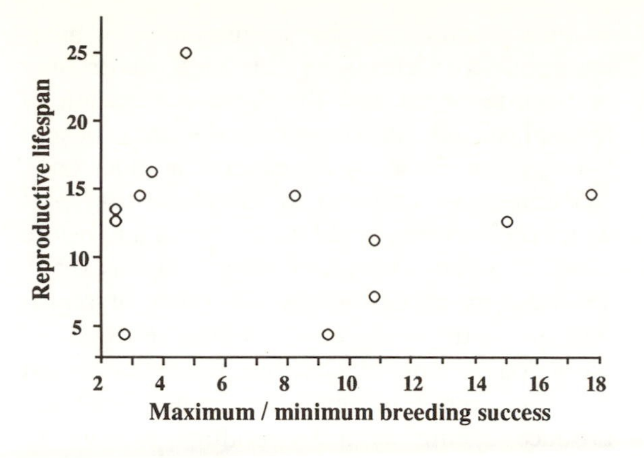

Figure 8.13 In flatfish, in contrast to the fish in Fig. 8.12, there is no correlation between length of reproductive lifespan and inter-year variation in breeding success (after Table 5 in Roff 1981*a*).

inter-year variation in reproductive success (the lower right-hand corner of the plot is occupied).

On the other hand, in a larger sample of flatfish species age at maturity was positively correlated with inter-year variation in reproductive success (Fig. 8.14). Roff concluded that variation in reproductive lifespan in flatfish is not a response to variation in reproductive success but a correlate of variation in age at maturity.

We do not know what the correlation would have been in Fig. 8.13 if the sample used had been the same as in Fig. 8.14, and we do not yet understand the potential response to selection of

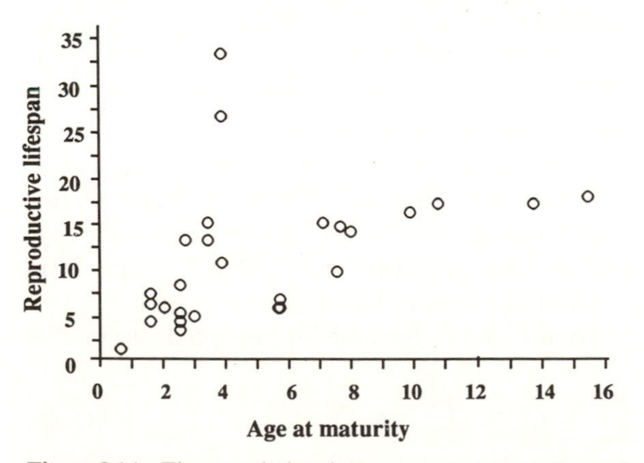

Figure 8.14 The correlation between age at maturity and reproductive lifespan in a sample of flatfish larger than that used in Fig. 8.13 is significantly positive ($r = 0.44$, $p < 0.05$) (after Roff 1981*a*).

the different life history traits of flatfish. However, Roff's cautionary remark is certainly valid.

Summary of the evidence

There is no good experimental evidence on the impact of environmental variation on the evolution of iteroparity. The correlational evidence is decidedly mixed. Anyone attempting to test these ideas should try to separate the effects of changes in mean mortality rates from the effects of changes in the variation of mortality rates, for both changes in means and changes in variances should affect reproductive lifespan. It is hard to arrange conditions under which variances change independently of means, but until that is done the status of ideas on the evolution of semelparity and iteroparity will not be nearly as secure as the status of ideas on the evolution of age and size at maturity and reproductive effort.

INTRINSIC EFFECTS LIMITING THE LIFESPAN OF THE DISPOSABLE SOMA: AGEING

Introduction

Mankind has long been haunted by mortality. Why do we become more susceptible to death as we grow older? Why do we not live forever if accidents and violence are avoided? Why are their intrinsic effects on our mortality rates? Evolutionary theory has greatly clarified the conceptual foundations of research on ageing.

Here I do not try to contrast the gerontological approach to ageing, with its biochemical and molecular emphasis, with the evolutionary one. That has been done by Rose (1991), whose recent book excuses the brevity of this treatment. The data produced by the two approaches are consistent with each other; the evolutionary interpretation has greater predictive and explanatory power.

A necessary condition for the evolution of ageing is organisms with a separation of germ line and soma, for the first of two key ideas in the theory, that the force of selection declines with age, is only true of the soma, not of the germ line. Weismann (1885) saw that ageing is a property of the soma, while Fisher (1930) and Haldane (1941) knew that the force of selection declines with age, although they expressed the notion in somewhat different ways—Fisher as a decline in reproductive value, Haldane as the evolutionary irrelevance of phenomena occurring in post-reproductive individuals. This is particularly clear when we take r as the definition of fitness, but it must also be true at some point in life, perhaps in older age classes than would be the case for r, no matter what the definition of fitness.

The second key idea is that the intrinsic properties of the organism interact with the declining force of selection to produce age-specific effects on all aspects of organismal function—physiology, biochemistry, molecular biology (cf. Fig. 8.1). These intrinsic effects degrade reproduction and survival as the organism ages. Medawar (1946, 1952) first suggested that this must be the case. One of the mechanisms he suggested, the one for which he is primarily remembered, was mutations with age-specific effects. If deleterious mutations are held in the population by a mutation–selection balance, if the strength of selection declines with age, and if there are mutations with age-specific deleterious effects, they must increase in frequency as the age-class that they affect becomes older. This is now known as the *mutation accumulation hypothesis.*

While Medawar mentioned genes with pleiotropic effects, Williams (1957) developed that idea fully. Consider an allele that has positive effects early in life, during the juvenile or early reproductive period, and deleterious effects late in life, during the late or post-reproductive period. Its net effect on fitness will be positive, because the younger age classes make a large contribution to fitness than the older ones (cf. Chapter 2). Such mutations will spread and, if they remain polymorphic, they will maintain a negative genetic correlation between life history traits occurring early and late in life. This idea is now known as the *antagonistic pleiotropy hypothesis.*

Thus the evolutionary answer to the question 'Why do we age?' has three parts. First, we have a clear separation of germ line and soma. Second, the force of selection declines with age until past some point, determined by factors discussed under the evolution of reproductive lifespan, older organisms are irrelevant to evolution. Third, under these conditions two sorts of genetic effects become possible, (a) the accumulation of more mutations with effects on older age classes than on younger ones and (b) the accumulation of antagonistically pleiotropic genes that benefit younger age classes at the expense of older ones. Maturation is the point in the life history past which these effects should start to occur and before which they should never be seen (Williams 1957). Thus ageing should follow the onset of reproduction with widespread, diffuse erosion of physiological and biochemical functions caused by many genes of relatively small effect that produce ageing as a by-product, not as an adaptation. Ageing caused by major genes with large effects is not ruled out but is not expected to be the usual case.

The evolutionary theory

Fisher's (1930) reproductive value eventually always declines past some age, but it is not the most reliable indicator of age-specific selection pressures (Hamilton 1966; Caswell 1989*a*). A better indicator is the age-specific sensitivity of fitness to changes in reproduction or survival rates (Hamilton 1966; Emlen 1970, 1973; Charlesworth and Williamson 1975; Charlesworth 1980). Hamilton showed that if r is the measure of fitness, then the age-specific sensitivity of fitness to a change in the log of the survival rate from age a to age $a + 1$ (i.e. $\ln(p_a)$) is given by

$$\frac{\partial r}{\partial \ln(p_a)} = \left(\sum_{a+1}^{\infty} e^{-rx} l_x m_x\right) \Bigg/ \left(\sum_{1}^{\infty} x\, e^{-rx} l_x m_x\right). \tag{8.24}$$

An example is given in Fig. 8.15.

The second sum is the same for all age classes; it is the first that changes. After maturation this sum must decrease with age because fewer age classes are included in it. Details vary from population to population and over time.

Hamilton did the same for age-specific fecundity:

$$\frac{\partial r}{\partial m_a} = e^{-ra} l_a \Bigg/ \left(\sum_{1}^{\infty} x\, e^{-rx} l_x m_x\right). \tag{8.25}$$

For one population of humans, the sensitivity of fitness to changes in fecundity did not change much across the age classes (Fig. 8.16).

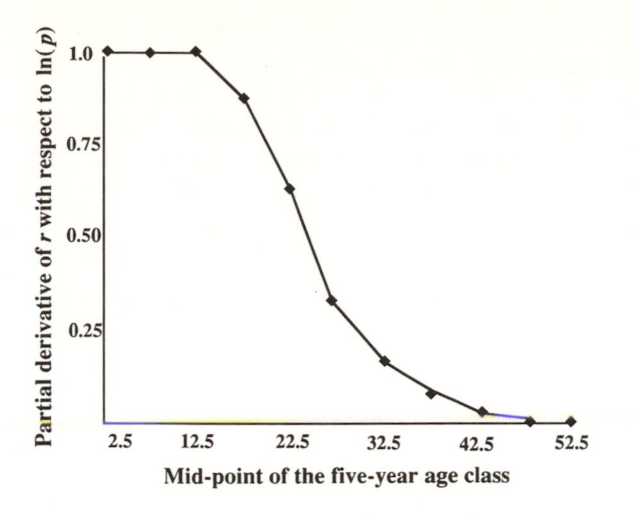

Figure 8.15 Age-specific partial derivatives of r with respect to $\ln(p_x)$ for the population of the United States about 1940 (after Charlesworth and Williamson 1975).

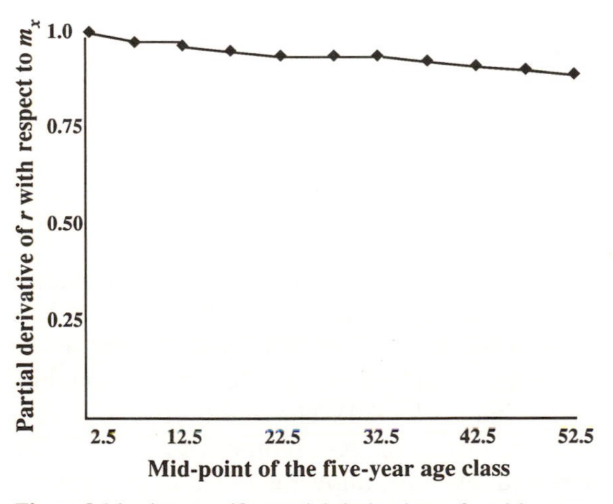

Figure 8.16 Age-specific partial derivatives of r with respect to m_x for the population of the United States about 1940 (after Charlesworth and Williamson 1975).

How much does the age-dependent sensitivity of fitness to changes in reproduction and survival depend on the choice of r as a fitness measure? The fitness measure makes a quantitative but not a qualitative difference. To see this, consider an extreme case, a forest tree with an age at maturity of 100 years and a reproductive lifespan of 500 years living in an ecosystem dominated by patch dynamics. Assume it is at evolutionary equilibrium. This means that each individual replaces itself, on average, with one surviving offspring over the course of 400 reproductive seasons. Now divide the life in half. The chances that the surviving and reproducing offspring will have been produced in the first 200 years must be greater than the chances for the second 200 years because even if the parent does not senesce, accidents happen, and fewer parents survive to reproduce during the second half of life than during the first. 'There is therefore little advantage to be gained from investment in potential somatic immortality when in practice the return from this investment may not be realised' (Kirkwood 1987). This holds no matter what sort of fitness measure is appropriate to the ecological situation, which may include all sorts of complex density- and frequency-dependent effects in populations far from demographic equilibrium. The force of selection always declines with age, albeit sometimes pretty slowly.

An immediate consequence is that fitness must be maximized at some level of investment in repair which is less than would be required for indefinite somatic survival (Fig. 8.17, Kirkwood 1985).

Kirkwood's disposable soma theory establishes a connection between the evolutionary theory of ageing and the processes traditionally studied by gerontologists. It is important for the link it makes to a long tradition and a large database. It predicts that 'ageing should be the result of the accumulation of unrepaired somatic damage . . . [and] that species with different longevities should exhibit corresponding differences in their levels of somatic maintenance and repair' (Kirkwood 1985, p. 38).

Repair is known to be costly. More than 2 per cent of the energy budget of cells is spent on DNA repair and proofreading, on processes that determine accuracy in protein synthesis, on protein turnover, and on the scavenging of oxygen radicals (Kirkwood, pers. comm.).

The genes that affect repair processes should have antagonistically pleiotropic effects. Figure 8.18, in which it is assumed that increased investment in repair means decreased investment in growth and reproduction, indicates the life history consequences of intrinsic changes in allocations to repair and maintenance. Conversely,

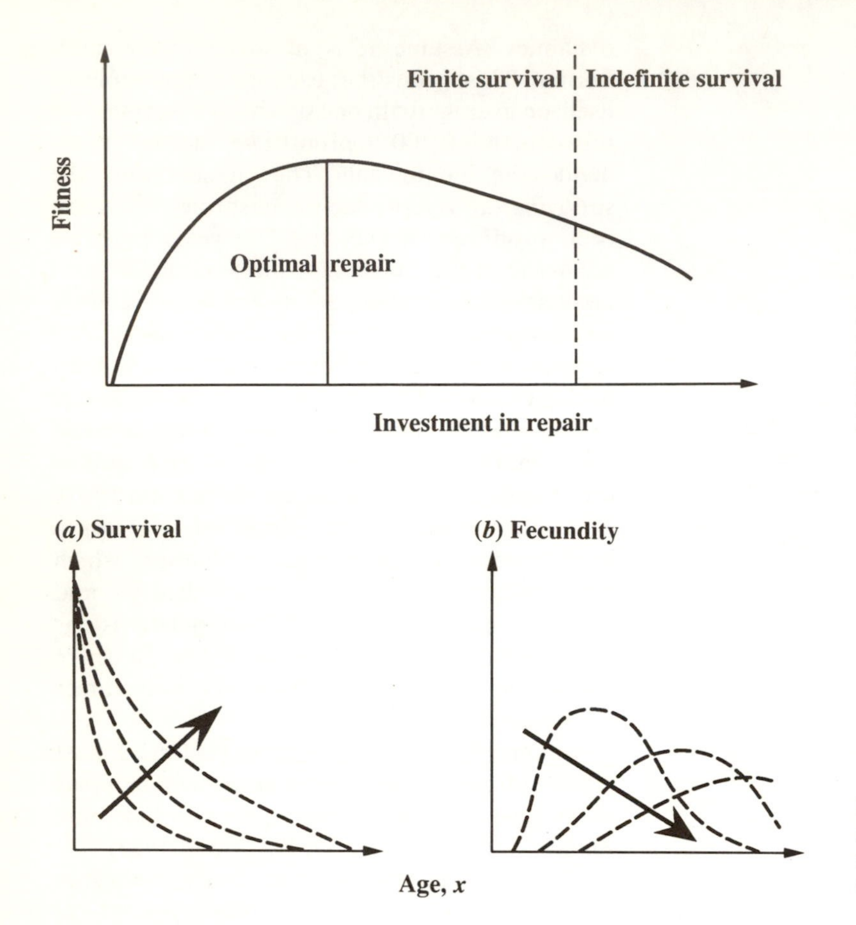

Figure 8.17 Disposable soma theory predicts that the optimal level of investment in repair will be below that required for indefinite somatic survival (after Kirkwood 1985).

Figure 8.18 The effects on survival and reproduction of increasing the investment in somatic repair and maintenance (after Kirkwood 1987).

and not depicted, when the external sources of adult mortality increase, then investment in repair is devalued, and we should see shifts to less repair and higher reproductive investment.

Experimental data

The extensive experimental tradition in the evolutionary biology of ageing has penetrated to as great a depth as the other outstanding experimental tradition in life history evolution, that on clutch size. One of the early clues was provided by Maynard Smith's (1959) work on the 'grandchildless' mutant in *Drosophila*, where homozygous daughters lack ovaries and live much longer than normal females. This suggested a link between reproduction and longevity that is still being unravelled.

Laboratory selection on lifespan

The main experimental designs were suggested by Edney and Gill (1968). First, in three laboratories large-scale experiments have been done in which fruit flies were forced to reproduce late in life: those of Wattiaux (1968), Rose (Rose 1984*a*, *b*; Hutchinson and Rose 1990), and Luckinbill (Luckinbill *et al.* 1984). The results are consistent: when the force of selection on older age classes is increased by only allowing older organisms to reproduce, ageing is postponed. One table (Table 8.7) can be taken as representative.

Partridge (1987) suggested that there is no direct evidence that early reproduction causes a delayed effect on late survival. Service (1989) showed that the differences between flies forced to reproduce early or late in life does not depend on their reproductive activity, for selected flies kept virgin throughout their lives still exhibit the difference in lifespan. This suggests that the physiology of the flies has been altered through selection, but it is not yet clear which aspect of physiology is more important—the pre-adult

Table 8.7 Mean lifespans in days survived from eclosion for populations of *Drosophila melanogaster* that had been forced to reproduce either early or late in life for 25 generations. From Hutchinson and Rose (1990)

Late-reproduced populations	Early-reproduced populations	Difference
71.3	60.7	10.6
55.1	46.7	8.4
64.4	49.5	14.9
59.6	47.2	12.4
62.2	46.3	15.9

period during which 80 per cent of adult protein is synthesized and not later replaced (Maynard Smith *et al.* 1970) or repair mechanisms operating in the adult flies—about 20 per cent of the total protein in adult male *D. subobscura* turns over with a half-life of about 10 days (Maynard Smith *et al.* 1970).

The second type of experiment, in which the effective lifespan is shortened in treatments that are then compared with controls maintained with a normal laboratory lifespan, has been done by Sokal (1970) and Mertz (1975) on flour beetles (*Tribolium*) and by Mueller (1987) on *Drosophila.* The results are not so consistent as those from experiments that forced a delay in reproduction. In Sokal's experiment, one sex responded as expected in each pair of selected and control lines. In Mertz's experiment, no significant difference was found, but the results are difficult to interpret because of significant genotype × environment interactions in the perpetual virgins used in the assay (Rose 1991). In Mueller's experiment, flies forced to reproduce early for more than 100 generations had reduced fecundity late in life, and unpublished work indicates that they also have reduced longevity (Rose 1991).

Many other experiments on fruit flies have reported inconsistent results, but careful work, both experimental and analytical, by Rose and Service (1985), Clare and Luckinbill (1985), Reznick *et al.* (1986), Clark (1987), and Service *et al.* (1988) has shown that all the experiments that apparently contradict the evolutionary theory have been afflicted by one of the following problems. (a) The experiments must be done at high, not at low, larval densities, for only under these conditions do the genetic differences in larval growth and survival crucial for correlated responses in lifespan and fecundity appear to be expressed. (b) The flies must be outbred, not inbred. (c) The flies must have been maintained in the laboratory long enough to allow them to come into evolutionary equilibrium with laboratory culture, i.e. more than 10 generations. Otherwise the patterns observed could result from the genotype × environment interactions elicited by the difference between laboratory and field and from the temporary linkage disequilibrium induced in the population by the strong selection associated with the move to the laboratory. (d) The assays must be done on organisms that are allowed to reproduce normally, not on virgins, for only under these conditions are the genetic differences in lifespan reliably elicited.

When these considerations are taken into account, the evidence can be summarized rather optimistically. Lifespan responds rapidly to selection in the laboratory in a manner consistent with evolutionary theory.

The population genetic mechanisms involved

That selection to increase late reproduction produces a correlated response of decreased early reproduction is evidence that antagonistic pleiotropy may be involved in at least some of the experiments. However, not all laboratories report that flies selected for increased lifespan show reduced early fertility. Partridge and Fowler (1991) have found that increased longevity is achieved through slower growth and larger size at eclosion, a correlate of which is greater lifetime fecundity. It is not pleiotropy that is at issue—pleiotropy is implicated in the results of almost all carefully done experiments—but which traits are linked by the pleiotropy. Given the nature of insect development and physiology, it is likely that selection on many adult traits, longevity included, can only be achieved by changes in larval physiology and behaviour.

The cost of exposure to males in reduced lifespan is well known in wild-type flies (Partridge *et al.* 1987) and has been shown to consist primarily of the cost of mating (Fowler and

Partridge 1989). Further demonstrations where the effects of single, identified genes are known are the *dunce* allele in fruit flies (cf. Rose 1991) and the *age-1* mutant in the nematode *Caenorhabdites elegans* (Friedman and Johnson 1988). Flies with *dunce* have poor memory because the allele reduces or abolishes cAMP phosphodiesterase activity. Females homozygous for *dunce* mate about twice as often as normal females and their lifespan is only 50 per cent of normal when they are kept with males. Nematodes with the *age-1* allele live 60 per cent longer at 20°C than nematodes without it, but their fertility is decreased by about 75 per cent. These are identified genes with antagonistic pleiotropic effects on reproduction and survival.

The best evidence for the mutation accumulation hypothesis comes from the response to relaxation of selection for increased longevity. Both Rose *et al.* (1987) and Service *et al.* (1988) report that when selection is relaxed, not all characters exhibit the reversed response expected under the antagonistic pleiotropy hypothesis. Some differences between selected stocks and their controls persist in the absence of selection, as would be expected under mutation accumulation. Other differences respond as would be expected under antagonistic pleiotropy. It is not surprising that both mechanisms are involved, for they are not mutually exclusive (Rose 1991).

Physiological mechanisms

In the experimental tradition on clutch size (Chapter 7), a great deal is known about the phenotypic consequences of manipulating clutch sizes up or down, and much interest is now concentrated on the physiological mechanisms that mediate the response. In the experimental tradition on ageing, a great deal is known about the genetic consequences of manipulating early and late reproduction up and down, and much interest is now concentrated on the physiological mechanisms that mediate the response. Physiology is moving closer to the centre of life history evolution. This return to an attitude widely held two decades ago and never totally abandoned will give greater insight into phenomena already understood at other levels. It is diversity among lineages in physiology and development that hold the key to understanding the comparative patterns of life history variation. The theory of population genetics is general. The theory of quantitative genetics is general. The theory of demography is general. If we want to find the source of differences among taxa, we must look where lineage-specificity dominates. That is the level of physiology and development.

The physiological mechanisms underlying ageing are particularly well understood in fruit flies, for which recent reviews are available in Partridge (1990) and Rose (1991). Maynard Smith (1958) showed that simple rate-of-living theories do not work, for flies that start at a low temperature, spend a period at a higher temperature, and then return to the low temperature rapidly reacquire the mortality schedule characteristic of uninterrupted life at the low temperature. The effects on mortality of the higher temperature are not permanent. Partridge and her colleagues have demonstrated that mating reduces lifespan for both males and females and that laying eggs reduces female lifespan (Partridge and Farquhar 1981; Partridge and Andrews 1985; Partridge *et al.* 1986, 1987). The last-listed paper describes an experiment with a design similar to that used by Maynard Smith (1958): the effects of egg-laying were not permanent either.

Flies selected to live longer have greater flight endurance (Graves *et al.* 1988; Graves and Rose 1990). Diversion of energy into fat stores is one mechanism of stress resistance in flies selected for greater lifespan (Service 1987). This is consistent with Zwaan *et al.* (1991), who found that larval crowding affects adult fat content, which in turn affects both starvation resistance and longevity. Dietary restriction prolongs lifespan in males and females selected for both short and long life, and the effect on females is even greater in the females selected for long life than it is in the females selected for short life (Rose, pers. comm.). All this points to the physiological processes involved in the reaction to selection to change lifespan.

The number of genes involved

Both the Rose and the Luckinbill laboratories (Hutchinson and Rose 1990; Luckinbill *et al.* 1989) suggest that there are many loci on all three major chromosomes that contribute to the differences in lifespan of the stocks selected in both groups. Whereas single genes of major effect on lifespan are known as mutants in some organisms (e.g., *age-1* in *Caenorhabdites*), no one has demonstrated that the response of lifespan to artificial selection is based on such genes, and evolutionary theory suggests that it is unlikely that it would be.

Summary of experiments on ageing in fruit flies

Flies selected for longer life do live longer. They have either lower fecundity early in life, lower larval survival, or slower larval growth, larger size at eclosion, and higher total lifetime fecundity. They also have better stamina in flight and better resistance to ethanol.

CHAPTER SUMMARY

The evolution of the length of the reproductive lifespan is controlled by the relationships of the mean and variance of adult and juvenile mortality. Increases in the mean and variance of adult mortality should decrease the reproductive lifespan, eventually producing semelparity. Increases in the mean and variance of juvenile mortality can lengthen the reproductive lifespan, eventually producing very long-lived organisms with high investments in somatic structures and their repair. However, not all types of environmental variance select for longer reproductive lifespan. The evidence on the evolution of reproductive lifespan is not strong and not experimental. The most thoroughly explored case study is consistent with the explanations outlined.

Even without ageing, reproductive lifespans would vary because extrinsic mortality rates vary. The key evidence for ageing is an increase with age in the intrinsic component of age-specific mortality rates. Identifying ageing requires a method for distinguishing between intrinsic and extrinsic causes of mortality.

The evolutionary theory of ageing assumes that the force of selection must decline with age. This permits two effects: the accumulation of more deleterious mutations having effects on older age classes, and the accumulation of genes that have antagonistically pleiotropic effects, improving the young while damaging the old. The experimental tradition testing these ideas is strong but limited to a few species. Both mutation accumulation and antagonistic pleiotropy seem to affect the ageing process in fruit flies. Mating and laying eggs shorten life. Fat reserves figure in the physiology of senescence and are strongly influenced by the larval environment. Physiological data on other organisms is available, but comparable genetic data are lacking.

RECOMMENDED READING

Bell, G. (1988). *Sex and death in protozoa.* (Cambridge University Press, Cambridge).

Charnov, E. L. and Schaffer, W. M. (1973). Life history consequences of natural selection: Cole's result revisited. *Am. Nat.* **107**, 791–3.

Medawar, P. B. (1952). *An unsolved problem of biology.* (H. K. Lewis, London).

Rose, M. R. (1991). *Evolutionary biology of ageing.* (Oxford University Press, Oxford).

Seger, J. and Brockmann, H. J. (1987). What is bet-hedging? *Oxford Surv. Evol. Biol.* **4**, 182–211.

Williams, G. C. (1957). Pleiotropy, natural selection, and the evolution of senescence. *Evolution* **11**, 398–411.

Young, T. (1990). Evolution of semelparity in Mount Kenya lobelias. *Evol. Ecol.* **4**, 157–72.

PROBLEM

1. Orzack and Tuljapurkar (1989) is one of the most sophisticated theoretical treatments of the evolution of iteroparity currently available. Grosberg (1988) and Young (1990) are among the best empirical studies in which semelparity and iteroparity are explicitly compared. What sort of data would one have to gather over how long a period of time in either system to test the ideas presented by Orzack and Tuljapurkar?

DISCUSSION OF PART II

LIFE HISTORY STRATEGIES

Sherlock Holmes knew the significance of the dog that did not bark. This book did not dwell on life history strategies. The word 'strategy' means simply 'complex adaptation'; it refers to the coordinated evolution of all the life history traits together. Early on there were hopes that only a few syndromes would be found into which most traits fell, but those hopes were premature. It was first necessary to understand the evolution of one trait at a time. This book summarizes that attempt. The problem was more complex than it seemed at first, and it was necessary to reject some of the early claims, analyse the foundations, and build the analysis up trait by trait. This does not mean that life history strategies are unimportant or uninteresting. They have great potential importance. However, their analysis must be based on and consistent with the evolution of the individual traits.

Rejection of *r* and *K*-selection

In the 1960s and 1970s interest in life history evolution was stimulated by the identification of a dichotomy between species that matured early, had many small offspring, made a large reproductive effort, and died young, and species that matured later, had a few large offspring, made a small reproductive effort, and lived a long time (MacArthur and Wilson 1967; Pianka 1970). The explanation offered was microevolutionary and based on differences in mode of population regulation. It argued that short-lived species with high reproductive rates had evolved under density-independent conditions and called them '*r*-selected', implying that such circumstances selected for a high intrinsic rate of increase (*r*). Long-lived species with low reproductive rates were thought to have evolved under density-dependent conditions and were called '*K*-selected', implying that such circumstances selected for ability to withstand high densities of conspecifics (*K* represents saturation density).

This explanation was suggestive and influential but incorrect. First, it was couched at the level of population regulation rather than demographic mechanism and confused the statistical description of population processes with the selection pressures that act on individual organisms.

Second, even the best studies that appeared to support it (e.g. Solbrig and Simpson 1974; McNaughton 1975; Law *et al.* 1977) did not establish the mode of population regulation. They documented patterns of reproduction consistent with '*r* and *K*-selection' but did not exclude hypotheses based on age-specific models.

Third, even as a method of just classifying life histories, the *r*–*K* dichotomy fails for about 50 per cent of the species for which we have reliable data (Wilbur *et al.* 1974; Stearns 1977).

Fourth, it has been tested in artificial selection experiments on fruit flies (Taylor and Condra 1980; Barclay and Gregory 1981; Mueller and Ayala 1981) and on protozoa (Luckinbill 1979). In three cases, the results were not consistent with the predictions of *r* and *K*-selection but were consistent with age-specific models. In the fourth, Mueller and Ayala (1981) did not find that populations selected for rapid growth have lower saturation densities; lines that grew rapidly at one density also grew rapidly at other densities. Recently Mueller (pers. comm.) has found that the major difference in lines selected under conditions of high larval density is a change in larval feeding and pupation behaviour—they forage deeper into the medium and have less of a tendency to pupate on the surface of the larval medium. Thus one assumption of *r* and *K*-selection, a trade-off between rapid population growth at low densities and replacement efficiency at high densities, does not hold for Mueller's flies.

Fifth, there are several types of age-specific selection pressures contained within the categories 'density independent' and 'density dependent' (Schaffer 1979; Charlesworth 1980). Those differences are crucial to accurate predictions. For example, *density-dependent* mortality applied equally to all age classes selects for increased reproduction early in life, which is supposed to be '*r*-selected' by *density-independent* mortality regimes (Charlesworth 1980). Clearly, we must analyse the age-specific effects of the mortality as well as the mode of population regulation that generates the mortality.

Sixth, even if we take '*r*-selected' or '*K*-selected' only as labels for sets of reproductive traits, they are inappropriate because they conceal important exceptions and imply a type of causation that has never been established. Although we have known since the work of MacArthur (1962), Roughgarden (1971), and Charlesworth (1980) that *K* is an appropriate fitness measure under density-dependent population regulation, we do not yet know what changes in life history traits to expect under *K*-selection. One reason is that we do not have equations that relate changes in life history traits to changes in *K* that hold under density-dependence. Remedying this deficiency is a challenge for future research on life history theory.

Why list the reasons for preferring age-specific models to *r* and *K*-selection? *r* and *K*-selection enjoyed great initial success because it was convenient, simple, and in accord with a popular explanation of population regulation. It became widespread in the literature and continued to structure research programmes long after its deficiencies were recognized. In the five years starting with 1977, there were on average 42 references to *r* and *K*-selection per year in the set of papers accessible to the BIOSIS literature search service. In the five years starting with 1984, there were on average 16 such references. Interest persists but is declining.

Some recent papers that explicitly discuss the problems with *r* and *K*-selection and try to save its strengths while discarding its weaknesses are constructive. The lack of a comprehensive theory of how life histories should evolve under density dependence is a minor scandal; responsible discussion of the strengths of *K*-selection is one approach to creating such a theory. Enough people have found it a useful framework in which to interpret their observations that it must contain an element of truth. The problem is to identify that element. Other papers use *r* and *K*-selection as a convenient description of a pattern without having any idea of what caused it. That is a misleading waste of time.

I discussed the deficiencies of *r* and *K*-selection to direct energy towards age-specific models, where success has been demonstrated, and to draw attention to the need for density-dependent life history theory. *r* and *K*-selection was a helpful initial generalization with a part worth saving: the focus on density dependence.

Habitats do not map directly on to life histories

Good natural historians have noticed that habitat, lifestyle, and life history are correlated. For example, Skutch (1949) and Dobzhansky (1950) suggested that in any lineage different life histories are found in tropical and temperate habitats *because* the tropics are more predictable. In *r* and *K*-selection, this idea reappeared in the statement that 'stable' or 'predictable' habitats should be '*K*-selecting', whereas 'fluctuating' or 'unpredictable' habitats should be '*r*-selecting'.

These ideas have not had much success for several reasons. First, the dichotomies between tropical and temperate, stable and fluctuating, predictable and unpredictable conceal within each term several types of habitats whose differences are critical to successful predictions. The problem arises because the mechanisms linking habitats to life histories are not specified, and when one does specify them, one finds that one has to consider many different types of unpredictability. Is the habitat unpredictable with respect to food supply? Adult mortality rates? Juvenile mortality rates? Unpredictable in what sense?

Second, how one classifies a diverse set of habitats depends on the lineage because the definition of the environment changes with the organisms under study. If one were to ask various biologists to submit classifications of habitats that were best suited to the study of their respective organisms, those classifications would differ considerably. A habitat that is stable for one organism is fluctuating for another. If we want generality in our explanations, they must be placed in a different framework. It is simpler and more general to deal with explanations couched in terms of mortality schedules rather than habitats.

Nevertheless, well-qualified opinion asserts that it is worth trying to relate habitat to life history (e.g. Southwood 1977, 1988; Grime 1977; Alerstam and Högstedt 1982; Taylor *et al.* 1990). These authors want to explain patterns relating certain lineages to certain habitats. To satisfy, such explanations should demonstrate a mechanism that links habitats to life histories. According to this book, one candidate is the impact of habitats on age- and size-specific fecundity and mortality schedules. Thus we seek to understand not habitat → life history but habitat → mortality regime → life history. Another candidate is the long-term association of certain lineages with certain habitats—an historical explanation with a strong comparative component.

I am more optimistic about this line of work than I was several years ago, for Promislow and Harvey (1990) have shown how to bring mortality rates into broad comparative analyses, and Charnov and Berrigan (1990) have uncovered patterns suggesting that mortality rates affect all organisms of a given basic design in much the same way. If we put these two insights together with previous work on habitat as a template for life history evolution (Southwood 1977, 1988), they suggest that the association of habitats with mortality regimes may produce sets of patterns relating habitat to life history with one pattern for each major design class of organisms, and with quite a bit of variation in patterns among design classes (e.g. insects, birds, herbs, trees). We may have to reconcile ourselves to dealing with sets of patterns rather than single patterns, but sets of patterns could prove useful in future attempts to relate life history evolution to community ecology.

ARE ANY GENERAL MODELS USEFUL?

There are virtually no general predictions in life history theory because some organism can always be found with a tricky and unexpected trade-off that violates the assumptions, because the predictions themselves depend on the state of the life history, and because we do not yet have reliable source laws to predict the critical trade-offs. Thus it is more sensible to treat the theory as a general framework that tells us which questions need to be answered when building a model for some particular organism, than it is to try to use the predictions of general models. If you are interested in testing life history theory, then collaborate with a theoretician and build a model of your particular organism, testing both assumptions and predictions against your data. There are few predictions extant in the literature general enough to be convincingly and fairly tested on some randomly chosen organism without modification.

This does not mean that general models are not useful. They are often the only models whose qualitative results are simple enough to remember and thus the ones that have given most workers in the field the insights used in daily conversation. For example, the reproductive effort model of Chapter 7 has been repeatedly confirmed in many qualitative features. However, to make quantitative predictions of changes in reproductive effort would require tailoring the model to a specific organism, and that would result in a degree of complexity that would be hard to remember and inappropriate to apply to other organisms.

ARE THERE ANY GENETIC CONSTRAINTS AT ALL?

The answer one gives to this question is to a certain degree a matter of intellectual taste and background, and opinions are diverse. I do not believe there are any genetic constraints on the equilibrium state of a population, only constraints imposed by the structure of the organisms in which the genes are expressed. Many geneticists would disagree with me. You might want to consider whether they were confusing the language in which they expressed their measurements for the causes underlying the phenomena measured.

Genetic variation does limit the rate of response to selection—that is not at issue. And genetics can be a useful tool for getting at the critical features of organismal structure that produce constraints on the equilibrium state. But it is not the only such tool, and it is not necessarily the tool that should be chosen first. Often more insight may be gained from a combination of functional morphology, developmental biology, and careful comparisons of closely related species (cf. Chapter 5).

Single-locus genes with known physiological effects are much more informative objects to analyse than genetic covariances between two polygenic traits, for the meaning of a genetic covariance must normally be hedged about with so many qualifying statements that it would not be useful to consider them the causes of anything (cf. Chapter 3). Quantitative genetics is a method of phenomenological description with major problems when applied to causation (Lewontin 1970).

That does not mean that the application of quantitative genetics to life history evolution has been worthless. Quite the contrary—it has helped a lot. By focusing attention on genetic covariances between traits, it has given us one clear technical definition of how traits are linked together and has clarified the meaning of microevolutionary trade-offs and how they differ from physiological trade-offs. In so doing, it has put the question: 'What is responsible for the linkages between traits?' To answer this question, however, quantitative genetics is only one of several tools available for probing physiology and development, and it is not necessarily the most efficient method for measuring trade-offs and constraints. We need a method of predicting the relative importance and quantitative impact of major trade-offs independent of demographic and quantitative genetic measurements, i.e. from first principles of physiology and development.

EXTENSIONS OF LIFE HISTORY THEORY

Complex life histories

The simple version of life history theory developed in this book says nothing directly about the evolution of complex life cycles, a subject significantly advanced by Istock (1967, 1983) and Wilbur (1980). Despite those contributions, the problem remains a fertile field. Caswell's (1989*a*) life cycle graphs seem to be an obvious tool. The subject encompasses not only the two-stage larval → adult life cycles of many amphibians, insects, and marine invertebrates, but the complex life cycles of parasites, coelenterates, and algae in which there may be changes between sexual and asexual phases, between haploidy and diploidy, between sessile and motile forms, and between parasitic and free-living forms, coupled in some cases with radical morphological change.

Modular organisms

The lack of discussion of the life histories of modular organisms is the weakness of this book that I most regret. My lack of familiarity with the appropriate literature and the time it would have taken to master it caused the omission. A classic reference for plants is Harper (1977), and entries to the recent botanical literature can be made through Watkinson and White (1985), Schmid *et al.* (1988), and Cain (1990). Recent references for clonal animals include Buss (1987) and Hughes (1989). Modular organisms *are* different. Their biology raises the central questions of how more inclusive units of selection arise and how conflicts of interest between levels of selection are resolved.

Frequency dependence and ESS models

The role of frequency dependence in maintaining alternative reproductive phenotypes within populations of bluegill sunfish, blue-headed wrasses, and pandalid shrimp provides some of the best evidence available for the interaction of life histories and behaviour in evolution. Only the need to restrict the length of this book and the excuse provided by the excellent coverage given the topic elsewhere (e.g. Gross and Charnov 1980; Charnov 1982; Maynard Smith 1982; Warner 1984) has kept me from giving it the extensive coverage it deserves. Kawecki (pers. comm.) has recently shown that the ESS age and size at maturity of organisms with ecology like parasitoids is considerably older and larger than what one would predict from optimality theory. Such new applications of ESSs should steadily change the status of predictions.

POTENTIAL APPLICATIONS OF LIFE HISTORY THEORY

Fisheries management

Fisheries biology gave to life history theory the concept of reproductive compensation—when stocks are reduced by fishing, fish have more food per capita, grow faster, reach larger sizes, and have higher fecundities. The higher fecundities compensate for the smaller population size and help maintain the population in the face of fishing pressure. In the language of life history theory, reproductive compensation describes a reaction norm for fecundity as a function of population density, and it suggests that the evolution of reaction norms for birth and death rates as functions of population density will form a central element of future models used in fisheries management.

While considerable work has already been done on the consequences of density-dependent effects on birth and death rates for population dynamics, that work needs to be extended to include genetic variation for the reaction norms of birth and death rates and the evolution of the population mean reaction norms. Such models of the *evolution* of single- or multi-species population dynamics could form the basis for research on the next point.

Community ecology

One test of the usefulness of a theory is its ability to organize and clarify another field. A relatively unexplored application of life history theory is the reduction of community ecology to the interaction of the reaction norms of the birth and death rates as functions of the densities and age distributions of the interacting species. Life history theory could predict the evolution of the reaction norms of the vital rates of two or more interacting species. This method might yield insights that would extend those already achieved in community ecology. Essential background for anyone interested in this approach can be found in Tilman (1982) and Clark (1990).

Parasite–host, predator–prey, and plant–herbivore interactions

The concept of trade-off is central in life history evolution, and it is starting to be applied to the evolution of parasite–host, predator–prey, and plant–herbivore relationships. One key question is whether resistance has a cost expressed in decreased reproductive performance. Such effects have been found for induced morphological defences in *Daphnia* (see Dodson 1989 for a review) and barnacles (Lively 1986*a*, *b*) and for induced chemical defences in tobacco (Baldwin *et al.* 1990). They have not yet been adequately investigated in the evolution of drug resistance and pesticide resistance, where they could be of major significance. For example, if the evolution of resistance to drugs has a cost in reproductive performance, then when the drug is withdrawn, resistance should disappear, and a cyclical pattern of drug treatments could prove to be more effective than continuous use of one drug and as effective as successive replacement of one drug by another (Koella, pers. comm.).

Behaviour

Life history theory and behavioural ecology have had an off-and-on courtship for the last twenty years. The two fields share a common concern for workable definitions of fitness and there is a formal similarity in modelling approaches, for optimal control theory is applied in both fields. While life history evolution limits the behaviour that can evolve, these limits are not surprising enough to catalyse interest. For example, interactions among three generations only occur in species selected for iteroparity, and parent–offspring conflict only occurs in populations in which the parents live long enough after the birth of their offspring to provide parental care. These statements neither surprise nor help.

In two other areas the linkage between the evolution of life histories and behaviour is so strong and natural that research proceeds as a healthy mixture of methods of both fields. The first is the relation of sex allocation to mating pattern and parental care—monogamy, polygyny, and so forth (cf. Charnov 1982, Clutton-Brock 1988). The second is predicting optimal clutch size when parents and offspring are in conflict over parental investment and offspring independence (cf. Charnov and Skinner 1984; Godfray 1987*b*; Parker and Mock 1987).

PARADOXES

Some observations on life histories seem to be at odds with theory. Here are a few; there are doubtless more.

Organisms that gain fecundity with size and have high fecundity at maturity should be iteroparous. Why then is semelparity associated with very high fecundity in catadramous eels (Bell 1980)?

Trees deal with spatial and temporal variation in juvenile mortality by dispersing their seeds and having a long reproductive lifespan (Stearns and Crandall 1981*b*). This decreases the variance over generations in lifetime reproductive success and increases geometric mean fitness. Why then is there a semelparous species of large canopy tree (*Tachigalia*) in the neotropical forest (Foster 1977)? *Tachigalia* should not exist, but it does.

Why do bamboos reproduce asexually for many years, even attaining the size of a small forest, then switch to sexual reproduction and die (Janzen 1976)?

Why are there any cases in which growth stops before maturity, most dramatic in albatrosses and condors, but in principle in all birds (Charlesworth 1980, pp. 258–9)? Can no bird grow and fly at the same time?

Are dung beetles really semelparous and if they are, how do they survive with such low fecundity (cf. Chapter 7)?

Why do baleen whales have no parental care and no reproductive senescence whereas the odontocete whales have parental care and strong reproductive senescence (Promislow 1991)? Are parental care and senescence associated in other groups?

Why do so many seabirds delay maturity and have a clutch of one egg although the growth rates of nestlings vary threefold within the group and adults can successfully rear enlarged clutches (Ricklefs 1990)?

Why are seed size and offspring size so often relatively constant within species despite heterogeneous environments?

PROBLEMS

1. Why do Kozlowski and Wiegert's predictions of reaction norms for age and size at maturity (his switching curves) differ from Stearns and Koella's? Is it because the former uses R_O and the latter use r for fitness?
2. By doing microscopic simulations of organisms with age structure and trade-offs among life history traits, construct a set of predictions about the the evolution of life history traits in which no assumptions about fitness are made. One lets the simulated organisms reproduce and die, then follows the distribution of traits over time. Put in some dependence on resources, then some predators, diseases, or parasites, and see what difference it makes.
3. Consider the interaction of reaction norms for life history traits with population dynamics and community ecology. How would you include resource dynamics? How much genetic variation for these reaction norms will be maintained by variation in population density? Does the genetic variation in reaction norms for birth and death rates that is maintained by changes in population density itself stabilize interspecific interactions?

APPENDIX 1: GENETIC CORRELATIONS OF LIFE HISTORY TRAITS

(a) Genetic correlations between life history traits measured from the resemblance of relatives (after Bell and Koufopanou 1986)

Organism	Correlation	Material	Reference
Between longevity and fecundity			
Goose	+	Small flock	Merritt (1962)
Fowl	+	Small flock	Dempster and Lowry (1952)
Mouse	+	Inbred lines	Roderick and Storer (1961)
Drosophila	≈0		Tantawy and Rakha (1964)
Drosophila	+	Inbred and outbred	Gowen and Johnson (1946)
Drosophila	+	Inbred lines	Rose (1984*a*)
Drosophila	+		Temin (1966)
Drosophila	⩾0	Outbred	Murphy *et al.* (1983)
Drosophila	⩽0	Mixed results	Scheiner *et al.* (1989)
Tribolium	−		Soliman (1982)
Oncopeltus	+	Half-sibs from wild	Dingle and Hegmann (1982)
Barley	−	New mutations	Gaul (1961)
Daphnia	+	Clones	Lynch (1984)
Between longevity and early fecundity			
Drosophila	≈0		Tantawy and El-Helw (1966)
Drosophila	+	Inbred and hybrid	Giesel (1979)
Drosophila	+	Inbred and hybrid	Giesel *et al.* (1982)
Drosophila	+	Inbred and hybrid	Giesel and Zettler (1980)
Drosophila	−	Extracted recombinants	Hughes and Clark (1988)
Between longevity and late fecundity			
Drosophila	+	Inbred and hybrid	Giesel (1979)
Drosophila	≈0	Inbred and hybrid	Giesel and Zettler (1980)
Between early fecundity and late fecundity			
Oncopeltus	+	Half-sibs from wild	Dingle and Hegmann (1982)
Drosophila	+	Inbred and hybrid	Giesel *et al.* (1982)
Drosophila	−	Large outbred cage	Rose and Charlesworth (1981*a*)
Drosophila	+	Inbred and hybrid	Giesel and Zettler (1980)
Drosophila	+	Inbred lines	Rose (1984*a, b*)
Drosophila	−	At both 19 and 25°C	Scheiner *et al.* (1989)
Daphnia	+	Clones	Bell (1984*b*)
Daphnia	+	Clones	Lynch (1984)
Drosophila	≈0	Extracted recombinants	Hughes and Clark (1988)

(b) Correlated responses to selection on life history traits (from Bell and Koufopanou 1986)

Organism	Correlated response	Gens*	Reference
Response of longevity to upward selection on early fecundity			
Chicken	Upward (+ −)	4	Nordskog and Festing (1962)
Chicken	None	12	Morris (1963)
Goose	Downward	10	Merritt (1962)
Tribolium	Downward	40	Sokal (1970)
Tribolium	None	12	Mertz (1975)
Drosophila	None	3	Rose and Charlesworth (1981*b*)
Drosophila	Downward	15	Rose (1984*b*)
Response of longevity to upward selection on late fecundity			
Drosophila	Downward	15	Rose (1984*b*)
Response of late fecundity to upward selection on early fecundity			
Chicken	Downward	6	Lerner (1958)
Chicken	Downward	5	Erasmus (1962)
Chicken	Downward	12	Morris (1963)
Mouse	Downward	23	Wallinga and Bakker (1978)
Tribolium	Downward	12	Mertz (1975)
Drosophila	None	3	Rose and Charlesworth (1981*b*)
Drosophila	Downward	17	Luckinbill and Clare (1985)
Drosophila	Downward	120	Mueller (1986)
Response of early fecundity to upward selection on late fecundity			
Drosophila	Downward		Wattiaux (1968)
Drosophila	Downward	3	Rose and Charlesworth (1981*b*)
Drosophila	Downward	17	Luckinbill and Clare (1985)
Drosophila	Downward	15	Rose (1984*b*)

* Gens = number of generations in which selection was practised.

APPENDIX 2: EVIDENCE ON TRADE-OFFS

The evidence presented in these tables has not been systematically screened for admissibility according to explicit criteria. It is therefore of variable quality and should be taken as such.

(a) The effect of mating on longevity in arthropods, nematodes, and mammals. (From Bell and Koufapanou 1986)

Organism	Sex	Effect	Reference
Arthropods and nematodes			
Agonum	F	V > C	Murdoch (1966)
Agromyza	F	V > C	Quiring and MacNeil (1984)
Artemia	F, M	V ⩾ C	Browne (1982)
Corixa	F	V ⩾ C	Calow (1973)
Diplogasteritus	F	V > C	Woombs and Laybourne-Parry (1984)
Drosophila	F, M	V > C	Bilewicz (1953)
Drosophila	F	V > C	Maynard Smith (1958)
Drosophila	F, M	V > C	Malick and Kidwell (1966)
Drosophila	F, M	V ⩾ C	Kidwell and Malick (1967)
Drosophila	M	V > C > CC	Partridge and Farquhar (1981)
Drosophila	F	C > CC	Turner and Anderson (1983)
Drosophila	F	C > CC	Fowler and Partridge (1989)
Dysdercus	F	V ⩾ C	Clarke and Sardesai (1959)
Ephestia	F	V > C	Norris (1933)
Fumea	F	V > C	Matthes (1951)
Melanoplus	F	V > C	Dean (1981)
Mesocyclops	F	V > C	Feifarek *et al.* (1983)
Panagrellus	F, M	V ⩾ C	Abdulrhaman and Samoiloff (1975)
Paroigolaimella	F	V > C	Woombs and Laybourne-Parry (1984)
Periplaneta	F	V > C	Griffiths and Tauber (1942)
Rhabditis	F	V > C	Woombs and Laybourne-Parry (1984)
Tribolium	F	V > C	Sonleitner (1961)
Tribolium	F	V > C	Mertz (1975)
Trogoderma	F	V ⩾ C	Loschiavo (1968)
Mammals			
Mus	F, M	V < C	Agduhr (1939)
Mus	F	V > C	Mühlbock (1959)
Rattus	M	V ⩽ C	Slonaker (1924)
Rattus	M	V < C	Drori and Folman (1969)

Symbols: F, female; M, male; V, virgin; C, allowed to copulate; CC, allowed more copulations than C. V > C = virgins live longer than mated individuals. V < C = virgins die younger than mated individuals. Where V ⩾ C, virgins live longer under some treatments but not under others.

(b) Phenotypic correlations between fecundity and parental survival (after Calow 1973; Stearns 1976; Bell 1984*a*; Warner 1984; Bell and Koufopanou 1986; Lindén and Møller 1989)

Organism	Common name	Sex	Correlation	Reference
In laboratory culture				
Aelosoma	aquatic oligochaete		≈0	Bell (1984*b*)
Agromyza	a leaf-mining fly		>0	Quiring and McNeil (1984)
Asplanchna	a predatory rotifer		<0	Snell and King (1977)
Philodina	asexual benthic rotifer		<0	Bell (1984*b*)
Platyias	planktonic rotifer		≈0	Bell (1984*a*)
Cypridopsis	ostracod		≈0	Bell (1984*b*)
Daphnia	water flea		≈0	Bell (1984*b*)
Drosophila	fruit fly		>0	Kidwell and Malick (1967)
Drosophila			>0	Murphy *et al.* (1983)
Carabidae spp.	ground beetles		≈0	van Dijk (1979)
Tribolium	flour beetle		≈0	Mertz (1975)
Chorthippus	grasshopper		<0	Souza Santos and Begon (1987)
Melanoplus	grasshopper		≈0	Dean (1981)
Gargaphia	hemipteran bug		≈0	Tallamy and Denno (1982)
Dysdercus	hemipteran bug		<0	Clarke and Sardesai (1959)
Panolis	moth		>0	Leather and Burnand (1987)
Oryzias	medaka fish		<0	Hirshfield (1980)
Pimephales	minnow		<0	Markus (1934)
Gallus	domestic chicken		<0	Hall and Marble (1931)
Gallus			>0	Dempster and Lowry (1952)
Mesocyclops	cyclopoid copepod		≈0	Feifarek *et al.* (1983)
Pristina	aquatic oligochaete		≈0	Bell (1984*b*)
In unmanipulated field populations				
Astrocaryum	a tropical palm	F	<0	Pinero *et al.* (1982)
Poa	meadow grass	F	<0	Law *et al.* (1979)
Podophyllum	mayapple	F	<0	Sohn and Policansky (1977)
Senecio	giant Kenyan senecio	F	<0	Smith and Young (1982)
Shaskyus	littoral gastropod	H	<0	Fotheringham (1971)
Ocenebra	littoral gastropod	H	<0	Fotheringham (1971)
Armidillidium	pillbug	F	<0	Paris and Pitelka (1962)
Pandalus	pandalid shrimp	F	<0	Allen (1959)
Agonum	ground beetle	F	<0	Murdoch (1966)
Tetraopes	milkweed beetle	M	>0	McCauley (1983)
Thalassoma	bluehead wrasse	M	≈0	Warner (1984)
Delichon	housemartin	F	<0	Bryant (1979)
Delichon		M	≈0	Bryant (1979)
Pica	magpie	F	>0	Högstedt (1981)
Passer	house sparrow	F, M	<0	Summer-Smith (1956)
Turdus	blackbird	F, M	<0	Snow (1958)
Parus	willow tit	F, M	<0	Ekman and Askenmo (1986)
Parus	crested tit	F, M	<0	Ekman and Askenmo (1986)
Parus	great tit	F, M	≈0	Den Boer-Hazewinkel (1987)

(b) Phenotypic correlations between fecundity and parental survival (*continued*)

Organism	Common name	Sex	Correlation	Reference
In unmanipulated field populations (*continued*)				
Melospiza	song sparrow	F	>0	Smith (1981)
Microtus	meadow vole	F	<0	Clough (1965)
Cervus	red deer	F	≈0	Clutton-Brock *et al.* (1983)
Cervus		M	>0	Clutton-Brock (1984)
Ovis	mountain sheep	M	<0	Geist (1971)
Papio	olive baboon	F, M	<0	Berger (1972)
In manipulated field populations				
Falco	kestrel	M	<0	Dijkstra *et al.* (1990)
Aegolius	Tengmalm's owl	M, F	≈0	Korpimäki (1988)
Larus	glaucous-winged gull	M, F	<0	Reid (1987)
Creagus	swallow-tailed gull	M, F	≈0	Harris (1970)
Ficedula	pied flycatcher	M	<0	Askenmo (1979)
Ficedula	collared flycatcher	M, F	≈0	Gustaffson and Sutherland (1988)
Iridoprocne	true swallow	F	≈0	DeSteven (1980)
Parus	great tit	F	<0	Kluyver (1951)
Parus	great tit	M, F	≈0	Boyce and Perrins (1987); Pettifor *et al.* (1988)
Parus	great tit	M, F	≈0	Tinbergen (1987)
Parus	great tit	M, F	<0	Lindén 1988)
Parus	blue tit	F	<0	Nur (1984*a*, 1988)*
Parus	blue tit	M	≈0	Nur (1984*a*, 1988)*
Parus	coal tit	M, F	≈0	Orell and Koivula (1988)
Passer	house sparrow	M, F	≈0	Hegner and Wingfield (1987)
Corvus	rook	M, F	≈0	Røskaft (1985)

* But see Pettifor (1990). Symbols: M = male, F = female, H = hermaphrodite

(c) Phenotypic correlations between early and late fecundity (after Warner 1984; Bell and Koufonapou 1986)

Organism	Common name	Sex	Correlation	Reference
In laboratory culture under good conditions				
Aelosoma	aquatic oligochaete		⩾0	Bell (1984*b*)
Biomphalaria	snail		<0	Minchella and Loverde (1981)
Cyprodopsis	ostracod		>0	Bell (1984*b*)
Daphnia	water flea		⩾0	Bell (1984*b*)
Daphnia	water flea		>0	Lynch (1984)
Gallus	domestic chicken		>0	Jull (1928)
Gallus	domestic chicken		>0	Harris and Lewis (1922)
Gallus	domestic chicken		>0	Hall and Marble (1931)
Gargaphia	hemipteran bug		⩾0	Tallamy and Denno (1982)

(c) Phenotypic correlations between early and late fecundity (*continued*)

Organism	Common name	Sex	Correlation	Reference
In laboratory culture under good conditions (*continued*)				
Philodina	asexual benthic rotifer		0	Bell (1984*b*)
Platyias	planktonic rotifer		>0	Bell (1984*b*)
Pristina	aquatic oligochaete		<0	Bell (1984*b*)
Tribolium	flour beetle		<0	Boyer (1978)
In unmanipulated field populations				
Poa	meadow grass	F	<0	Law *et al.* (1979)
Senecio	giant Kenyan senecio	F	<0	Smith and Young (1982)
Ficedula	pied flycatcher	F	>0	Harvey *et al.* (1985)
Passer	house sparrow	F	<0	McGillivray (1983)
Cervus	red deer	F	<0	Clutton-Brock *et al.* (1983)
In manipulated field populations				
Parus	great tit	M, F	<0	Slagsvold (1984)
Parus	great tit	M, F	<0	Tinbergen (1987)
Parus	great tit	M, F	<0	Lindén (1988)
Parus	great tit	M, F	≈0	Pettifor *et al.* (1988)
Parus	blue tit	M, F	<0	Nur (1988)
Troglodytes	house wren	F	≈0	Finke *et al.* (1987)
Corvus	rook	M, F	<0	Røskaft (1985)

(d) The effect of reproduction on growth (from Stearns 1976; Harper 1977; Warner 1984)

Organism	Common name	Correlation	Reference
Field populations			
Elminius	barnacle	<0	Crisp and Patel (1961)
Balanus	barnacle	<0	Barnes (1962)
Fagus	beech trees	<0	Rohmeder (1967)
Pseudotsuga	Douglas fir	<0	Eis *et al.* (1965)
Abies	grand fir	<0	Eis *et al.* (1965)
Pinus	western white pine	<0	Eis *et al.* (1965)
Betula	white birch	<0	Gross (1972)
Betula	yellow birch	<0	Gross (1972)
Picea	spruce trees	<0	Danilow (1953)
Iridoprocne	tree swallow	≈0	DeSteven (1980)
Thalassoma	blue-headed wrasse	<0	Warner (1984)
Laboratory studies			
Agropyron	grass	0	Reekie and Bazzaz (1987*c*)
Oryzias	medaka fish	<0	Hirshfield (1980)

(e) A summary of studies in which the trade-off between offspring size or quality and offspring number was measured

Organism	Common name	Correlation	Reference
Among individuals within populations			
Lomatium	umbellifer	0	Thompson (1984)
Phaseolus	field bean	<0	Adams (1967)
Avena	wheat	0	Puckridge and Donald (1967)
Chorthippus	field grasshopper	<0	Kriegsbaum (1988)
Mesocyclops	copepod	<0	Allan (1984)
Daphnia	cladoceran	<0	Lynch (1984)
		>0	Lynch (1984)*
Barbaorula	frog	0	Salthe and Duellman (1973)
Bufo	toad	0	Salthe and Duellman (1973)
Pachymedusa	frog	0	Salthe and Duellman (1973)
Agalychnis	frog	0	Salthe and Duellman (1973)
Rana	frog	0	Salthe and Duellman (1973)
Parus	great tit	<0	Henrich (unpub.)
Parus	great tit	<0	Smith *et al.* (1989)
Parus	blue tit	<0	Nur (1984*b*)
Agelaius	red-winged blackbird	<0	Cronmiller and Thompson (1980)
Homo	man	<0	
Among populations within species			
Gambusia	mosquito fish	0 lab	Stearns (1983*c*)
Gambusia	mosquito fish	<0 field	Stearns (1983*b*)
Among closely related species			
Solidago	goldenrod	<0	Werner and Platt (1976)
	salamanders	<0	Salthe (1969)

* Whether the trade-off is observed in *Daphnia* depends on the clone and the clutch (first, second, . . .) in which it is measured

APPENDIX 3: ELEMENTARY ALLOMETRY

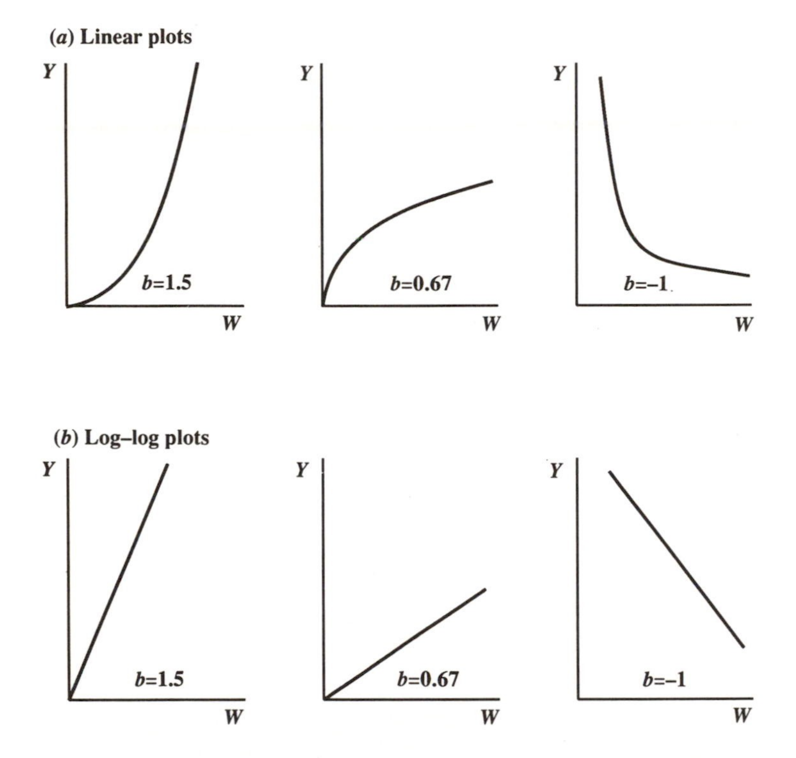

Figure A3.1 Many traits are related to body size or to one another by power functions. These are the general shapes of power functions on linear and log–log plots.

The most familiar example of an allometric argument concerns surface area to volume ratios. If shape is preserved as size increases, then the increase in the surface area of the organism, or any organ in it, increases as the square of a linear measure. The volume of the organism, or any organ in it, increases as the cube of a linear measure. If we let x be the linear measure—length of a bone, for example, S be surface area, V be volume, and a, b, c, d be constants of proportionality, then

$$S = ax^2 \quad \text{and} \quad V = bx^3$$

from which we get

$$x = cV^{1/3} \quad \text{and} \quad S = dV^{2/3}.$$

This means that surface area increases as the two-thirds power of volume. A proportionate increase in surface area with volume would imply an exponent of 1. Since 2/3 is less than 1, surface area is increasing less rapidly than volume. Thus the ratio of surface area to volume is getting smaller and smaller as volume increases. Volume is proportional to weight, to biomass, to the amount of tissue that needs to be supplied with oxygen and nutrients. The absorptive power of a lung or intestine is proportional to its surface area. Therefore we expect the length of intestines and the area of lungs to increase more rapidly than weight as an animal grows to preserve the proportions necessary for physiological function.

Given that two traits are related by a power

function, $Y = aW^b$,

(1) if $b > 1$, the ratio of Y to W increases as W increases;
(2) if $b = 1$, the ratio stays the same as W changes;
(3) if $b < 1$, the ratio decreases as W increases.
(4) The exponent in the power function is the slope of the straight line relating the two variables on a log–log graph. This slope can be estimated from the log-transformed data by doing a linear regression.

Peters (1983, Ch. 2) and McMahon and Bonner (1983, Ch. 2) give good reviews of the mathematics of power functions and allometric relations. Pagel and Harvey (1988) show how changes in the slope of an allometric relation estimated at different taxonomic levels—say among families as opposed to among genera—can arise as an artefact of the analysis. This effect should be accounted for before deciding on the pattern that needs explanation.

GLOSSARY

additive genetic variance: The variance among individuals in *breeding values* (*see* breeding value). It can be measured through the degree of resemblance among relatives.

altricial: Organisms born at an early stage of development, usually with their eyes closed, and incapable of temperature regulation. Compared to precocial organisms, they are relatively dependent on their parents.

allometry: The study of the proportions of the traits, principally as an exponential function of body size.

antagonistic pleiotropy: Pleiotropy occurs when one allele has effects on two or more traits. Antagonistic pleiotropy occurs when the effects increase fitness through one trait and decrease it through another.

average effect: The average effect of an allele is the average deviation from the population mean of those individuals that have received the allele from one parent, when the allele received from the other parent is a random sample of the population. The average effect of an allele A_2 can also be defined as the change that would occur in the population mean if all A_1 alleles were changed into A_2 alleles. Average effects are measured in terms of genotypic values (*see* genotypic value).

breeding value: The breeding value of an individual is twice the deviation of the mean of its progeny from the population mean when it is mated at random with a sample of the population. The breeding value of an individual is also equal to the average effects of the alleles that it carries, summed across all segregating loci.

broad-sense heritability: Broad-sense heritability is the proportion of total phenotypic variance among individuals in a population that is accounted for by genetic effects in general. It can include maternal effects as well as effects of dominance and epistasis.

capital breeder: An organism that uses stored energy for reproduction.

cladistics: A synonym for 'phylogenetic systematics', a method of determining the evolutionary relationships of species that relies on the information contained in shared derived states of traits.

cohort: All the organisms in a population that were born within a defined time interval and belong to a single age class.

demography: The analysis of the effects of age structure on population dynamics and on natural selection.

determinate growth: Growth that ceases with maturation.

dominance variance: The portion of phenotypic variation that can be attributed to deviations from the breeding value caused by dominance effects, i.e. to the interactions between the two alleles at given loci.

elasticity: Like sensitivity, a measure of the amount one thing changes when another thing is changed. Elasticities are dimensionless sensitivities normalized for comparison among traits measured in different units.

environmental value: The mean phenotype of all genotypes in that environment, used as a sort of bioassay to get the organisms to define how good that environment is, and often used as the x-axis on reaction norm plots.

epistasis: The interaction of alleles at two or more loci in affecting the state of a single trait where the combined effect differs from the sum of the individual loci.

fitness: The expected contribution of an allele, genotype, or phenotype to future generations. The fitness of genes and organisms is always relative to the other genes and organisms that are present in the population. It is also a function of the environment in which it is measured.

fitness profile: The mapping of values of a trait on to fitness, expressed as a plot with fitness on the y-axis and a range of values of the trait on the x-axis.

General Life History Problem: A model, investigated many times, that predicts the optimal reproductive effort based on residual reproductive value, i.e. on the assumption that increased current reproduction decreases future reproduction either through direct effects on future reproduction or through effects on parental survival. A synonym for the Reproductive Effort Model.

generation time: The average age of mothers giving birth in a population in stable age distribution.

genetic correlation: The portion of a phenotypic correlation between two traits in a population that can be attributed to additive genetic effects.

genetic covariance: The covariance between two traits in a population that can be attributed to additive genetic effects.

genotype: 'Genotype' is used in at least two different senses in evolutionary biology. In Johanssen's distinction between genotype and phenotype, genotype refers to the total genetic content of the organism—all the genes on all the chromosomes plus those in the mitochondria, in the centrosome, and, in plants, in the chloroplasts. In population genetics, 'genotype' is also used to refer to the diploid combination of two alleles at one locus.

genotype × environment interaction: A technical term (G × E) used in quantitative genetics to refer to the case where different genotypes interact with the environment in different ways. A synonym is non-parallel reaction norms.

genotype × environment covariance: Different frequency distributions of genotypes over environments. *Not* G × E interaction.

genotypic value: If we cloned individual genotypes and reared them in a constant environment until maternal effects had disappeared, the effect of the environment on their deviations from the population mean would be zero, and the deviation of a clone from the mean of the clones would be its genotypic value.

heritability: Narrow-sense heritability is the proportion of total phenotypic variance among individuals in a population that is accounted for by additive genetic variance. The rate of evolution of a trait depends on its narrow-sense heritability, not on its broad-sense heritability, which usually overestimates the amount of genetic variation present that can respond to natural selection (*see* broad-sense heritability).

homeotherm: An organism that regulates its body temperature within a narrow range despite variation in the temperature of the environment.

imprimitive life history: A life history in which the age classes oscillate without ever converging to a stable age distribution.

income breeder: An organism that uses energy acquired during the reproductive period rather than stored energy for reproduction (compared to a capital breeder).

indeterminate growth: Growth that continues past maturation and may continue until the end of life.

interaction variance: The portion of the phenotypic variation that can be attributed to deviations from the breeding values caused by interaction effects among rather than within loci.

intrinsic rate of natural increase: In demography and population dynamics, the rate r at which an age-structured population increases in stable age distribution, i.e. with constant birth and death schedules.

iteroparity: More than one reproductive event per lifetime (*see* semelparity).

Lack clutch: The clutch that maximizes the number of offspring fledged under the assumption that fledging mortality is a function of clutch size.

lek: A mating place found in certain species, such as sage grouse and hammerhead bats, where males gather to compete for females through ritualized displays and direct combat, and where females gather to choose males and mate. Usually found in species that feed on widely scattered food. Thought to lead to great variation in male reproductive success.

Leslie matrix: A method of representing birth and survival rates that is convenient for analysing population dynamics and life history evolution. The first row contains the age-specific birth rates, the first subdiagonal the age-specific survival rates.

life history evolution: The evolution of the major features of a life cycle, principally the age distribution of birth and death rates, growth rates, and the size of offspring.

lineage-specific effect: A feature characteristic of a lineage that does not vary within that lineage and differs from others. Lineages are groups of species related by shared ancestry.

macroevolutionary trade-off: A trade-off defined by a negative correlation between two traits above the species level when both traits are fixed within species.

Malthusian parameter: In population genetics, the rate of increase of an allele at a locus that has age-specific effects on birth and death rates. One measure of fitness.

maternal effect: All phenotypic effects in the progeny that can be associated with the mother but not attributed to her breeding value. They often include both cytoplasmic effects and effects of environments shared by sibs.

Mendelian trait: A trait determined by a few genes of large effect which can, through crossing experiments, be shown to segregate in the classical ratios.

metric trait: A synonym for quantitative trait, both to be distinguished from Mendelian traits.

microevolutionary trade-off: A trade-off defined by a change in one trait in a population that increases fitness but is linked to a change in another trait that decreases fitness. Best measured by the correlated response to selection.

mutation accumulation: Where genes have age-specific effects, the amount of genetic variation maintained by a balance between mutation and selection increases with the age at which the genes have their effects, for the selection pressures on age classes decrease with age.

narrow-sense heritability: The proportion of phenotypic variability accounted for by variation in the breeding values of individuals (*see* heritability; broad-sense heritability).

natural selection: The process that takes place in a population of entities that vary among themselves with respect to reproduction and survival, whereby those with greater differential reproductive success have higher expected representation in future generations.

nidicolous: Birds that hatch at an early stage of

development, usually without feathers, with their eyes closed, and incapable of temperature regulation. An ornithological synonym for 'altricial'.

nidifugous: Birds that hatch at an advanced stage of development, fully feathered, with their eyes open, capable of temperature regulation and, in some cases, even flight. An ornithological synonym for 'precocial'.

phenotype: The phenotype is the expression of the genotype in interaction with the environment during the course of the growth and development of the organism. Strictly speaking, any part of the organism that is not DNA can be referred to as phenotypic. All aspects of the phenotype are the product of interactions between genes and environment. The life history of an organism is one of the more comprehensive examples of a phenotype, for it results from processes stretching from molecular biology to ecology.

phylogenetic autocorrelation: A method for determining the relative contributions of ancestry and current variation to the value of a trait.

phylogenetic subtraction: A method for determining the proportion of a correlation between two traits that is contributed by higher taxonomic levels, such as classes, orders, or families. The mean values of the traits are calculated for a given level, the correlations among these mean values are calculated, then the mean values for, say, the orders are subtracted from the mean values of the respective families, the correlations among the traits are recalculated for the residuals of the family means, and the difference to the first method of calculating the correlations is noted.

phylogenetic systematics: A method of determining the evolutionary relationships among species by relying on the information delivered by shared derived characters and ignoring other aspects of similarity. Also called cladistics or Hennigan systematics.

physiological trade-off: A trade-off caused by allocation decisions between two or more processes that compete directly with one another for limited resources within a single individual. *See* trade-off; microevolutionary trade-off.

pleiotropy: One gene affects two or more traits.

poikilotherm: An organism that allows its body temperature to vary with the environmental temperature.

precocial: Organisms born at an advanced stage of development with their eyes open and capable of temperature regulation (*see* nidifugous).

quantitative genetics: The genetics of traits that vary continuously, like body size and growth rate, and that are determined by large numbers of genes each with relatively small effects. It is a statistical method of inferring genetic influences from phenotypic variation. As such, it is fundamentally a phenotypic rather than genetic theory of evolution, and it is phenomenological rather than mechanistic.

quantitative trait: Synonym for metric trait, a trait determined by many genes of small effect.

reaction norm: A reaction norm is the mapping of the genotype onto the phenotype as a function of the environment. It is usually expressed as a plot of phenotypic values against environmental values. The reaction norm of a genotype is the full set of phenotypes that that genotype will express in interaction with the full set of environments in which it can survive.

Reproductive Effort Model: See General Life History Problem.

reproductive value: The number of offspring that an average organism in a particular age class can expect to have over the rest of its life under the conditions prevailing, discounted back to the present by the current population growth rate. So defined, reproductive value is a characteristic of age classes and not of individuals, but the reproductive value of its age class is an estimate of the reproductive value of an individual.

residual reproductive value: The number of offspring that an average organism in a particular age class can expect to have *after* the current reproductive event.

semelparity: One reproductive event per lifetime. 'Big bang' reproduction (*see* iteroparity).

sensitivity: The amount by which one thing changes when another thing is changed a standard amount.

selection gradient: The selection pressures acting on traits when they are considered in isolation from other traits or, equivalently, a vector containing entries for a series of traits and representing the sensitivity of fitness to changes in those traits.

senescence: An increase with age in that portion of the age-specific mortality rate that can be attributed to intrinsic changes in the organism.

stable age distribution: A distribution of organisms among age classes in which the *proportion* of organisms in each age class remains constant through time (*see* stationary age distribution).

stationary age distribution: A distribution of organisms among age classes in which the *number* of organisms in each age class remains constant through time (*see* stable age distribution).

surplus power: The energy left after the costs of standard and active metabolism have been met.

synapomorphy: A trait that is shared by two or more species in a derived state.

trade-off: A trade-off occurs when an increase in one thing implies a decrease in something else. Physiological trade-offs occur when two traits compete for materials and energy within a single organism. Microevolutionary trade-offs occur when selection on one trait decreases the value of a second trait.

REFERENCES

Abdulrahman, M. and Samoiloff, M. R. (1975). Sex-specific aging in the nematode *Pangrellus redivivus*. *Can. J. Zool.* **53**, 651–6.

Abrahamson, W. G. (1979). Patterns of resource allocation in wild flower populations of fields and woods. *Am. J. Bot.* **55**, 71–9.

Adams, M. W. (1967). Basis of yield component compensation in crop plants with special reference to the field bean *Phaseolus vulgaris*. *Crop Sci.* **7**, 505–10.

Agduhr, E. (1939). Internal secretion and resistance to injurious factors. *Acta Med. Scand.* **99**, 387.

Alberch, P. and Gale, E. A. (1985). A developmental analysis of an evolutionary trend: Digital reduction in amphibians. *Evolution* **39**, 8–23.

Alerstam, T. and Högstedt, G. (1982). Bird migration and reproduction in relation to habitats for survival and breeding. *Ornis. Scand.* **13**, 25–37.

Allan, J. D. (1984). Life history variation in a freshwater copepod: evidence from population crosses. *Evolution* **38**, 280–91.

Allen, J. A. (1959). On the biology of *Pandalus borealis* Kroyer with reference to a population off the Northumberland coast. *J. Mar. Biol. Assoc. U.K.* **38**, 189–220.

Alm, G. (1959). Connection between maturity, size, and age in fishes. *Rep. Inst. Freshw. Res., Drottningholm* **40**, 5–145.

Altman, P. A. and Dittmer, D. S. (eds). (1972). *Biology Data Book*, Vol. 1, 2nd edn. (Federation of American Societies for Experimental Biology, Bethesda, MD).

Anderson, D. J. (1990*a*). On the evolution of human brood size. *Evolution* **44**, 438–40.

Anderson, D. J. (1990*b*). Evolution of obligate siblicide in boobies. 1. A test of the insurance-egg hypothesis. *Am. Nat.* **135**, 334–50.

Andersson, M. (1976). Clutch size in the long-tailed skua *Stercorarius longicaudus*: some field experiments. *Ibis* **118**, 586–8.

Andersson, M. (1978). Natural selection of offspring numbers: Some possible intergeneration effects. *Am. Nat.* **112**, 762–6.

Ang, P. O., Jr. and De Wreede, R. E. (1990). Matrix models for algal life history stages. *Mar. Ecol. Prog. Ser.* **59**, 171–81.

Arnold, S. J. and Wade, M. J. (1984*a*). On the measurement of natural and sexual selection: Theory. *Evolution* **38**, 709–19.

Arnold, S. J. and Wade, M. J. (1984*b*). On the measurement of natural and sexual selection: Applications. *Evolution* **38**, 720–34.

Asbirk, S. (1979). The adaptive significance of reproductive pattern in the black guillemot *Cepphus grylle*. *Vid. Med. Dank. nat. Forening* **141**, 29–80.

Askenmo, C. (1977). Effects of addition and removal of nestlings on nestling weight, nestling survival and female weight loss in the pied flycatcher *Ficedula hypoleuca* (Pallas). *Ornis. Scand.* **8**, 1–8.

Askenmo, C. (1979). Reproductive effort and return rate of male pied flycatchers. *Am. Nat.* **114**, 748–53.

Atchley, W. R. (1984). Ontogeny, timing of development, and genetic variance–covariance structure. *Am. Nat.* **123**, 519–40.

Atchley, W. R. (1987). Developmental quantitative genetics and the evolution of ontogenies. *Evolution* **41**, 316–30.

Atchley, W. R., Plummer, A. A., and Riska, B. (1985). Genetic analysis of size-scaling patterns in the mouse mandible. *Genetics* **111**, 579–95.

Austad, S. N. and Sunquist, M. E. (1986). Sex-ratio manipulation in the common opossum. *Nature* **324**, 58–60.

Bakker, K. (1959). Feeding period, growth, and pupation in larvae of *Drosophila melanogaster*. *Ent. Exp. Appl.* **2**, 171–86.

Baldwin, I. T., Sims, C. L., and Kean, S. E. (1990). The reproductive consequences associated with inducible alkaloidal responses in wild tobacco. *Ecology* **71**, 252–62.

Barclay, H. J. and Gregory, P. T. (1981). An experimental test of models predicting life history characteristics. *Am. Nat.* **117**, 944–61.

Barnes, H. (1962). So-called anecdysis in *Balanus balanoides* and the effect of breeding upon the growth of the calcareous shell of some common barnacles. *Limnol. Oceanogr.* **7**, 462–73.

Barton, N. H. and Turelli, M. (1989). Evolutionary quantitative genetics: How little do we know? *Ann. Rev. Genet.* **23**, 337–70.

Baur, B. (1987). Can cannibalistic hatchlings of the land snail *Arianta arbustorum* distinguish between sib and non-sib eggs? *Behavior* **103**, 259–65.

Baur, B. (1991, in press). Cannibalism in gastropods. In *Cannibalism: ecology and evolution among diverse taxa* (eds M. Elgar and B. Crespi). (Oxford University Press, Oxford).

Becker, W. A. (1984). *Manual of procedures in quantitative genetics.* (Academic Enterprises, Pullman, WA).

Beer, A. E. and Quebbeman, J. F. (1982). The immunobiology and immunopathology of the maternal–fetal relationship. In *Physiopathology of hypophysical disturbances and diseases of reproduction* (eds A. de Nicola, J. Blaquier, and R. J. Soto). (Liss, New York).

Begon, M., Harper, J. L., and Townsend, C. R. (1990). *Ecology.* 2nd edn. (Blackwell Scientific, Oxford).

Bell, G. (1976). On breeding more than once. *Am. Nat.* **110**, 57–77.

Bell, G. (1980). The costs of reproduction and their consequences. *Am. Nat.* **116**, 45–76.

Bell, G. (1984*a*). Evolutionary and nonevolutionary theories of senescence. *Am. Nat.* **124**, 600–3.

Bell, G. (1984*b*). Measuring the cost of reproduction. I. The correlation structure of the life table of a plankton rotifer. *Evolution* **38**, 300–13.

Bell, G. (1984*c*). Measuring the cost of reproduction. II. The correlation structure of the life tables of five freshwater invertebrates. *Evolution* **38**, 314–26.

Bell, G. (1988). *Sex and death in protozoa.* (Cambridge University Press, Cambridge).

Bell, G. (1989). A comparative method. *Am. Nat.* **133**, 553–71.

Bell, G. (1990*a*). The ecology and genetics of fitness in *Chlamydomonas.* I. Genotype-by-environment interaction among pure strains. *Proc. Roy. Soc. Lond. B* **240**, 295–321.

Bell, G. (1990*b*). The ecology and genetics of fitness in *Chlamydomonas.* II. The properties of mixtures of strains. *Proc. Roy. Soc. Lond. B* **240**, 323–50.

Bell, G. (1991, in press). *The evolutionary biology of crop plants.* (Cambridge University Press, Cambridge).

Bell, G. and Koufopanou, V. (1986). The cost of reproduction. *Oxford Surv. Evol. Biol.* **3**, 83–131.

Bell, M., Baumgartner, J. V., and Olson, E. C. (1985). Patterns of temporal change in single morphological characters of a Miocene stickleback fish. *Paleobiology* **11**, 258–71.

Bellman, R. (1957). *Dynamic programming.* (Princeton University Press, Princeton).

Berger, M. E. (1972). Population structure of olive baboons (*Papio anubis*) (J. P. Fisher) in the Laikipia District of Kenya. *E. Afr. Wildlf. J.*, **10**, 159–64.

Berven, K. A. (1982). The genetic basis of altitudinal variation in the wood frog *Rana sylvatica.* I. An experimental analysis of life-history traits. *Evolution* **36**, 962–83.

Berven, K. A. (1987). The heritable basis of variation in larval developmental patterns within populations of the wood frog, *Rana sylvatica. Evolution* **41**, 1088–97.

Beverton, R. J. H. and Holt, S. J. (1959). A review of the lifespans and mortality rates of fish in nature and the relation to growth and other physiological characteristics. In *CIBA Foundation. Colloquia on ageing. The lifespan of animals* (pp. 142–77). (Churchill, London).

Bielak, A. T. and Power, G. (1986). Changes in mean weight, sea-age composition, and catch-per-unit-effort of Atlantic Salmon (*Salmo salar*) angled in the Godbout River, Quebec, 1859–1983. *Can. J. Fish. Aquat. Sci.* **43**, 281–7.

Bilewicz, S. (1953). Experiments on the effects of reproductive functions on the length of life of *Drosophila melanogaster. Folio Biol.* **1**, 177–94.

Birch, L. C., Dobzhansky, Th., Elliott, P. O., and Lewontin, R. C. (1963). Relative fitness of geographic races of *Drosophila serrata. Evolution* **17**, 72–83.

Black, J. N. (1957). The early vegetative growth of three strains of subterranean clover (*Trifolium subterraneum* L.) in relation to size of seed. *Aust. J. Agric. Res.* **8**, 1–14.

Blackmore, M. S. and Charnov, E. L. (1989). Adaptive variation in environmental sex determination in a nematode. *Am. Nat.* **134**, 817–23.

Boag, P. T. and Grant, P. R. (1978). The heritability of external morphology in Darwin's finches. *Nature* **274**, 793–4.

Boyce, M. S. (1984). Restitution of *r*- and *K*-selection as a model of density-dependent natural selection. *Ann. Rev. Ecol. Syst.* **15**, 427–48.

Boyce, M. S. and Perrins, C. M. (1987). Optimizing great tit clutch size in a fluctuating environment. *Ecology* **68**, 142–53.

Boyer, J. F. (1978). Reproductive compensation in *Tribolium casteneum. Evolution* **32**, 519–28.

Bradley, A. J., McDonald, I. R., and Lee, A. K. (1980). Stress and mortality in a small marsupial (*Antechinus stuartii*, Macleay). *Gen. Comp. Endocrinol.* **40**, 188–200.

Bradshaw, A. D. (1965). Evolutionary significance of phenotypic plasticity in plants. *Adv. Genetics* **13**, 115–55.

Brockelman, W. Y. (1975). Competition, the fitness of offspring, and optimal clutch size. *Am. Nat.* **107**, 677–99.

Brodie, E. D. III. (1989). Behavioral modification as a means of reducing the cost of reproduction. *Am. Nat.* **134**, 225–38.

Brough, C. N. and Dixon, A. F. G. (1989). Intraclonal trade-off between reproductive investment and size of fat body in the vetch aphid, *Megoura viciae. Funct. Ecol.* **3**, 747–52.

Brown, K. M. (1983). Do life history tactics exist at the intraspecific level? Data from freshwater snails. *Am. Nat.* **121**, 871–9.

Browne, R. A. (1982). The costs of reproduction in brine shrimp. *Ecology* **63**, 43–7.

Bryant, D. M. (1975). Breeding biology of house martins *Delichon urbica* in relation to aerial insect abundance. *Ibis* **117**, 180–216.

Bryant, D. M. (1979). Reproductive costs in the house martin (*Delichon urbica*). *J. Anim. Ecol.* **48**, 655–75.

Bryant, D. M. and Westerterp, K. R. (1983). Time and energy limits to brood size in house martins (*Delichon urbica*). *J. Anim. Ecol.* **52**, 905–25.

Bull, J. J. and Charnov, E. L. (1985). On irreversible evolution. *Evolution* **39**, 1149–55.

Bulmer, M. G. (1980). *The mathematical theory of population genetics.* (Oxford University Press, Oxford).

Bulmer, M. G. (1985). Selection of iteroparity in a variable environment. *Am. Nat.* **126**, 63–71.

Buss, L. W. (1987). *The evolution of individuality.* (Princeton University Press, Princeton).

Cain, M. L. (1990). Models of clonal growth in *Solidago altissima. J. Ecol.* **78**, 27–46.

Calder, W. A., III. (1979). The kiwi and egg design: evolution as a package deal. *Bioscience* **29**, 461–7.

Calder, W. A., III. (1984). *Size, function, and life history.* (Harvard University Press, Harvard).

Calow, P. (1973). The relationship between fecundity, phenology, and longevity: a systems approach. *Am. Nat.* **107**, 559–74.

Calow, P. and Woollhead, A. S. (1977). The relation between ration, reproductive effort and age-specific mortality in the evolution of life history strategies—some observations on freshwater triclads. *J. Anim. Ecol.* **46**, 765–82.

Case, T. J. (1978). On the evolution and adaptive significance of postnatal growth rates in the terrestrial vertebrates. *Q. Rev. Biol.* **53**, 243–82.

Casper, B. B. (1990). Seedling establishment from one- and two-seeded fruits of *Cryptantha flava*: a test of parent–offspring conflict. *Am. Nat.* **136**, 167–77.

Caswell, H. (1978). A general formula for the sensitivity of population growth rate to changes in life history parameters. *Theor. Pop. Biol.* **14**, 215–30.

Caswell, H. (1980). On the equivalence of maximizing reproductive value and maximizing fitness. *Ecology* **61**, 19–24.

Caswell, H. (1989*a*). *Matrix population models.* (Sinauer, Sunderland, MA).

Caswell, H. (1989*b*). Life-history strategies. In *Ecological concepts* (ed. J. M. Chevett). (Blackwell Scientific, Oxford).

Caswell, H. and Hastings, A. (1980). Fecundity, developmental time, and population growth rate: an analytical solution. *Theor. Pop. Biol.* **17**, 71–9.

Caughley, G. (1966). Mortality patterns in mammals. *Ecology* **47**, 906–18.

Charlesworth, B. (1973). Selection in populations with overlapping generations. V. Natural selection and life histories. *Am. Nat.* **107**, 303–11.

Charlesworth, B. (1980). *Evolution in age-structured populations.* (Cambridge University Press, Cambridge).

Charlesworth, B. (1987). The heritability of fitness. In *Sexual selection: testing the alternatives* (eds. J. W. Bradbury and M. B. Anderson). (John Wiley, New York).

Charlesworth, B. (1990*a*). Optimization models, quantitative genetics, and mutation. *Evolution* **44**, 520–38.

Charlesworth, B. (1990*b*). Evolution: life and times of the guppy. *Nature* **346**, 313–14.

Charlesworth, B. and Léon, J. A. (1976). The relation of reproductive effort to age. *Am. Nat.* **110**, 449–59.

Charlesworth, B. and Williamson, J. A. (1975). The probability of the survival of a mutant gene in an age-structured population and implications for the evolution of life-histories. *Genet. Res.* **26**, 1–10.

Charnov, E. L. (1979). Natural selection and sex change in pandalid shrimp: test of a life-history trait. *Am. Nat.* **113**, 715–34.

Charnov, E. L. (1982). *The theory of sex allocation.* (Princeton University Press, Princeton).

Charnov, E. L. (1986). Life history evolution in a "recruitment population": why are adult mortality rates constant? *Oikos* **47**, 129–34.

Charnov, E. L. (1989). Phenotypic evolution under Fisher's fundamental theorem of natural selection. *Heredity* **62**, 113–16.

Charnov, E. L. and Berrigan, D. (1990). Dimensionless numbers and life history evolution: age at maturity versus the adult lifespan. *Evol. Ecol.* **4**, 273–5.

Charnov, E. L., and Berrigan, D. (1991). Dimensionless numbers and the assembly rules for life histories. *Philos. Trans. Roy. Soc. B* **332**, 41–8.

Charnov, E. L. and Krebs, J. R. (1974). On clutch size and fitness. *Ibis* **116**, 217–19.

Charnov, E. L. and Schaffer, W. M. (1973). Life history consequences of natural selection: Cole's result revisited. *Am. Nat.* **107**, 791–3.

Charnov, E. L. and Skinner, S. W. (1984). Evolution of host selection and clutch size in parasitoid wasps. *Florida Entom.* **67**, 5–21.

Charnov, E. L. and Skinner, S. W. (1988). Clutch size in parasitoids: the egg production rate as a constraint. *Evol. Ecol.* **2**, 167–74.

Chesson, P. L. (1982). The stabilizing effect of a random environment. *J. Math. Biol.* **15**, 1–36.

Chesson, P. L. and Warner, R. R. (1981). Environmental variability promotes coexistence in lottery competitive systems. *Am. Nat.* **117**, 923–43.

Cheverud, J. M. (1988). A comparison of genetic and phenotypic correlations. *Evolution* **42**, 958–68.

Cheverud, J. M., Dow, M. M., and Leutenegger, W. (1985). The quantitative assessment of phylogenetic constraints in comparative analyses: Sexual dimorphism in body weight among primates. *Evolution* **39**, 1335–51.

Clare, M. J. and Luckinbill, L. S. (1985). The effects of gene environment interaction on the expression of longevity. *Heredity* **55**, 19–29.

Clark, A. G. (1987). Senescence and the genetic correlation hang-up. *Am. Nat.* **129**, 932–40.

Clark, A. G. and Keith, L. E. (1988). Variation among extracted lines of *Drosophila melanogaster* in triacylglyceral and carbohydrate storage. *Genetics* **119**, 595–607.

Clark, J. S. (1990). Integration of ecological levels: individual plant growth, population mortality and ecosystem processes. *J. Ecol.* **78**, 275–99.

Clarke, K. U. and Sardesai, J. B. (1959). An analysis of the effects of temperature upon growth and reproduction of *Dysducus fasciatus* Sign. (Hemiptera, Pyrrhocoridae). *Bull. Entomol. Res.* **50**, 387–405.

Clough, G. C. (1965). Viability in wild meadow voles under various conditions of population density, season, and reproductive activity. *Ecology* **46**, 119–34.

Clutton-Brock, T. H. (1984). Reproductive effort and terminal investment in iteroparous animals. *Am. Nat.* **123**, 212–29.

Clutton-Brock, T. H. (ed.) (1988). *Reproductive success: Studies of individual variation in contrasting breeding systems.* (University of Chicago Press, Chicago).

Clutton-Brock, T. H. (1991). *The evolution of parental care.* (Princeton University Press, Princeton).

Clutton-Brock, T. H. and Harvey, P. H. (1977). Primate ecology and social organisation. *J. Zool. Soc. Lond.* **183**, 1–39.

Clutton-Brock, T. H. and Iason, G. R. (1986). Sex ratio variation in mammals. *Q. Rev. Biol.* **61**, 339–74.

Clutton-Brock, T. H., Guinness, F. E., and Albon, S. D. (1982). *Red deer: behavior and ecology of two sexes.* (University of Chicago Press, Chicago).

Clutton-Brock, T. H., Guinness, F. E., and Albon, S. D. (1983). The cost of reproduction to red deer hinds. *J. Anim. Ecol.* **52**, 367–83.

Cockburn, A., Lee, A. K., and Martin, R. W. (1983). Macrogeographic variation in litter size in *Antechinus* (Marsupalia: Dasyuridae). *Evolution* **37**, 86–95.

Coddington, J. A. (1988). Cladistic tests of adaptational hypotheses. *Cladistics* **4**, 3–22.

Cohen, D. (1966). Optimizing reproduction in a randomly varying environment. *J. Theor. Biol.* **12**, 119–29.

Cohen, D. (1967). Optimizing reproduction in a randomly varying environment when a correlation may exist between the conditions at the time a choice has to be made and the subsequent outcome. *J. Theor. Biol.* **16**, 1–14.

Cohen, J. E. (1977). Ergodicity of age structure in populations with Markovian vital rates. III. Finite-state moments and growth rate; an illustration. *Adv. Appl. Prob.* **9**, 462–75.

Cole, L. C. (1954). The population consequences of life history phenomena. *Q. Rev. Biol.* **29**, 103–37.

Comfort, A. (1979). *The biology of senescence.* 3rd edn. (Churchill Livingstone, Edinburgh).

Comstock, R. E. and Moll, R. H. (1963). Genotype–environment interactions. In *Statistical genetics in plant breeding* (eds W. D. Hanson and H. F. Robinson). (NAS-NRC Publ. 982, Washington, DC).

Conaway, C. H., Sadler, K. C., and Hazelwood, D. H. (1974). Geographic variation in litter size and onset of breeding in cottontails. *J. Wildl. Mgmt.* **38**, 473–81.

Corkhill, P. (1973). Food and feeding ecology of puffins. *Bird Study* **20**, 207–20.

Crandall, R. E. and Stearns, S. C. (1982). Variational models of life-histories. *Theor. Pop. Biol.* **21**, 11–23.

Crespi, B. J. (1990). Measuring the effect of natural selection on phenotypic interaction systems. *Am. Nat.* **135**, 32–47.

Crisp, D. J. and Patel, B. S. (1961). The interaction between breeding and growth rate in the barnacle *Elminius modestus* Darwin. *Limnol. Oceanogr.* **6**, 105–15.

Cronmiller, J. R. and Thompson, C. F. (1980). Experimental manipulation of brood size in red-winged blackbirds. *Auk* **97**, 559–65.

Crossner, K. A. (1977). Natural selection and clutch size in the European starling. *Ecology* **58**, 885–92.

Curio, E. (1973). Towards a methodology of teleonomy. *Experientia* **29**, 1045–59.

Daan, S., Dijkstra, C., and Tinbergen, J. M. (1990). Family planning in the kestrel (*Falco tinnunculus*): The ultimate control of covariation of laying date and clutch size. *Behavior* **114**, 83–116.

Danilow, D. (1953). Einfluss der Samenerzeugung auf die Struktur der Jahrringe. *Allg. Forstz.* **8**, 454–5.

Darwin, C. (1859). *The origin of species.* (John Murray, London).

Darwin, C. (1871). *The descent of man and selection in relation to sex.* (John Murray, London).

Dean, J. (1981). The relationship between lifespan and reproduction in the grasshopper *Melanoplus*. *Oecologia* **48**, 365–8.

DeMartini, E. E. and Anderson, M. E. (1980). Comparative survivorship and life history of painted greenling (*Oxylebius pictus*) in Puget Sound, Washington, and Monterey Bay, California. *Env. Biol. Fish.* **5**, 33–47.

Dempster, E. R. (1955). Maintenance of genetic heterogeneity. *Cold Spring Harbor Symp. Quant. Biol.* **20**, 25–32.

Dempster, E. R. and Lowry, D. (1952). Continuous selection for egg production in poultry. *Genetics* **37**, 693–708.

Den Boer-Hazewinkel, J. (1987). On the costs of

reproduction: parental survival and production of second clutches in the great tit. *Ardea* **75**, 99–110.

DeSteven, D. (1980). Clutch size, breeding success and parental survival in the tree swallow (*Iridoprocne bicolor*). *Evolution* **34**, 278–91.

Dhondt, A. A., Adriaensen, F., Matthysen, E., and Kempenaers, B. (1990). Nonadaptive clutch sizes in tits. *Nature* **348**, 723–5.

Dijk, Th. S. van (1979). On the relationship between reproduction, age, and survival in two carabid beetles: *Calanthus melanocephala* L. and *Pterostichus coerulescens* L. (Coleoptera, Carabidae). *Oecologia* (*Berl.*) **40**, 63–80.

Dijkstra, C. (1988). Reproductive tactics in the kestrel, *Falco tinnunculus*. A study in evolutionary biology. *Ph.D. thesis*, University of Groningen.

Dijkstra, C., Bult, A., Bijlsma, S., Daan, S., Meijer, T., and Zijlstra, M. (1990). Brood size manipulations in the kestrel (*Falco tinnunculus*): effects on offspring and parent survival. *J. Anim. Ecol.* **59**, 269–86.

Dingle, H. and Hegmann, J. (eds) (1982). *Evolution and genetics of life histories*. (Springer, Berlin).

Dingle, H., Evans, K. E., and Palmer, J. O. (1988). Responses to selection among life-history traits in a nonmigratory population of milkweed bugs (*Oncopeltus fasciatus*). *Evolution* **42**, 79–92.

Dobzhansky, Th. (1950). Evolution in the tropics. *Am. Sci.* **38**, 209–21.

Dodson, S. L. (1989). Predator-induced reaction norms. *Bioscience* **39**, 447–52.

Draper, N. and Smith, H. (1981). *Applied regression analysis*, 2nd edn. (John Wiley, New York).

Drent, R. H. and Daan, S. (1980). The prudent parent: energetic adjustments in avian breeding. In *The integrated study of bird populations* (eds H. Klomp and J. W. Woldendorp). (North-Holland, Amsterdam).

Drori, D. and Folman, Y. (1969). The effect of mating on the longevity of male rats. *Exp. Gerontol.* **4**, 263–6.

Dudley, J. W. (1977). 76 generations of selection for oil and protein percentage in maize. In Pollak *et al.* (1977), pp. 459–73.

Dunham, A. D. and Miles, D. B. (1985). Patterns of covariation in life history traits of squamate reptiles: the effects of size and phylogeny reconsidered. *Am. Nat.* **126**, 231–57.

Ebert, D. (1991). The effect of size at birth, maturation threshold and genetic differences on the life-history of *Daphnia magna*. *Oecologia* (*Berlin*) **86**, 243–50.

Edney, E. B. and Gill, R. W. (1968). Evolution of senescence and specific longevity. *Nature* **220**, 281–2.

Edwards, T. C. Jr. and Collopy, M. W. (1983). Obligate and facultative brood reduction in eagles: an examination of factors that influence fratricide. *Auk* **100**, 630–5.

Efron, B. (1981). Nonparametric estimates of standard error. The jackknife, the bootstrap and other methods. *Biometrika* **68**, 589–99.

Eis, S., Garman, S. H., and Ebell, L. F. (1965). Relation between cone production and diameter increment in Douglas fir (*Pseudotsuga menziesii* (Mirb.) Franco), grand fir (*Abies grandis* (Dougl.) Lindl.) and western white pine (*Pinus monticola* Dougl.). *Can. J. Bot.* **43**, 1553–9.

Eisen, E. J. and Legates, J. E. (1966). Genotype × sex interaction and the genetic correlation between the sexes for body weight in *Mus musculus*. *Genetics* **54**, 611–23.

Eisenberg, J. F. (1981). *The mammalian radiations. An analysis of trends in evolution, adaptation, and behavior*. (University of Chicago Press, Chicago).

Ekbohm, G., Fagerström, T., and Agren, G. I. (1980). Natural selection for variation in offspring numbers: comments on a paper by J. H. Gillespie. *Am. Nat.* **115**, 445–7.

Ekman, J. and Askenmo, C. (1986). Reproductive cost, age-specific survival and a comparison of the reproductive strategy in two european tits (Genus *Parus*). *Evolution* **40**, 159–68.

Elgar, M. A. (1990). Evolutionary compromise between a few large and many small eggs: comparative evidence in teleost fish. *Oikos* **59**, 283–7.

Elinson, R. P. (1989). *Egg evolution*. In *Complex organismal functions: integration and evolution in vertebrates* (eds D. B. Wake and G. Roth). (John Wiley, New York).

Emerson, S. B. (1986). Heterochrony and frogs: the relationship of a life history trait to morphological form. *Am. Nat.* **127**, 167–83.

Emlen, J. M. (1970). Age specificity and ecological theory. *Ecology* **51**, 588–601.

Emlen, J. M. (1973). *Ecology: An evolutionary approach*. (Addison-Wesley, Reading, MA).

Emlen, J. M. (1984). *Population biology. The coevolution of population dynamics and behavior*. (Macmillan, New York).

Enfield, F. D. (1977). Selection experiments in tribolium designed to look at gene action issues. In Pollak *et al.* (1977) pp. 177–90.

Erasmus, J. E. (1962). Part-period selection for egg production. In *Proceedings of the 12th World Poultry Congress Sydney 1962*.

Euler, L. (1760). Recherches générales sur la mortalité: la multiplication du genre humain. *Mem. Acad. Sci., Berlin* **16**, 144–64.

Falconer, D. S. (1952). The problem of environment and selection. *Am. Nat.* **26**, 293–8.

Falconer, D. S. (1989). Introduction to quantitative genetics, 3rd edn. (Longman, London).

Feifarek, B. P., Wyngaard, A. C., and Allan, J. D. G.

(1983). The cost of reproduction in a freshwater copepod. *Oecologia* **56**, 166–8.

Felsenstein, J. (1976). The theoretical population genetics of variable selection and migration. *Ann. Rev. Genetics* **10**, 253–80.

Felsenstein, J. (1985). Phylogenies and the comparative method. *Am. Nat.* **125**, 1–15.

Felsenstein, J. (1988). Phylogenies from molecular sequences: inference and reliability. *Ann. Rev. Genetics* **22**, 521–55.

Festa-Bianchet, M. (1989). Individual differences, parasites, and the costs of reproduction for bighorn ewes (*Ovis canadensis*). *J. Anim. Ecol.* **58**, 785–96.

Finke, M. A., Milinkovitch, D. J., and Thompson, C. F. (1987). Evolution of clutch size: an experimental test in the house wren (*Troglodytes aedon*). *J. Anim. Ecol.* **56**, 99–114.

Fisher, R. A. (1918). The correlation between relatives on the supposition of Mendelian inheritance. *Trans. Roy. Soc. Edin.* **52**, 399–431.

Fisher, R. A. (1922). The mathematical foundations of theoretical statistics. *Phil. Trans. Roy. Soc. A* **222**, 306–68.

Fisher, R. A. (1930). *The genetical theory of natural selection.* (Dover, New York).

Fleming, T. H. and Rauscher, R. J. (1978). On the evolution of litter size in *Peromyscus leucopus*. *Evolution* **32**, 45–55.

Flower, S. S. (1938). Further notes on the duration of life in animals. IV. Birds. *Proc. Zool. Soc. Lond. A*, 195.

Fone, A. L. (1989). A comparative study of annual and perennial *Hypochoeris* (Asteraceae). *J. Ecol.* **77**, 495–508.

Foster, R. B. (1977). *Tachigalia versicolor* is a suicidal neotropical tree. *Nature* **268**, 624–6.

Fotheringham, N. (1971). Life history patterns of the littoral gastropods *Shaskyus festivus* (Hinds) and *Ocenebra poulsoni* (Carpenter) (Prosobranchia: Muricidae). *Ecology* **52**, 742–57.

Fowler, K. and Partridge, L. (1989). A cost of mating in female fruitflies. *Nature* **338**, 760–1.

Fox, G. A. (1989). Life tables and statistical inferences. *Bull. Ecol. Soc. Amer.* **70**, 229–30.

Frank, S. A. and Slatkin, M. (1990). Evolution in a variable environment. *Am. Nat.* **136**, 244–60.

Friedman, D. B. and Johnson, T. E. (1988). A mutation in the *age-1* gene in *Caenorhabditis elegans* lengthens life and reduces hermaphoditic fertility. *Genetics* **118**, 75–86.

Frisch, R. W. (1978). Population, food intake, and fertility. *Science* **199**, 22–30.

Gadgil, M. and Bossert, W. (1970). Life history consequences of natural selection. *Am. Nat.* **104**, 1–24.

Gaillard, J. M., Pontier, D., Allaine, D., Lebreton, J. D., Trouvilliez, J., and Clobert, J. (1989). An analysis of demographic tactics in birds and mammals. *Oikos* **56**, 59–76.

Gaines, M. S., Vogt, K. J., Hamrick, J. L., and Cadwell, J. (1974). Reproductive strategies and growth-patterns in sunflowers (*Helianthus*). *Am. Nat.* **108**, 889–94.

Gargett, V. (1978). Sibling aggression in the black eagle in the Matopos, Rhodesia. *Ostrich* **49**, 57–63.

Gaul, H. (1961). Use of induced mutations in seed-propagated species. In: *Mutations and plant breeding*, pp. 206–51. Natural Academy of Sciences. USA, Publication 891, Washington DC.

Gebhardt, M. D. (1989). The quantitative genetics of phenotypic plasticity of life history traits in *Drosophila*. *Ph.D. thesis*, University of Basle.

Gebhardt, M. D. and Stearns, S. C. (1988). Reaction norms for developmental time and weight at eclosion in *Drosophila mercatorum*. *J. evol. Biol.* **1**, 335–54.

Geist, V. (1971). *Mountain sheep: A study in behavior and evolution.* (University of Chicago Press, Chicago).

Ghiselin, M. T. (1985). A moveable eater. *Natural History* **94**, 54–61.

Giesel, J. T. (1979). Genetic co-variation of survivorship and other fitness indices in *Drosophila melanogaster*. *Exp. Gerontol.* **14**, 323–8.

Giesel, J. T. (1986). Genetic correlation structure of life history variables in outbred, wild *Drosophila melanogaster*: effects of photoperiod regimen. *Am. Nat.* **128**, 593–603.

Giesel, J. T. and Zettler, E. E. (1980). Genetic correlations of life-historical parameters and certain fitness indices in *Drosophila melanogaster*: rm, rs, and diet breadth. *Oecologia* **47**, 299–302.

Giesel, J. T., Murphy, P. A., and Manlove, M. N. (1982). The influence of temperature on genetic interrelationships of life history traits in a population of *Drosophila melanogaster*: what tangled data sets we weave. *Am. Nat.* **119**, 464–79.

Gillespie, J. H. (1977). Natural selection for variances in offspring numbers: a new evolutionary principle. *Am. Nat.* **111**, 1010–14.

Gillespie, J. H. and Turelli, M. (1989). Genotype–environment interactions and the maintenance of polygenic variation. *Genetics* **121**, 129–38.

Gingerich, P. D. (1983). Rates of evolution: effects of time and temporal scaling. *Science* **222**, 159–61.

Gittleman, J. L. (1986). Carnivore life history patterns: allometric, phylogenetic, and ecological associations. *Am. Nat.* **127**, 744–71.

Godfray, H. C. J. (1987*a*). The evolution of clutch size in parasitic wasps. *Am. Nat.* **129**, 221–33.

Godfray, H. C. J. (1987*b*). The evolution of clutch size in invertebrates. *Oxford Surv. Evol. Biol.* **4**, 117–54.

Godfray, H. C. J. and Parker, G. A. (1991). Clutch size, fecundity and parent-offspring conflict. *Philos. Trans. Roy. Soc. B* **332**, 67–79.

Goodman, D. (1982). Optimal life histories, optimal notation, and the value of reproductive value. *Am. Nat.* **119**, 803–23.

Goodman, D. (1984). Risk spreading as an adaptive strategy in iteroparous life histories. *Theor. Pop. Biol.* **25**, 1–20.

Gould, S. J. and Lewontin, R. C. (1979). The spandrels of San Marco and the Panglossian paradigm: a critique of the adaptationist program. *Proc. Roy. Soc. Lond. B* **205**, 581–98.

Goulden, C. E. and Hornig, L. L. (1980). Population oscillations and energy reserves in planktonic cladocera and their consequences to competition. *Proc. Natl. Acad. Sci. USA* **77**, 1716–20.

Gowen, W. and Johnson, L. E. (1946). On the mechanism of heterosis. I. Metabolic capacity of different races of *Drosophila melanogaster* for egg production. *Am. Nat.* **80**, 149–79.

Grainger, E. H. (1953). On the age, growth, migration, reproductive potential and feeding habits of the arctic char (*Salvelinus alpinus*) of Frobisher Bay, Baffin Island. *J. Fish. Res. Bd. Can.* **10**, 326–70.

Grant, P. R. (1986). *Ecology and evolution of Darwin's finches.* (Princeton University Press, Princeton).

Graves, J. L. Jr. and Rose, M. R. (1990). Flight duration of *Drosophila melanogaster* selected for postponed senescence. In *Genetic effects on aging II* (ed. D. E. Harrison). (Telford Press, Caldwell, NJ).

Graves, J. L. Jr., Luckinbill, L. S., and Nichols, A. (1988). Flight duration and wing beat frequency in long- and short-lived *Drosophila melanogaster. J. Insect Physiol.* **34**, 1021–6.

Green, J. (1954). Size and reproduction in *Daphnia magna* (Crustacea: Cladocera). *Proc. Zool. Soc. Lond.* **124**, 535–45.

Green, R. F. and Painter, P. R. (1975). Selection for fertility and development time. *Am. Nat.* **109**, 1–10.

Griffiths, J. T. and Tauber, O. E. (1942). Fecundity, longevity and parthenogenesis of the American roach, *Periplaneta americana* L. *Physiolog. Zool.* **15**, 196–206.

Grime, J. P. (1977). Evidence for the existence of three primary strategies in plants and its relevance to ecological and evolutionary theory. *Am. Nat.* **111**, 1169–94.

van Groenendael, J., de Kroon, H., and Caswell, H. (1988). Projection matrices in population biology. *TREE* **3**, 264–9.

Groeters, F. R. and Dingle, H. (1987). Genetic and maternal influences on life history plasticity in response to photoperiod by milkweed bugs (*Oncopeltus fasciatus*). *Am. Nat.* **129**, 332–46.

Grosberg, R. K. (1988). Life-history variation within a population of the colonial ascidian *Botryllus schlosseri.* I. The genetic and environmental control of seasonal variation. *Evolution* **42**, 900–20.

Gross, H. L. (1972). Crown deterioration and reduced growth associated with excessive seed production by birch. *Can. J. Bot.* **50**, 2431–7.

Gross, M. R. and Charnov, E. L. (1980). Alternative male life histories in bluegill sunfish. *Proc. Natl. Acad. Sci. USA* **77**, 6937–40.

Gupta, A. P. and Lewontin, R. C. (1982). A study of reaction norms in natural populations of *Drosophila pseudoobscura. Evolution* **36**, 934–48.

Gustafsson, L. (1986). Lifetime reproductive success and heritability: empirical support for Fisher's fundamental theorem. *Am. Nat.* **128**, 761–4.

Gustaffson, L. and Pärt, T. (1990). Acceleration of senescence in the collared flycatcher *Ficedula albicollis* by reproductive costs. *Nature* **347**, 279–81.

Gustaffson, L. and Sutherland, W. J. (1988). The costs of reproduction in the collared flycatcher *Ficedula albicollis. Nature* **335**, 813–15.

Hahn, W. E. and Tinkle, D. W. (1965). Fat body cycling and experimental evidence for its adaptive significance to ovarian follicle development in the lizard *Uta stansburiana. J. Exp. Zool.* **158**, 79–86.

Hails, C. J. and Bryant, D. M. (1979). Reproductive energetics of a freeliving bird. *J. Anim. Ecol.* **48**, 471–82.

Haldane, J. B. S. (1941). *New paths in genetics.* (Allen and Unwin, London).

Haldane, J. B. S. (1949). Suggestions as to the quantitative measurement of rates of evolution. *Evolution* **3**, 51–6.

Haldane, J. B. S. and Jayakar, S. D. (1963). Polymorphism due to selection of varying direction. *J. Genet.* **58**, 237–42.

Hall, G. O. and Marble, D. R. (1931). The relationship between the first year egg production and the egg production of later years. *Poultry Sci.* **10**, 194–203.

Hamilton, W. D. (1966). The moulding of senescence by natural selection. *J. Theor. Biol.* **12**, 12–45.

Harper, J. L. (1977). *Population biology of plants.* (Academic Press, London).

Harris, J. A. and Lewis, H. R. (1922). The correlation between first and second year egg production in the domestic fowl. *Genetics* **7**, 274–318.

Harris, M. P. (1966). Breeding biology of the Manx shearwater *Puffinus puffinus. Ibis* **108**, 17–33.

Harris, M. P. (1969). The biology of stormpetrels in the Galapagos Islands. *Proc. Cal. Acad. Sci.* **37**, 95–166.

Harris, M. P. (1970). Breeding ecology of the swallow-tailed gull *Creagus furcatus. Auk* **87**, 215–43.

Harris, M. P. and Plumb, W. J. (1965). Experiments on the ability of herring gulls *Larus argentatus* and lesser black-backed gulls *L. fuscus* to raise larger than normal broods. *Ibis* **107**, 256–7.

von Hartmann, L. (1954). Der Trauerfliegenschnäpper. III. Die Nahrungsbiologie. *Acta Zool. Fenn.* **83**, 1–196.

Harvey, P. H. and Clutton-Brock, T. H. (1985). Life history variation in primates. *Evolution* **39**, 559–81.

Harvey, P. H. and Mace, G. M. (1982). Comparisons between taxa and adaptive trends: Problems of methodology. In King's College Sociobiology Group (1982).

Harvey, P. H. and Pagel, M. (1991). *The comparative method in evolutionary biology*. (Oxford University Press, Oxford).

Harvey, P. H. and Zammuto, R. M. (1985). Patterns of mortality and age at first reproduction in natural populations of mammals. *Nature* **315**, 319–20.

Harvey, P. H., Stenning, M. J., and Campbell, B. (1985). Individual variation in seasonal breeding success of pied flycatchers (*Ficedula hypoleuca*). *J. Anim. Ecol.* **54**, 391–8.

Harvey, P. H., Read, A. F., and Promislow, D. E. L. (1989). Life history variation in placental mammals: unifying the data with theory. *Oxford Surv. Evol. Biol.* **6**, 13–32.

Hassall, M. and Grayson, F. W. L. (1988). The occurrence of an additional instar in the development of *Chorthippus bruneus* (Orthoptera: Gomphocerinae). *J. Nat. Hist.* **21**, 329–37.

Hastings, A. and Caswell, H. (1979). Role of environmental variability in the evolution and life history strategies. *Proc. Natl. Acad. Sci. USA* **76**, 4700–3.

Haukioja, E., Lemmetyinen, R., and Pikkola, M. (1989). Why are twins so rare in *Homo sapiens*? *Am. Nat.* **133**, 572–7.

Haymes, G. T. and Morris, R. D. (1977). Brood size manipulations in herring gulls. *Can. J. Zool.* **55**, 1762–6.

Hegner, R. E. and Wingfield, J. C. (1987). Effects of brood-size manipulations on parental investment, breeding success, and reproductive endocrinology of house sparrows. *Auk* **104**, 470–80.

Heller, J. (1990). Longevity in molluscs. *Malacologia* **31**, 259–95.

Hennig, W. (1950). *Grundzüge einer Theorie der phylogenetischen Systematik*. (Deutscher Zentralverlag, Berlin).

Henrich, S. (1988). Variation in offspring sizes of the poeciliid fish *Heterandria formosa* in relation to fitness. *Oikos* **51**, 13–18.

Hill, W. G. (1982). Predictions of response to artificial selection from new mutations. *Genet. Res. Camb.* **40**, 255–78.

Hirshfield, M. F. (1980). An experimental analysis of reproductive effort and cost in the Japanese medaka *Oryzias latipes*. *Ecology* **61**, 282–92.

Hirshfield, M. F. and Tinkle, D. W. (1975). Natural selection and the evolution of reproductive effort. *Proc. Natl. Acad. Sci. USA* **72**, 2227–31.

Hislop, J. R. G. (1984). A comparison of the reproductive tactics and strategies of cod, haddock, whiting and Norway pout in the North Sea. In *Fish reproduction* (eds G. W. Potts and R. J. Wootton). (Academic Press, New York).

Högstedt, G. (1980). Evolution of clutch size in birds: Adaptive variation in relation to territory quality. *Science* **210**, 1148–50.

Högstedt, G. (1981). Should there be a positive or negative correlation between survival of adults in a bird population and their clutch size? *Am. Nat.* **118**, 568–71.

Howe, H. F. (1978). Initial investment, clutch size, and brood reduction in the common grackle (*Quiscalus quiscula* L.). *Ecology* **59**, 1109–12.

Hubbell, S. P. (1979). Tree dispersion, abundance, and diversity in a tropical dry forest. *Science* **203**, 1299–309.

Huey, R. B. and Hertz, P. E. (1984). Is a jack-of-all-temperatures a master of none? *Evolution* **38**, 441–4.

Huey, R. B. and Stevenson, R. D. (1979). Integrating thermal physiology and ecology of ectotherms: A discussion of approaches. *Amer. Zool.* **19**, 357–66.

Hughes, D. M. and Clark, A. G. (1988). Analysis of the genetic structure of life history of *Drosophila melanogaster* using recombinant extracted lines. *Evolution* **42**, 1309–20.

Hughes, R. N. (1989). *A functional biology of clonal animals*. (Chapman and Hall, London).

Hughes, T. P. and Connell, J. H. (1987). Population dynamics based on size or age? A reef-coral analysis. *Am. Nat.* **129**, 818–29.

Hussell, D. J. T. (1972). Factors affecting clutch size in arctic passerines. *Ecol. Monogr.* **42**, 317–64.

Hutchinson, E. W. and Rose, M. R. (1990). Quantitative genetic analysis of postponed aging in *Drosophila melanogaster*. In *Genetic effects on aging II* (ed. D. E. Harrison). (Telford Press, Caldwell, NJ).

Hutchinson, G. E. (1978). *An introduction to population ecology*. (Yale University Press, Yale).

Innes, D. G. L. (1978). A reexamination of litter size in some North American microtines. *Can. J. Zool.* **56**, 1488–96.

Intriligator, M. D. (1971). *Mathematical optimization and economic theory*. (Prentice-Hall, Englewood Cliffs, NJ).

Istock, C. A. (1967). The evolution of complex life cycle phenomena: An ecological perspective. *Evolution* **21**, 592–605.

Istock, C. A. (1983). Boundaries to life history variation and evolution. In *Population biology. Retrospect and prospect* (eds C. E. King and P. S. Dawson). (Columbia University Press, New York).

Istock, C. A., Zisfein, J., and Vaura, K. J. (1976). Ecology and evolution of the pitcher-plant mosquito. 2. The substructure of fitness. *Evolution* **30**, 535–47.

Jain, S. K. (1978). Inheritance of phenotypic plasticity in soft chess, *Bromus mollis* L. (Gramineae). *Experientia* **34**, 835–6.

Janzen, D. H. (1976). Why bamboos wait so long to flower. *Ann. Rev. Ecol. Syst.* **7**, 347–92.

Jarvis, M. J. F. (1974). The ecological significance of clutch size in the South African gannet (*Sula capensis*), (Liechtenstein). *J. Anim. Ecol.* **43**, 1–17.

Johannsen, W. (1909). *Elemente der Exakten Erblichkeitslehre.* (Gustav Fischer, Jena).

Jones, A. (1974). Sexual maturity, fecundity and growth of the turbot *Scophthalmus maximus* L. *J. Mar. Biol. Assoc. UK* **54**, 109–25.

de Jong, G. (1989). Phenotypically plastic characters in isolated populations. In *Evolutionary biology of transient unstable populations* (ed. A. Fontdevila). (Springer, Berlin).

de Jong, G. (1990*a*). Genotype by environment interaction and the genetic covariance between environments: Multilocus genetics. *Genetica* **81**, 171–7.

de Jong, G. (1990*b*). Quantitative genetics of reaction norms. *J. evol. Biol.* **3**, 447–68.

Jouventin, P. and Weimerskirch, H. (1990). Satellite tracking of wandering albatrosses. *Nature* **343**, 746–8.

Jull, M. A. (1928). Second year egg production in relation to first year egg production in the domestic fowl. *Poultry Sci.* **7**, 276–86.

Kaplan, R. H. and Cooper, W. S. (1984). The evolution of developmental plasticity in reproductive characteristics: an application of the "adaptive coin-flipping" principle. *Am. Nat.* **123**, 393–410.

Kenagy, G. J., Masman, D., Sharbaugh, S. M., and Nagy, K. A. (1990). Energy expenditure during lactation to litter size in free-living golden-mantled ground squirrels. *J. Anim. Ecol.* **59**, 73–88.

Keyfitz, N. (1968). *Introduction to the mathematics of population.* (Addison-Wesley, Reading, MA).

Kidwell, J. F. and Malick, L. (1967). The effect of genotype, mating status, weight and egg production on longevity in *Drosophila melanogaster*. *J. Hered.* **58**, 169–72.

King's College Sociobiology Group (ed.) (1982). *Current problems in sociobiology.* Cambridge University Press, Cambridge.

Kirkwood, T. B. L. (1981). Repair and its evolution: survival versus reproduction. In *Physiological ecology: an evolutionary approach to resource use* (eds A. R. R. Townsend and P. Calow). (Blackwell, Oxford).

Kirkwood, T. B. L. (1985). Comparative and evolutionary aspects of longevity. In *Handbook of the biology of aging*, 2nd edn (eds C. E. Finch and E. L. Schneider). (Van Nostrand Reinhold, New York).

Kirkwood, T. B. L. (1987). Immortality of the germ line versus disposability of the soma. In *Evolution of longevity in animals* (eds. A. D. H. Woodhead and K. H. Thompson). (Plenum, New York).

Klomp, H. (1970). Clutch size in birds. *Ardea* **58**, 1–121.

Klomp, H. and Teerink, B. J. (1967). The significance of oviposition rates in the egg parasite, *Trichogamma evanescens*. *Arch. Neerl. Zool.* **17**, 350–75.

Kluyver, H. N. (1951). The population ecology of the Great Tit, *Parus m. major*. *Ardea* **39**, 1–135.

Kohn, L. A. P. and Atchley, W. R. (1988). How similar are genetic correlation structures? Data from mice and rats. *Evolution* **42**, 467–81.

Korpimäki, E. (1988). Costs of reproduction and success of manipulated broods under varying food conditions in Tengmalm's owl. *J. Anim. Ecol.* **57**, 1027–39.

Koufopanou, V. and Bell, G. (1984). Measuring the cost of reproduction. IV. Predation experiments with *Daphnia pulex*. *Oecologia* **64**, 81–6.

Kozlowski, J. (1980). Density dependence, the logistic equation, and *r* selection and *K* selection—a critique and an alternative approach. *Evol. Theory* **5**, 89–102.

Kozlowski, J. and Stearns, S. C. (1989). Hypotheses for the production of excess zygotes: models of bet-hedging and selective abortion. *Evolution* **43**, 1369–77.

Kozlowski, J. and Uchmanski, J. (1987). Optimal individual growth and reproduction in perennial species with indeterminate growth. *Evol. Ecol.* **1**, 214–30.

Kozlowski, J. and Wiegert, R. G. (1987). Optimal age and size at maturity in annuals and perennials with determinate growth. *Evol. Ecol.* **1**, 231–44.

Kozlowski, J. and Ziòlko, M. (1988). Gradual transition from vegetative to reproductive growth is optimal when the maximum rate of reproductive growth is limited. *Theor. Pop. Biol.* **34**, 118–29.

Kriegsbaum, H. (1988). Untersuchungen zur "Lebensgeschichte" von Feldhäuschrecken (Acrididae, Gomphocerinae): Fortpflanzungsstrategie und akustisches Verhalten im natürlichen Habitat. *Ph.D. thesis*, University of Erlangen.

Kuroda, N. (1959). Field studies on the grey starling *Sturnus cineraeus* Temminck. 2. Breeding biology (pt. 3). *Misc. rep. Yamashima Inst.* **13**, 535–52.

Lack, D. (1947). The significance of clutch size. *Ibis* **89**, 302–52.

Lack, D. (1954). *The natural regulation of animal numbers.* (Oxford University Press, Oxford).

Lack, D. (1966). *Population studies of birds.* (Oxford University Press, Oxford).

Lack, D. (1968). *Ecological adaptations for breeding in birds.* (Methuen, London).

Lande, R. (1975). The maintenance of genetic variability by mutation in a polygenic character with linked loci. *Genet. Res.* **26**, 221–35.

Lande, R. (1979). Quantitative genetic analysis of multivariate evolution applied to brain:body size allometry. *Evolution* **34**, 402–16.

Lande, R. (1982*a*). A quantitative genetic theory of life history evolution. *Ecology* **63**, 607–15.

Lande, R. (1982*b*). Elements of a quantitative genetic

model of life history evolution. In *Evolution and genetics of life histories* (eds H. Dingle and J. Hegmann). (Springer, Berlin).

Lande, R. and Arnold, S. J. (1983). The measurement of selection on correlated characters. *Evolution* **37**, 1210–26.

Lauder, G. V. (1986). Homology, analogy, and the evolution of behavior. In *Evolution of animal behaviour* (eds M. H. Nitecki and J. A. Kitchell). (Oxford University Press, Oxford).

Lauder, G. V. and Liem, K. (1989). The role of historical factors in the evolution of complex organismal functions. In *Complex organismal functions: integration and evolution in vertebrates* (eds D. B. Wake and G. Roth). (John Wiley, New York).

Law, R., Bradshaw, A. D., and Putwain, P. D. (1977). Life-history variation in *Poa annuum*. *Evolution* **31**, 233–46.

Law, R., Bradshaw, A. D., and Putwain, P. D. (1979*a*). The cost of reproduction in annual meadow grass. *Am. Nat.* **113**, 3–16.

Law, R., Bradshaw, A. D., and Putwain, P. D. (1979*b*). Optimal life histories under age-specific predation. *Am. Nat.* **114**, 399–417.

Lazarus, J. and Inglis, I. (1986). Shared and unshared parental investment, parent-offspring conflict, and brood size. *Anim. Behav.* **34**, 1791–804.

Leather, S. R. and Burnand, A. C. (1987). Factors affecting life-history parameters of the pine beauty moth, *Panolis flammea* (D&S): the hidden costs of reproduction. *Funct. Ecol.* **1**, 331–8.

Lee, E. T. (1980). *Statistical methods for survival data analysis.* (Lifetime Learning, Belmont, California, USA).

Lefkovitch, L. P. (1965). The population growth of organisms grouped by stages. *Biometrics* **21**, 1–18.

Leggett, W. C. and Carscadden, J. E. (1978). Latitudinal variation in reproductive characteristics of American shad (*Alosa sapidissima*): Evidence for population-specific life history strategies in fish. *J. Fish. Res. Bd. Canada* **35**, 1469–78.

Lenski, R. E. and Service, P. M. (1982). The statistical analysis of population growth rates calculated from schedules of survivorship and fecundity. *Ecology* **63**, 655–62.

Léon, J. A. (1976). Life histories as adaptive strategies. *J. Theor. Biol.* **60**, 301–35.

Leopold, A. C. (1961). Senescence in plant development. *Science* **134**, 1727–32.

Lerner, I. M. (1958). *The genetic basis of selection.* (John Wiley, New York).

Leslie, P. H. (1945). On the use of matrices in certain population mathematics. *Biometrika* **33**, 183–212.

Leslie, P. H. (1948). Some further notes on the use of matrices in population mathematics. *Biometrika* **35**, 213–45.

Lessells, C. M. (1986). Brood size in Canada geese: a manipulation experiment. *J. Anim. Ecol.* **55**, 669–90.

Lessios, H. A. (1990). Adaptation and phylogeny as determinants of egg size in echinoderms from the two sides of the Isthmus of Panama. *Am. Nat.* **135**, 1–13.

Leutenegger, W. (1979). Evolution of litter size in primates. *Am. Nat.* **114**, 525–31.

Levene, H. (1953). Genetic equilibrium when more than one ecological niche is available. *Am. Nat.* **87**, 311–13.

Leverich, W. J. and Levin, D. A. (1979). Age-specific survivorship and reproduction in *Phlox drummondii*. *Am. Nat.* **113**, 881–903.

Levins, R. (1968). *Evolution in changing environments.* (Princeton University Press, Princeton).

Lewontin, R. C. (1965). Selection for colonizing ability. In *The genetics of colonizing species* (eds H. G. Baker and G. L. Stebbins). (Academic Press, New York).

Lewontin, R. C. (1970). The analysis of variance and the analysis of causes. *Am. J. Hum. Genet.* **26**, 400–11.

Lindén, M. (1988). Reproductive trade-off between first and second clutches in the great tit *Parus major*: an experimental study. *Oikos* **51**, 285–90.

Lindén, M. (1990). Reproductive investment and its fitness consequences in the great tit *Parus major*. *Ph.D. thesis*, University of Uppsala.

Lindén, M. and Møller, A. P. (1989). Cost of reproduction and covariation of life history traits in birds. *TREE* **4**, 367–71.

Lively, C. M. (1986*a*). Predator-induced shell dimorphism in the acorn barnacle *Chthamalus anisopoma*. *Evolution* **40**, 232–42.

Lively, C. M. (1986*b*). Competition, comparative life histories, and maintenance of shell dimorphism in a barnacle. *Ecology* **67**, 858–64.

Lloyd, C. S. (1977). The ability of the razorbill *Alca torda* to raise an additional chick to fledging. *Ornis Scand.* **8**, 155–9.

Lloyd, D. G. (1980). Sexual strategies in plants. I. An hypothesis of serial adjustment of maternal investment during one reproductive session. *New Phytol.* **86**, 69–79.

Lloyd, D. G. (1987). Selection of offspring size at independence and other size-versus-number strategies. *Am. Nat.* **129**, 800–17.

Lloyd, D. G. (1988). A general principle for the allocation of limited resources. *Evol. Ecol.* **2**, 175–87.

Lofsvold, D. (1986). Quantitative genetics of morphological differentiation in *Peromyscus*. I. Tests of the homogeneity of genetic covariance structure among species and subspecies. *Evolution* **40**, 559–73.

Loman, J. (1980). Brood size optimization and adaptation among hooded crows *Corvus corone*. *Ibis* **122**, 494–500.

Loschiavo, R. S. (1968). Effect of oviposition on egg production and longevity in *Trogoderma parabile* (Coleoptera: Dermestidae). *Can. Entomol.* **100**, 86–9.

Lotka, A. J. (1907). Studies on the mode of growth of material aggregates. *Am. J. Sci.* **24**, 199–216; 375–6.

Lotz, L. A. P. (1990). The relation between age and size at first flowering of *Plantago major* in various habitats. *J. Ecol.* **78**, 757–71.

Luckinbill, L. S. (1979). Selection of the *r/K* continuum in experimental populations of protozoa. *Am. Nat.* **113**, 427–37.

Luckinbill, L. S. and Clare, M. J. (1985). Selection for life span in *Drosophila melanogaster. Heredity* **55**, 9–18.

Luckinbill, L. S., Arking, R., Clare, M. J., Cirocco, W. C., and Buck, S. A. (1984). Selection for delayed senescence in *Drosophila melanogaster. Evolution* **38**, 996–1003.

Luckinbill, L. S., Grudzien, T. A., Rhine, S., and Weisman, G. (1989). The genetic basis of adaptation to selection for longevity in *Drosophila melanogaster. Evol. Ecol.* **3**, 31–9.

Lynch, M. J. (1984). The limits to life history evolution in *Daphnia. Evolution* **38**, 465–82.

Lynch, M. J. (1985). Spontaneous mutations for life-history characters in an obligate parthenogen. *Evolution* **39**, 804–18.

Lynch, M. J. and Gabriel, W. (1987). Environmental tolerance. *Am. Nat.* **129**, 283–303.

Lynch, M. J., Spitze, K., and Crease, T. (1989). The distribution of life-history evariation in the *Daphnia pulex* complex. *Evolution* **43**, 1724–36.

MacArthur, R. H. (1962). Some generalized theorems of natural selection. *Proc. Natl. Acad. Sci. USA* **48**, 1893–7.

MacArthur, R. H. and Wilson, E. O. (1967). *Theory of island biogeography.* (Princeton University Press, Princeton).

McCauley, D. E. (1983). An estimate of the relative opportunities for natural and sexual selection in a population of milkweed beetles. *Evolution* **37**, 701–7.

McGillivray, W. B. (1983). Intraseasonal reproductive costs for the house sparrow (*Passer domesticus*). *Auk* **100**, 25–32.

McGinley, M. A. and Charnov, E. L. (1988). Multiple resources and the optimal balance between size and number of offspring. *Evol. Ecol.* **2**, 77–84.

McGinley, M. A., Temme, D. H., and Geber, M. A. (1987). Parental investment in offspring in variable environments: theoretical and empirical considerations. *Am. Nat.* **130**, 370–98.

MacKay, T. F. C. (1981). Genetic variation in varying environments. *Genet. Res.* **37**, 79–93.

McLennan, D. A., Brooks, D. R., and McPhail, J. D. (1988). The benefits of communication between comparative ethology and phylogenetic systematics: a case study using gasterosteid fishes. *Can. J. Zool.* **66**, 2177–90.

McMahon, T. A. and Bonner, J. T. *On size and life.* (Scientific American Books, New York).

McNaughton, S. J. (1975). *r*- and *K*-selection in *Typha. Am. Nat.* **109**, 251–61.

Maddison, D. (1986). *MacClade User Manual.* (D. and W. Maddison, Cambridge, MA).

Maddison, W. D. (1990). A method for testing the correlated evolution of two binary characters: Are gains or losses concentrated on certain branches of a phylogenetic tree? *Evolution* **44**, 539–57.

Malick, L. E. and Kidwell, J. F. (1966). The effect of mating status, sex and genotype on longevity in *Drosophila melanogaster. Genetics* **54**, 203–9.

Mangel, M. and Clark, C. W. (1988). *Dynamic modeling in behavioral ecology.* (Princeton University Press, Princeton).

Mann, R. H. K. (1973). Observations on the age, growth, reproduction and food of the roach, *Rutilus rutilus* (L.) in two rivers in southern England. *J. Fish. Biol.* **5**, 707–36.

Mann, R. H. K. (1974). Observations on the age, growth, reproduction and food of the dace, *Leuciscus leuciscus* (L.) in two rivers in southern England. *J. Fish. Biol.* **6**, 237–53.

Mann, R. H. K. (1976). Observations on the age, growth, reproduction and food of the pike, *Esox lucius* (L.) in two rivers in southern England. *J. Fish. Biol.* **8**, 179–97.

Mann, R. H. K. (1980). Observations on the age, growth, reproduction and food of the gudgeon in two rivers in southern England. *J. Fish. Biol.* **17**, 163–76.

Mann, R. H. K. and Mills, C. A. (1979). Demographic aspects of fish fecundity. In *Fish phenology* (ed. P. J. Miller). (Academic Press, New York).

Markus, H. C. (1934). Life history of the black-headed minnow, *Pimephales promelas. Copeia* **1934**, 116–22.

Marshall, D. L., Fowler, M. L., and Levin, D. A. (1985). Plasticity in yield components in natural populations of three species of *Sesbania. Ecology* **66**, 753–61.

Marshall, D. L., Levin, D. A., and Fowler, N. L. (1986). Plasticity of yield components in response to stress in *Sesbania macrocarpa* and *Sesbania vesicaria* (Leguminosae). *Am. Nat.* **127**, 508–21.

Masman, D., Dijkstra, C., Daan, S., and Bult, A. (1989). Energetic limitation of avian parental effort: Field experiments in the kestrel (*Falco tinninculus*). *J. evol. Biol.* **2**, 435–56.

Mather, K. and Jinks, J. L. (1982). *Biometrical genetics.* (Chapman and Hall, London).

Matthes, E. (1951). Der Einfluss der Fortpflanzung auf die Lebensdauer eines Schmetterlings (*Fumea crassiorella*). *Zeit. Verg. Physiol.* **33**, 1.

Maynard Smith, J. (1958). The effect of temperature and of egg laying on the longevity of *Drosophila subobscura. J. Exp. Biol.* **35**, 832–42.

Maynard Smith, J. (1959). Sex-limited inheritance of longevity in *Drosophila subobscura*. *J. Genet.* **56**, 1–9.

Maynard Smith, J. (1982). *Evolution and the theory of games*. (Cambridge University Press, Cambridge).

Maynard Smith, J. (1989). *Evolutionary genetics*. (Oxford University Press, Oxford).

Maynard Smith, J., Bozcuk, A. N., and Tebbutt, S. (1970). Protein turnover in adult *Drosophila*. *J. Insect Physiol.* **16**, 601–13.

Maynard Smith, J., Burian, R., Kauffman, S., Alberch, P., Campbell, J., Goodwin, B. *et al.* (1985). Developmental constraints and evolution. *Q. Rev. Biol.* **60**, 265–87.

Meats, A. (1971). The relative importance to population increase of fluctuations in mortality, fecundity, and the time variables of the reproductive schedule. *Oecologia* **6**, 223–37.

Medawar, P. B. (1946). Old age and natural death. *Modern Quarterly* **1**, 30–56.

Medawar, P. B. (1952). *An unsolved problem of biology*. (H. K. Lewis, London).

Mellors, W. K. (1975). Selective predation of ephippial *Daphnia* and the resistance of ephippial eggs to digestion. *Ecology* **56**, 974–80.

Merritt, E. S. (1962). Selection for egg production in geese. In *Proceedings of the 12th World Poultry Congress, Sydney*.

Mertz, D. B. (1970). Notes on methods used in life history studies. In *Readings in ecology and ecological genetics* (eds J. H. Connell, D. B. Mertz, and W. W. Murdoch). (Harper & Row, New York).

Mertz, D. B. (1975). Senescent decline in flour beetle strains selected for early adult fitness. *Physiol. Zool.* **48**, 1–23.

Michod, R. E. (1979). Evolution of life histories in response to age-specific mortality factors. *Am. Nat.*. **113**, 531–50.

Michod, R. E. and Levin, B. (eds) (1987). *The evolution of sex*. (Sinauer, Sunderland, MA).

Milthorpe, F. L. and Moorby, J. (1979). An introduction to crop physiology, 2nd edn. (Cambridge University Press, Cambridge).

Minchella, D. J. (1985). Host life-history variation in response to parasitism. *Parasitology* **90**, 205–16.

Minchella, D. J. and Loverde, P. T. (1981). A cost of increased early reproductive effort in the snail *Biomphalaria glabrata*. *Am. Nat.* **118**, 876–81.

Mitchell-Olds, T. and Rutledge, J. J. (1986). Quantitative genetics in natural plant populations: a review of the theory. *Am. Nat.* **127**, 379–402.

Møller, A. P. (1990). Effects of parasitism by the haematophagous mite *Ornithonyssus bursa* on reproduction in the barn swallow *Hirundo rustica*. *Ecology* **71**, 2345–57.

Montague, J. R., Mangan, R. L., and Starmer, W. T. (1981). Reproductive allocation in the Hawaiian Drosophilidae: egg size and number. *Am. Nat.* **118**, 865–71.

Morris, J. A. (1963). Continuous selection for egg production using short-term records. *Aust. J. Agric. Res.* **14**, 909–25.

Mousseau, T. A. and Roff, D. A. (1987). Natural selection and the heritability of fitness components. *Heredity* **59**, 181–97.

Mühlbock, O. (1959). Factors influencing the lifespan of inbred mice. *Gerontologia* **3**, 177–83.

Mueller, L. D. (1986). The evolution of semelparity in laboratory populations of *Drosophila*. *Am. Nat.* **128**, 282–93.

Mueller, L. D. (1987). Evolution of accelerated senescence in laboratory populations of *Drosophila*. *Proc. Natl. Acad. Sci. USA* **84**, 1974–7.

Mueller, L. D. and Ayala, F. J. (1981). Trade-off between *r* selection and *K* selection in *Drosophila melanogaster* populations. *Proc. Natl. Acad. Sci. USA* **78**, 1303–5.

Müller, G. B. (1989). Ancestral patterns in bird limb development: A new look at Hampe's experiment. *J. Evol. Biol.* **3**, 31–48.

Murdoch, W. W. (1966). Population stability and life history phenomena. *Am. Nat.* **100**, 45–51.

Murphy, G. I. (1968). Pattern in life history and the environment. *Am. Nat.* **102**, 390–404.

Murphy, P. A., Giesel, J. T., and Manlove, M. N. (1983). Temperature effects on life history variation in *Drosophila simulans*. *Evolution* **37**, 1181–92.

Murray, B. G. Jr. (1990). Population dynamics, genetic change, and the measurement of fitness. *Oikos* **59**, 189–99.

Murton, R. K., Westwood, N. J., and Isaacson, A. J. (1974). Factors affecting egg-weight, body-weight and moult of the woodpigeon *Columba palumbus*. *Ibis* **116**, 52–73.

Myers, P. (1978). Sexual dimorphism in size of vespertilionid bats. *Am. Nat.* **112**, 701–11.

Nei, M. (1987). *Molecular evolutionary genetics*. (Columbia University Press, New York).

Nelson, J. B. (1966). Clutch size in the sulidae. *Nature* **210**, 435–6.

Nettleship, D. N. (1972). Breeding success of the common puffin (*Fratercula arctica* L.) on different habitats at Great Island, Newfoundland. *Ecol. Monogr.* **42**, 239–68.

Newman, R. A. (1988*a*). Genetic variation for larval anuran (*Scaphiopus couchii*) development time in an uncertain environment. *Evolution* **42**, 763–73.

Newman, R. A. (1988*b*). Adaptive plasticity in development of *Scaphiopus couchii* tadpoles in desert ponds. *Evolution* **42**, 774–83.

Noodén, L. D. (1988). Whole plant senescence. In *Senescence and aging in plants* (eds K. V. Thiman and A. C. Leopold). (Academic Press, San Diego).

van Noordwijk, A. J. (1989). Reaction norms in genetical ecology. *Bioscience* **39**, 453–9.

van Noordwijk, A. J. and Gebhardt, M. (1987). Reflections on the genetics of quantitative traits with continuous environmental variation. In *Genetic constraints on evolution* (ed. V. Loeschke). (Springer, Berlin).

van Noordwijk, A. J. and de Jong, G. (1986). Acquisition and allocation of resources: their influence on variation in life history tactics. *Am. Nat.* **128**, 137–42.

van Noordwijk, A. J., van Balen, J. H., and Scharloo, W. (1981). Genetic and environmental variation in clutch size of the great tit. *Neth. J. Zool.* **31**, 342–72.

Nordskog, A. W. and Festing, M. (1962). Selection and correlated responses in the fowl. In *Proceedings of the 12th World Poultry Congress Sydney*.

Norman, F. I. and Gottish, M. D. (1969). Artificial twinning in the short-tailed shearwater *Puffinus tenuirostris*. *Ibis* **111**, 391–3.

Norris, M. J. (1933). Contributions towards the study of insect fertility. II. Experiments on the factors influencing fertility in *Ephestia kuhniella*. *Proc. Zool. Soc. Lond.* 903.

Norton, H. T. J. (1928). Natural selection and mendelian variation. *Proc. Lond. Math. Soc.* **28**, 1–45.

Nur, N. (1984*a*). The consequences of brood size for breeding blue tits. I. Adult survival, weight change and the cost of reproduction. *J. Anim. Ecol.* **53**, 479–96.

Nur, N. (1984*b*). The consequences of brood size for breeding blue tits. II. Nestling weight, offspring survival and optimal brood size. *J. Anim. Ecol.* **53**, 497–517.

Nur, N. (1988). The consequences of brood size for breeding blue tits. III. Measuring the cost of reproduction: survival, future fecundity, and differential dispersal. *Evolution* **42**, 351–62.

O'Connor, R. J. (1978). Brood reduction in birds: Selection for fratricide, infanticide, and suicide? *Anim. Behav.* **26**, 79–96.

Orell, M. and Koivula, K. (1988). Cost of reproduction: parental survival and production of recruits in the willow tit *Parus montanus*. *Oecologia* **77**, 423–32.

Orzack, S. H. and Tuljapurkar, S. (1989). Population dynamics in variable environments. VII. The demography and evolution of iteroparity. *Am. Nat.* **133**, 901–23.

Oster, G. F. and Alberch, P. (1982). Evolution and bifurcation of developmental programs. *Evolution* **36**, 444–59.

Oster, G. F., Shubin, N., Murray, J. D., and Alberch, P. (1988). Evolution and morphogenetic rules: The shape of the vertebrate limb in ontogeny and phylogeny. *Evolution* **42**, 862–84.

Otto, C. and Nilsson, L. M. (1981). Why do beech and oak trees retain leaves until spring? *Oikos* **37**, 387–90.

Pagel, M. and Harvey, P. H. (1988). Recent developments in the analysis of comparative data. *Q. Rev. Biol.* **63**, 413–40.

Pani, S. N. and Lasley, J. F. (1972). *Genotype × environment interactions in animals*. (Missouri Agricultural Experiment Station Research Bull. 992, Columbia, MO).

Papageorgiou, N. K. (1979). The length-weight relationship, age, growth and reproduction of the roach *Rutilus rutilus* in Lake Volvi. *J. Fish. Biol.* **14**, 529–38.

Paris, O. A. and Pitelka, F. H. (1962). Population characteristics of the terrestrial isopod *Armidillidium vulgare* in California grassland. *Ecology* **43**, 229–48.

Parker, G. A. and Begon, M. (1986). Optimal egg size and clutch size: effects of environment and maternal phenotype. *Am. Nat.* **128**, 573–92.

Parker, G. A. and Courtney, S. P. (1984). Models of clutch size in insect oviposition. *Theor. Pop. Biol.* **26**, 27–48.

Parker, G. and Maynard Smith, J. (1990). Optimality theory in evolutionary biology. *Nature* **348**, 27–33.

Parker, G. and Mock, D. W. (1987). Parent–offspring conflict over clutch size. *Evol. Ecol.* **1**, 161–74.

Partridge, L. (1987). Is accelerated senescence a cost of reproduction? *Funct. Ecol.* **1**, 317–20.

Partridge, L. (1990). An experimentalist's approach to the role of costs of reproduction in the evolution of life-histories. In *Towards a more exact ecology* (eds P. J. Grubb and I. Whittaker). (Blackwell Scientific, Oxford).

Partridge, L. and Andrews, R. (1985). The effect of reproductive activity on the longevity of male *Drosophila melanogaster* is not caused by an acceleration of aging. *J. Insect Physiol.* **31**, 393–5.

Partridge, L. and Farquhar, M. (1981). Sexual activity reduces lifespan of male fruitflies. *Nature* **294**, 580–2.

Partridge, L. and Fowler, K. (1991, in press). Direct and correlated responses to selection on age at reproduction in *Drosophila melanogaster*. *Evolution*.

Partridge, L. and Harvey, P. (1985). Costs of reproduction. *Nature* **316**, 20–1.

Partridge, L., Fowler, K., Trevitt, S., and Sharp, W. (1986). An examination of the effects of males on the survival and egg-production rates of female *Drosophila melanogaster*. *J. Insect Physiol.* **32**, 925–9.

Partridge, L., Green, A., and Fowler, K. (1987). Effects of egg-production and of exposure to males on female survival in *Drosophila melanogaster*. *J. Insect Physiol.* **33**, 745–9.

Pease, C. M. and Bull, J. J. (1988). A critique of methods for measuring life history trade-offs. *J. evol. Biol.* **1**, 293–304.

Pederson, D. G. (1972). A comparison of four different experimental designs for the estimation of heritability. *Theor. Appt. Genet.* **42**, 371–7.

Perrins, C. M. (1964). Survival of young swifts in relation to brood-size. *Nature* **201**, 1147–8.

Peters, R. H. (1983). *The ecological implications of body size.* (Cambridge University Press, Cambridge).

Pettifor, R. A., Perrins, C. M., and McCleery, R. H. (1988). Inidividual optimization of clutch size in great tits. *Nature* **336**, 160–2.

Phillipi, T. and Seger, J. (1989). Hedging one's evolutionary bets, revisited. *TREE* **4**, 41–4.

Pianka, E. R. (1970). On "*r*" and "*K*" selection. *Am. Nat.* **104**, 592–7.

Pianka, E. R. and Parker, W. S. (1975). Age-specific reproductive tactics. *Am. Nat.* **109**, 453–64.

Pinero, D., Sarukhan, J., and Alberdi, P. (1982). The costs of reproduction in a tropical palm, *Astrocaryum mexicanum*. *J. Ecol.* **70**, 473–81.

Pitelka, L. F. (1977). Energy allocation in annual and perennial lupines (*Lupinus*: Leguminosae). *Ecology* **58**, 1055–65.

Plumb, W. J. (1965). Observations on the breeding biology of the razorbill. *British Birds* **48**, 449–56.

Pollak, E., Kempthorne, O., and Bailey, T. B. (1977). *Proceedings of the international conference on quantitative genetics.* (Iowa State University Press, Ames, Iowa).

Power, H. W., Kennedy, E. D., Romagnano, L. C., Lombardo, M. P., Hoffenberg, A. S., Stouffer, P. C., and McGuire, T. R. (1989). The parasitism insurance hypothesis: Why starlings leave space for parasitic eggs. *Condor* **91**, 753–65.

Price, T. D. and Grant, P. R. (1984). Life history traits and natural selection for small body size in a population of Darwin's finches. *Evolution* **38**, 483–94.

Promislow, D. E. L. (1991). Senescence in natural populations of mammals: A comparative study. *Evolution* **45**, 1869–87.

Promislow, D. E. L. and Harvey, P. H. (1990). Living fast and dying young: a comparative analysis of life history variation among mammals. *J. Zool.* **220**, 417–37.

Prothero, J. and Jürgens, K. D. (1987). Scaling of maximum life span in mammals: A review. *Basic Life Sci.* **42**, 49–74.

Prout, T. (1980). Some relationships between density-independent selection and density dependent population growth. *Evol. Biol.* **13**, 1–68.

Provine, W. B. (1986). *Sewall Wright and evolutionary biology.* (University of Chicago Press, Chicago).

Puckridge, D. W. and Donald, C. M. (1967). Competition among wheat plants sown at a wide range of densities. *Aust. J. Agric. Res.* **18**, 193–211.

Quiring, D. T. and McNeil, J. N. (1984). Influence of intraspecific larval competition and mating on the longevity and reproductive performance of females on the leaf miner *Agromyza frontella* (Rondani) (Diptera: Agromyzidae). *Can. J. Zool.* **62**, 2197–200.

Read, A. F. and Harvey, P. (1989). Life history differences among the eutherian radiations. *J. Zool.* **219**, 329–53.

Reekie, E. G. and Bazzaz, F. A. (1987*a*). Reproductive effort in plants. 1. Carbon allocation to reproduction. *Am. Nat.* **129**, 876–96.

Reekie, E. G. and Bazzaz, F. A. (1987*b*). Reproductive effort in plants. 2. Does carbon reflect the allocation of other resources? *Am. Nat.* **129**, 897–906.

Reekie, E. G. and Bazzaz, F. A. (1987*c*). Reproductive effort in plants. 3. Effect of reproduction on vegetative activity. *Am. Nat.* **129**, 907–19.

Reid, W. V. (1987). The cost of reproduction in the glaucous-winged gull. *Oecologia* **74**, 458–67.

Remane, A. (1952). *Die Grundlagen des natürlichen Systems, der vergleichenden Anatomie und der Phylogenetik.* (Akademische Verlagsgesellschaft, Leipzig).

Rensch, B. (1938). Einwirkung des Klimas bei der Ausprägung von Vogelrassen, mit besonderer Berücksichtigung der Flügelform und der Eizahl. In *Proceedings of the 8th International Ornithological Congress Oxford.*

Reznick, D. N. (1982*a*). The impact of predation on life history evolution in Trinidadian guppies: Genetic basis of observed life history patterns. *Evolution* **36**, 1236–50.

Reznick, D. N. (1982*b*). Genetic determination of offspring size in the guppy (*Poecilia reticulata*). *Am. Nat.* **120**, 181–8.

Reznick, D. N. (1983). The structure of guppy life histories: the tradeoff between growth and reproduction. *Ecology* **64**, 862–73.

Reznick, D. N. (1985). Cost of reproduction: An evaluation of the empirical evidence. *Oikos* **44**, 257–67.

Reznick, D. N. (1989). Life-history evolution in guppies: 2. Repeatability of field observations and the effects of season on life histories. *Evolution* **43**, 1285–97.

Reznick, D. N. (1990). Plasticity in age and size at maturity in male guppies (*Poecilia reticulata*): An experimental evaluation of alternative models of development. *J. Evol. Biol.* **3**, 185–204.

Reznick, D. N. and Bryga, H. (1987). Life-history evolution in guppies (*Poecilia reticulata*): 1. Phenotypic and genetic changes in an introduction experiment. *Evolution* **41**, 1370–85.

Reznick, D. N. and Endler, J. A. (1982). The impact of predation on life history evolution in Trinidadian guppies (*Poecilia reticulata*). *Evolution* **36**, 160–77.

Reznick, D. N., Perry, E., and Travis, J. (1986). Measuring the cost of reproduction: A comment on papers by Bell. *Evolution* **40**, 1338–44.

Reznick, D. N., Bryga, H., and Endler, J. A. (1990). Experimentally induced life-history evolution in a natural population. *Nature* **346**, 357–9.

Rice, D. W. and Kenyon, K. W. (1962). Breeding cycles and behavior of Laysan and black-footed albatrosses. *Auk* **79**, 517–67.

Ricklefs, R. E. (1977). On the evolution of reproductive strategies in birds: reproductive effort. *Am. Nat.* **111**, 453–78.

Ricklefs, R. E. (1990). Seabird life histories and the marine environment: Some speculations. *Colonial Waterbirds* **13**, 1–6.

Ridley, M. (1983). *The explanation of organic diversity*. (Oxford University Press, Oxford).

Ridley, M. (1986). The number of males in a primate troop. *Anim. Behav.* **34**, 1848–58.

Roach, D. A. (1986). Life history variation in *Geranium carolinianum*. 1. Covariation between characters at different stages of the life cycle. *Am. Nat.* **128**, 47–57.

Robertson, A. (1955). Selection in animals: Synthesis. *Cold Spr. Harb. Symp. Quant. Biol.* **20**, 225–9.

Robertson, F. W. (1957). Studies in quantitative inheritance XI. Genetic and environmental correlation between body size and egg production in *Drosophila melanogaster*. *J. Genet.* **55**, 428–43.

Robertson, O. H. (1961). Prolongation of the life span of Kokanee salmon (*Onchorhynchus nerka kennerlyi*) by castration before beginning of gonad development. *Proc. Natl. Acad. Sci. USA* **47**, 609–21.

Roderick, B. J. and Storer, J. B. (1961). Correlation between mean litter size and mean lifespan among 12 inbred strains of mice. *Science* **134**, 48–9.

Roff, D. A. (1981*a*). Reproductive uncertainty and the evolution of iteroparity—why don't flatfish put all their eggs in 1 basket? *Can. J. Fish. Aquat. Sci.* **38**, 968–77.

Roff, D. A. (1981*b*). On being the right size. *Am. Nat.* **118**, 405–22.

Roff, D. A. (1982). Reproductive strategies in flatfish: a first synthesis. *Can. J. Fish. Aquat. Sci.* **39**, 1686–98.

Roff, D. A. (1984). The evolution of life history parameters in teleosts. *Can. J. Fish. Aquat. Sci.* **41**, 989–1000.

Roff, D. A. (1986). Predicting body size with life history models. *Bioscience* **36**, 316–23.

Roff, D. A. and Mousseau, T. A. (1987). Quantitative genetics and fitness: lessons from *Drosophila*. *Heredity* **58**, 103–18.

Rohder, K. (1985). Increased viviparity of marine parasites at high latitudes. *Hydrobiologica* **127**, 197–202.

Rohmeder, E. (1967). Beziehungen zwischen Frucht- bzw. Samenerzeugung und Holzerzeugung der Waldbäume. *Allg. Forstzeitschr.* **22**, 33–9.

Rose, M. (1983). Theories of life-history evolution. *Amer. Zool.* **23**, 15–24.

Rose, M. (1984*a*). Genetic covariation in *Drosophila* life history: untangling the data. *Am. Nat.* **123**, 565–9.

Rose, M. (1984*b*). Artificial selection on a fitness-component in *Drosophila melanogaster*. *Evolution* **38**, 516–26.

Rose, M. (1991). *Evolutionary biology of aging*. (Oxford University Press, Oxford).

Rose, M. and Charlesworth, B. (1981*a*). Genetics of life history in *Drosophila melanogaster*. I. Sib analysis of adult females. *Genetics* **97**, 173–86.

Rose, M. and Charlesworth, B. (1981*b*). Genetics of life history in *Drosophila melanogaster*. II. Exploratory selection experiments. *Genetics* **97**, 187–96.

Rose, M. R. and Service, P. M. (1985). Evolution of aging. *Rev. Biol. Res. Aging* **2**, 85–98.

Rose, M. R., Service, P. M., and Hutchinson, E. W. (1987). Three approaches to tradeoffs in life-history evolution. In *Genetic constraints on evolution* (ed. V. Loeschke). (Springer, Berlin).

Røskaft, E. (1985). The effect of enlarged brood size on the future reproductive potential of the rook. *J. Anim. Ecol.* **54**, 255–60.

Roth, V. (1991). Homology and hierarchies: Problems solved and unsolved. *J. evol. Biol.* **4**, 167–94.

Roth, G. and Wake, D. B. (1985). Trends in the functional morphology and sensorimotor control of feeding behavior in salamanders: an example of the role of internal dynamics in evolution. *Acta Biotheor.* **34**, 175–92.

Roughgarden, J. (1971). Density-dependent natural selection. *Ecology* **5**, 453–68.

Roughgarden, J. (1979). *Theory of population genetics and evolutionary ecology: An introduction*. (Macmillan, New York).

Saether, B.-E. (1987). The influence of body weight on the covariation between reproductive traits in European birds. *Oikos* **48**, 79–88.

Salthe, S. N. (1969). Reproductive modes and the number and size of ova in the urodeles. *Am. Midl. Nat.* **81**, 467–90.

Salthe, S. N. and Duellman, W. E. (1973). Quantitative constraints associated with reproductive mode in anurans. In *Evolutionary biology of the anurans* (ed. J. L. Vials). (University of Missouri Press, Columbia).

Samson, D. A. and Werk, K. S. (1986). Size-dependent effects in the analysis of reproductive effort in plants. *Am. Nat.* **127**, 667–80.

Sang, J. H. (1950). Population growth in *Drosophila* cultures. *Biol. Rev.* **25**, 188–217.

Sano, Y. and Morishima, H. (1982). Variation in resource allocation and adaptive strategy of a wild rice, *Oryza perennis* Muench. *Bot. Gaz.* **143**, 518–23.

Sano, Y., Morishima, H., and Oka, H.-I. (1980). Intermediate perennial-annual populations of *Oryza perennis* found in Thailand and their evolutionary significance. *Bot. Mag. Tokyo* **93**, 291–305.

Sargent, R. C., Taylor, P. D., and Gross, M. R. (1987). Parental care and the evolution of egg size in fishes. *Am. Nat.* **129**, 32–46.

Satou, S. (1988). A fitness function for optimal life history in relation to density effects, competition, predation and stability of the environment. *Ecol. Res.* **3**, 145–61.

Schaffer, V. (1969). *The year of the whale.* (Charles Scribner, New York).

Schaffer, W. M. (1974*a*). Selection for optimal life histories: The effects of age structure. *Ecology* **5**, 291–303.

Schaffer, W. M. (1974*b*). Optimal reproductive effort in fluctuating environments. *Am. Nat.* **108**, 783–90.

Schaffer, W. M. (1979). Equivalence of maximizing reproductive strategies. *Proc. Natl. Acad. Sci. USA* **76**, 3567–9.

Schaffer, W. M. (1981). On reproductive value and fitness. *Ecology* **62**, 1683–5.

Schaffer, W. M. (1983). The application of optimal control theory to the general life history problem. *Am. Nat.* **121**, 418–31.

Schaffer, W. M. and Reed, C. A. (1972). The co-evolution of social behavior and cranial morphology in sheep and goats (Bovidae, Caprini). *Fieldiana (Zoology)* **61**, 1–88.

Scheinberg, E. (1973). The design and analysis of genetic experiments to modify genotype-environment interaction. I. Single pair matings and selection using one simple quantitative attribute measured in two environments. *Can. J. Genet. Cytol.* **15**, 635–45.

Scheiner, S. M. and Goodnight, C. J. (1984). The comparison of phenotypic plasticity and genetic variation in populations of the grass *Danthonia spicata. Evolution* **38**, 845–55.

Scheiner, S. M. and Lyman, R. F. (1989). The genetics of phenotypic plasticity. I. Heritability. *J. evol. Biol.* **2**, 95–108.

Scheiner, S. M. and Lyman, R. F. (1991). The genetics of phenotypic plasticity. II. The response to selection. *J. evol. Biol.* **4**, 23–50.

Scheiner, S. M., Caplan, R. L., and Lyman, R. F. (1991). The genetics of phenotypic plasticity. III. Genetic correlations and fluctuating asymmetries. *J. evol. Biol.* **4**, 51–68.

Schifferli, L. (1978). Experimental modification of brood size among house sparrows *Passer domesticus. Ibis* **120**, 365–9.

Schlichting, C. D. (1986). The evolution of phenotypic plasticity in plants. *Ann. Rev. Ecol. Syst.* **17**, 667–93.

Schlichting, C. D. (1989*a*). Phenotypic plasticity in *Phlox*. II. Plasticity of character correlations. *Oecologia* **78**, 496–501.

Schlichting, C. D. (1989*b*). Phenotypic integration and environmental change. *Bioscience* **39**, 460–4.

Schlichting, C. D. and Levin, D. A. (1986). Phenotypic plasticity: An evolving character. *Biol. J. Linn. Soc.* **29**, 37–47.

Schlichting, C. D. and Levin, D. A. (1988). Phenotypic plasticity in *Phlox*. I. Wild and cultivated populations of *Phlox drummondii. Am. J. Bot.* **75**, 161–9.

Schlichting, C. D. and Levin, D. A. (1990). Phenotypic plasticity in *Phlox*. III. Variation among natural populations of *P. drummondii. J. evol. Biol.* **3**, 411–28.

Schluter, D. (1988). Estimating the form of natural selection on a quantitative trait. *Evolution* **42**, 849–61.

Schmalhausen, I. I. (1949). *Factors of evolution: The theory of stabilizing selection.* (Blakiston, Philadelphia, reprinted 1986 by University of Chicago Press).

Schmid, B., Puttick, G. M., Burgess, K. H., and Bazzaz, F. A. (1988). Correlation between genet architecture and some life history features in three species of *Solidago. Oecologia* **75**, 459–64.

Seger, J. and Brockmann, H. J. (1987). What is bet-hedging? *Oxford Surv. Evol. Biol.* **4**, 182–211.

Service, P. M. (1987). Physiological mechanisms of increased stress resistance in *Drosophila melanogaster* selected for postponed senescence. *Physiol. Zool.* **60**, 321–6.

Service, P. M. (1989). The effect of mating status on lifespan, egg laying, and starvation resistance in *Drosophila melanogaster* in relation to selection on longevity. *J. Insect Physiol.* **35**, 447–52.

Service, P. M. and Lenski, R. E. (1982). Aphid genotypes, plant phenotypes, and genetic diversity: a demographic analysis of experimental data. *Evolution* **36**, 1276–82.

Service, P. M. and Rose, M. R. (1985). Genetic covariation among life-history components: The effect of novel environments. *Evolution* **39**, 943–4.

Service, P. M., Hutchinson, E. W., and Rose, M. R. (1988). Multiple genetic mechanisms for the evolution of senescence in *Drosophila melanogaster. Evolution* **42**, 708–16.

Shaw, R. G. (1987). Maximum-likelihood approaches to quantitative genetics of natural populations. *Evolution* **41**, 812–26.

Shine, R. (1978). Propagule size and parental care: the "safe harbor" hypothesis. *J. Theor. Biol.* **75**, 417–24.

Shine, R. (1980). Costs of reproduction in reptiles. *Oecologia* **46**, 92–100.

Shine, R. (1988). Constraints on reproductive investment: a comparison between aquatic and terrestrial snakes. *Evolution* **42**, 17–27.

Shorrocks, B. (1972). *Drosophila.* (Ginn, London).

Sibly, R. M. and Calow, P. (1986). *Physiological ecology of animals.* (Blackwell Scientific, Oxford).

Sibly, R. M. and Calow, P. (1989). A life-cycle theory of responses to stress. *Biol. J. Linn. Soc.* **37**, 101–16.

Simmons, R. (1986). Food provisioning, nestling growth and experimental manipulation of brood size in the

African redbreasted sparrowhawk *Accipiter rufiventris*. *Ornis Scand.* **17**, 31–40.

Sinervo, B. (1990). The evolution of maternal investment in lizards: An experimental and comparative analysis of egg size and its effects on offspring performance. *Evolution* **44**, 279–94.

Sinervo, B. and Huey, R. B. (1990). Allometric engineering: an experimental test of the causes of interpopulational differences in performance. *Science* **248**, 1106–9.

Sinervo, B. and McEdward, L. R. (1988). Developmental consequences of an evolutionary change in egg size: an experimental test. *Evolution* **42**, 885–9.

Skelly, D. K. and Werner, E. E. (1990). Behavioral and life-historical responses of larval American toads to an odonate predator. *Ecology* **71**, 2313–22.

Skinner, S. W. (1985). Clutch size as an optimal foraging problem for insects. *Behav. Ecol. Sociobiol.* **17**, 231–8.

Skutch, A. F. (1949). Do tropical birds rear as many young as they can nourish? *Ibis* **91**, 430–55.

Slagsvold, T. (1982). Clutch size, nest size, and hatching asynchrony in birds: Experiments with the fieldfare (*Turdus pilaris*). *Ecology* **63**, 1389–99.

Slagsvold, T. (1984). Clutch size variation of birds in relation to nest predation: on the cost of reproduction. *J. Anim. Ecol.* **53**, 945–53.

Slatkin, M. (1987). Gene flow and the geographic structure of natural populations. *Science* **236**, 787–92.

Slonaker, J. R. (1924). The effect of copulation, pregnancy, pseudo-pregnancy and lactation on the voluntary activity and food consumption of the albino rat. *Am. J. Physiol.* **71**, 362.

Smith, A. P. and Young, T. P. (1982). The costs of reproduction in *Senecio leriodendron*, a giant rosette species of Mt Kenya. *Oecologia* **55**, 243–7.

Smith, C. C. and Fretwell, S. D. (1974). The optimal balance between the size and number of offspring. *Am. Nat.* **108**, 499–506.

Smith, H. G., Källander, H., and Nilsson, J.-A. (1987). Effects of experimentally altered brood size on frequency and timing of second clutches in the great tit. *Auk* **104**, 700–6.

Smith, H. G., Källander, H., and Nilsson, J.-A. (1989). The trade-off between offspring number and quality in the great tit *Parus major*. *J. Anim. Ecol.* **58**, 383–402.

Smith, J. N. M. (1986). Does high fecundity reduce survival in song sparrows? *Evolution* **35**, 1142–8.

Snell, T. W. (1978). Fecundity, developmental time, and population growth rate. *Oecologia* **32**, 119–25.

Snell, T. W. and King, C. E. (1977). Lifespan and fecundity patterns in rotifers: the cost of reproduction. *Evolution* **31**, 882–90.

Snow, D. (1958). *A study of blackbirds.* (Allen and Unwin, London).

Sohn, J. J. and Policansky, D. (1977). The costs of reproduction in the mayapple *Podophyllum peltatum* (Berberidaceae). *Ecology* **58**, 1366–74.

Sokal, R. R. (1970). Senescence and genetic load: evidence from *Tribolium*. *Science* **167**, 1733–4.

Solbrig, O. T. and Simpson, B. B. (1974). Components of regulation of a population of dandelions in Michigan. *J. Ecol.* **63**, 473–86.

Soliman, M. H. (1982). Directional and stabilizing selection for developmental time and correlated response in reproductive fitness in *Tribolium castaneum*. *Theor. Appl. Gen.* **63**, 111–16.

Soller, M., Brody, T., Eiton, Y., Agursky, T., and Wexler, C. (1984). Minimum weight for onset of sexual maturity in female chickens: Heritability and phenotypic and genetic correlations with early growth rate. *Poultry Sci.* **63**, 2103–13.

Sonleitner, F. J. (1961). Factors affecting egg cannibalism and fecundity in populations of adult *Tribolium castaneum* Herbst. *Physiol. Zool.* **34**, 233–5.

Southwood, T. R. E. (1977). Habitat, the templet for ecological strategies? *J. Anim. Ecol.* **46**, 337–66.

Southwood, T. R. E. (1988). Tactics, strategies and templets. *Oikos* **52**, 3–18.

Souza Santos, P. de Jr. and Begon, M. (1987). Survival costs of reproduction in grasshoppers. *Funct. Ecol.* **1**, 215–22.

Spira, T. P. and Pollak, O. D. (1986). Comparative reproductive biology of alpine biennial and perennial gentians (*Gentiana*: Gentianaceae) in California. *Amer. J. Bot.* **73**, 39–79.

Springer, S. (1948). Oviphagous embryos of the sand shark, *Carcharias taurus*. *Copeia* 430–6.

Stanton, M. L. (1984). Seed variation in wild radish: effect of seed size on components of seedling and adult fitness. *Ecology* **65**, 1105–12.

Staples, D. J. (1975). Production biology of the upland bully *Philypnodon breviceps* Stokell in a small New Zealand lake. I. Life history, food, feeding and activity rhythms. *J. Fish. Biol.* **7**, 1–24.

Stearns, S. C. (1976). Life history tactics: A review of the ideas. *Q. Rev. Biol.* **51**, 3–47.

Stearns, S. C. (1977). The evolution of life history traits. A critique of the theory and a review of the data. *Ann. Rev. Ecol. Syst.* **8**, 145–71.

Stearns, S. C. (1980). A new view of life-history evolution. *Oikos* **35**, 266–81.

Stearns, S. C. (1983*a*). The evolution of life-history traits in mosquitofish since their introduction to Hawaii in 1905: rate of evolution, heritabilities, and developmental plasticity. *Amer. Zool.* **23**, 65–76.

Stearns, S. C. (1983*b*). A natural experiment in life-history evolution: field data on the introduction of mosquitofish, *Gambusia affinis*, to Hawaii. *Evolution* **37**, 601–17.

Stearns, S. C. (1983*c*). The genetic basis of differences in life-history traits among six stocks of mosquitofish that shared ancestors in 1905. *Evolution* **37**, 618–27.

Stearns, S. C. (1983*d*). The impact of size and phylogeny on patterns of covariation in the life history traits of mammals. *Oikos* **41**, 173–87.

Stearns, S. C. (1987). The selection arena hypothesis. In *The evolution of sex and its consequences* (ed. S. C. Stearns). (Birkhäuser, Basel).

Stearns, S. C. (1989*a*). Comparative and experimental approaches to the evolutionary ecology of development. *Geobios Sp. Mem.* **12**, 349–55.

Stearns, S. C. (1989*b*). Tradeoffs in life history evolution. *Funct. Ecol.* **3**, 259–68.

Stearns, S. C. (1989*c*). The evolutionary significance of reaction norms. *Bioscience* **39**, 436–46.

Stearns, S. C. and Crandall, R. E. (1981*a*). Quantitative predictions of delayed maturity. *Evolution* **35**, 455–63.

Stearns, S. C. and Crandall, R. E. (1981*b*). Bet-hedging and persistence as adaptations of colonizers. In *Evolution today* (eds G. G. E. Scudder and J. L. Reveal). (Hunt Institute, Philadelphia, PA).

Stearns, S. C. and Crandall, R. E. (1984). Plasticity for age and size at sexual maturity: A life-history adaptation to unavoidable stress. In *Fish reproduction* (eds G. Potts and R. Wootton). (Academic Press, New York).

Stearns, S. C. and Koella, J. (1986). The evolution of phenotypic plasticity in life-history traits: Predictions for norms of reaction for age- and size-at-maturity. *Evolution* **40**, 893–913.

Stearns, S. C. and Sage, R. D. (1980). Maladaptation in a marginal population of mosquitofish, *Gambusia affinis*. *Evolution* **34**, 65–75.

Stearns, S. C., de Jong, G., and Newman, R. (1991). The effects of phenotypic plasticity on genetic correlations. *TREE* **6**, 122–6.

Strathmann, R. R. (1977). Egg size, larval development, and juvenile size in benthic marine invertebrates. *Am. Nat.* **111**, 373–6.

Strathmann, R. R. (1985). Feeding and nonfeeding larval development and life-history evolution in marine invertebrates. *Ann. Rev. Ecol. Syst.* **16**, 339–61.

Strathmann, R. R. (1990). Why life histories evolve differently in the sea. *Amer. Zool.* **30**, 197–207.

Struik, G. O. (1965). Growth patterns in some native annual and perennial herbs in southern Wisconsin. *Ecology* **46**, 401–20.

Summers-Smith, D. (1956). Mortality of the house sparrow. *Bird Study* **3**, 265–70.

Suzuki, D. T., Griffiths, A. J. F., Miller, J. H., and Lewontin, R. C. (1986). *Introduction to genetic analysis*, 3rd edn. (W. H. Freeman, New York).

Tallamy, D. W. and Denno, R. F. (1982). Life-history trade-offs in *Gargaphia solani* (Hemiptera: Tingidae): the cost of reproduction. *Ecology* **63**, 616–20.

Tantawy, A. O. and El-Helw, M. R. (1966). Studies on natural populations of *Drosophila melanogaster*. V. Correlated response to selection in *Drosophila melanogaster*. *Genetics* **53**, 97–110.

Tantawy, A. O. and Rakha, F. A. (1964). Studies on natural populations of *Drosophila*. IV. Genetic variances of and correlation between four characters in *Drosophila melanogaster* and *D. simulans*. *Genetics* **50**, 1349–55.

Tarburton, M. K. (1987). An experimental manipulation of clutch and brood size of white-rumped swiftlets: *Aerodromus spodiopygius* of Fiji. *Ibis* **129**, 107–14.

Taylor, C. E. and Condra, C. (1980). *r* selection and *K* selection in *Drosophila pseudoobscura*. *Evolution* **34**, 1183–93.

Taylor, D. R., Aarssen, L. W., and Loehle, C. (1990). On the relationship between *r/K* selection and environmental carrying capacity: A new habitat templet for plant life history strategies. *Oikos* **58**, 239–50.

Taylor, F. (1981). Ecology and evolution of physiological time in insects. *Am. Nat.* **117**, 1–23.

Taylor, H. M., Gourley, R. S., Lawrence, C. E., and Kaplan, R. S. (1974). Natural selection of life history attributes: an analytical approach. *Theor. Pop. Biol.* **5**, 104–22.

Temin, R. G. (1966). Homozygous viability and fertility loads in *Drosophila melanogaster*. *Genetics* **53**, 27–56.

Temme, D. H. and Charnov, E. L. (1987). Brood size adjustment in birds: economical tracking in a temporally varying environment. *J. Theor. Biol.* **126**, 137–47.

Thompson, J. N. (1984). Variation among individual seed masses in *Lomatium greyi* (Umbelliferae) under controlled conditions: magnitude and partitioning of the variance. *Ecology* **65**, 626–31.

Thompson, J. N. and Pellmyr, O. (1989). Origins of variance in seed number and mass: interaction of sex expression and herbivory in *Lomatium salmoniflorum*. *Oecologia* **79**, 395–402.

Thorson, G. (1950). Reproductive and larval ecology of marine bottom invertebrates. *Biol. Rev.* **25**, 1–45.

Tilman, D. (1982). *Resource competition and community dynamics*. (Princeton University Press, Princeton).

Tinbergen, J. M. (1987). Costs of reproduction in the great tit: intraseasonal costs associated with brood size. *Ardea* **75**, 111–22.

Tinkle, D. W. and Ballinger, R. E. (1972). *Sceloporus undulatus*: a study of the intraspecific comparative demography of a lizard. *Ecology* **53**, 570–84.

Travis, J. (1984). Anuran size at metamorphosis: experimental test of a model based on intraspecific competition. *Ecology* **65**, 1155–60.

Trendall, J. T. (1981). Covariation of life-history traits in the mosquitofish, *Gambusia affinis*. *Am. Nat.* **119**, 774–83.

Tuomi, J. (1990). On clutch size and parental survival. *Oikos* **58**, 387–9.

Tuomi, J., Hakala, T., and Haukioja, E. (1983). Alternative concepts of reproductive effort, costs of reproduction, and selection in life-history evolution. *Amer. Zool.* **23**, 25–34.

Turelli, M. (1978). Does environmental variability limit niche overlap? *Proc. Natl. Acad. Sci. USA* **75**, 5065–89.

Turelli, M. (1988). Phenotypic evolution, constant covariances, and the maintenance of additive variance. *Evolution* **42**, 1342–7.

Turner, C. L. (1937). Reproductive cycles and superfetation in poeciliid fishes. *Biol. Bull.* **72**, 145–64.

Tuttle, M. D. and Ryan, M. J. (1981). Bat predation and the evolution of frog vocalizations in the Neotropics. *Science* **214**, 677–8.

Tweddle, D. and Turner, J. L. (1977). Age, growth, and natural mortality rates of some cichlid fishes of Lake Malawi. *J. Fish. Biol.* **10**, 385–98.

Venable, D. L. and Burquez M., A. (1989). Quantitative genetics of size, shape, life-history, and fruit characteristics of the seed-heteromorphic composite *Heterosperma pinnatum*. I. Variation within and among populations. *Evolution* **43**, 113–24.

Vermeer, K. (1963). The breeding biology of the glaucous-winged gull *Larus glaucescens* on Mandarte Island. *Occ. Pap. Brit. Col. Prov. Mus.* **13**, 1–104.

Via, S. (1984). The quantitative genetics of polyphagy in an insect herbivore. II. Genetic correlations in larval performance within and across host plants. *Evolution* **38**, 896–905.

Via, S. and Lande, R. (1985). Genotype–environment interaction and the evolution of phenotypic plasticity. *Evolution* **39**, 505–22.

Via, S. and Lande, R. (1987). Evolution of genetic variability in a spatially heterogeneous environment: Effects of genotype–environment interaction. *Genet. Res.* **49**, 147–56.

Vogel, S. (1988). *Life's devices: the physical world of plants and animals.* (Princeton University Press, Princeton).

Waage, J. K. and Godfray, H. C. J. (1985). Reproductive strategies and population ecology of insect parasitoids. In *Behavioral ecology* (eds R. M. Sibly and R. H. Smith). (Blackwell Scientific, Oxford).

Waage, J. K. and Ng, S. M. (1984). The reproductive strategy of a parasitic wasp. I. Optimal progeny and sex allocation in *Trichogamma evanescens*. *J. Anim. Ecol.* **53**, 401–16.

Wagner, G. P. (1989). The biological homology concept. *Ann. Rev. Ecol. Syst.* **20**, 51–70.

Wake, D. B. and Larson, A. (1987). Multidimensional analysis of an evolving lineage. *Science* **238**, 42–8.

Wake, D. B. and Roth, G. (1989*a*). *Complex organismal functions: integration and evolution in vertebrates.* (John Wiley, New York).

Wake, D. B. and Roth, G. (1989*b*). The linkage between ontogeny and phylogeny in the evolution of complex systems. In *Complex organismal functions: integration and evolution in vertebrates* (eds D. B. Wake and G. Roth). (John Wiley, New York).

Wake, M. H. 1982. Diversity within a framework of constraints. Amphibian reproductive modes. In *Environmental adaptation and evolution* (eds D. Mossakowski and G. Roth). (G. Fischer, Stuttgart).

Wallinga, J. H. and Bakker, H. (1978). Effect of long-term selection for litter size in mice on lifetime reproduction. *J. Anim. Sci.* **46**, 1563–71.

Wanntorp, H.-E. (1983). Historical constraints in adaptation theory: traits and non-traits. *Oikos* **41**, 157–60.

Ward, J. G. (1973). Reproductive success, food supply, and the evolution of clutch size in the glaucous-winged gull. Ph.D. thesis, University of British Columbia.

Ward, P. (1965). The breeding biology of the black-faced dioch *Quelea quelea* in Nigeria. *Ibis* **107**, 326–49.

Ware, D. M. (1975). Relation between egg size, growth, and natural mortality of larval fish. *J. Fish. Res. Bd. Can.* **32**, 2503–12.

Ware, D. M. (1980). Bioenergetics of stock and recruitment. *Can. J. Fish. Aquat. Sci.* **37**, 1012–24.

Ware, D. M. (1982). Power and evolutionary fitness of teleosts. *Can. J. Fish. Aquat. Sci.* **39**, 3–13.

Warner, R. R. (1984). Deferred reproduction as a response to sexual selection in a coral reef fish: a test of the life historical consequences. *Evolution* **38**, 148–62.

Watkinson, A. R. and White, J. (1985). Some life-history consequences of modular construction in plants. *Phil. Trans. R. Soc. Lond. B* **313**, 31–51.

Wattiaux, J. M. (1968). Cumulative parental age effects in *Drosophila subobscura*. *Evolution* **22**, 406–21.

Weis, A. E. and Gorman, W. L. (1990). Measuring selection on reaction norms: an exploration of the *Eurosta–Solidago* system. *Evolution* **44**, 820–31.

Weismann, A. (1885). *Die Kontinuität des Keimplasmas als Grundlage einer Theorie der Vererbung.* (Gustav Fischer, Jena).

Werner, P. A. and Platt, W. J. (1976). Ecological relationship of co-occurring goldenrods (Solidago: Compositae). *Am. Nat.* **110**, 959–71.

Wilbur, H. M. (1977). Propagule size, number, and dispersion patterns in *Ambystoma* and *Asclepias*. *Am. Nat.* **111**, 43–68.

Wilbur, H. M. (1980). Complex life cycles. *Ann. Rev. Ecol. Syst.* **11**, 67–94.

Wilbur, H. M., Tinkle, D. W., and Collins, J. P. (1974). Environmental certainty, trophic level, and resource availability in life history evolution. *Am. Nat.* **108**, 805–17.

Wiley, E. O. (1981). *Phylogenetics*. (John Wiley, New York).

Wiley, R. H. (1974*a*). Evolution of social organization and life-history patterns among grouse. *Q. Rev. Biol.* **49**, 201–27.

Wiley, R. H. (1974*b*). Effects of delayed maturation on survival, fecundity, and the rate of population increase. *Am. Nat.* **108**, 705–9.

Williams, G. C. (1957). Pleiotropy, natural selection, and the evolution of senescence. *Evolution* **11**, 398–411.

Williams, G. C. (1966*a*). *Adaptation and natural selection*. (Princeton University Press, Princeton).

Williams, G. C. (1966*b*). Natural selection, the cost of reproduction, and a refinement of Lack's principle. *Am. Nat.* **100**, 687–90.

Williams, G. C. (1979). The question of adaptive sex ratio in outcrossed vertebrates. *Proc. Roy. Soc. Lond. B.* **205**, 567–80.

Willson, M. F. (1983). *Plant reproductive ecology*. (John Wiley, New York).

Winfield, I. J. and Townsend, C. R. (1983). The cost of copepod reproduction: increased susceptibility to fish predation. *Oecologia* **60**, 406–11.

Winkler, D. W. and Wallin, K. (1987). Offspring size and number: a life history model linking effort per offspring and total effort. *Am. Nat.* **129**, 708–20.

Wodinsky, J. (1977). Hormonal inhibitions of feeding and death in *Octopus*: control by optic gland secretion. *Science* **198**, 948–51.

Woltereck, R. (1909). Weitere experimentelle Untersuchungen über Artveränderung, speziell über das Wesen quantitativer Artunterschiede bei Daphniden. *Verh. Deutsch. Zool. Gesell.* **1909**, 110–72.

Woombs, M. and Laybourn-Parry, J. (1984). Growth, reproduction and longevity in nematodes from sewage-treatment plants. *Oecologia* **64**, 168–72.

Wootton, J. T. (1987). The effects of body mass, phylogeny, habitat, and trophic level on mammalian age at first reproduction. *Evolution* **41**, 732–49.

Wootton, R. J. (1973). The effect of food ration on egg production in the female three-spined stickleback, *Gasterosteus aculeatus* L. *J. Fish. Biol.* **5**, 89–96.

Wourms, J. P. (1972). The developmental biology of annual fishes. III. Pre-embryonic and embryonic diapause of variable duration in the eggs of annual fishes. *J. Exp. Zool.* **182**, 389–414.

Wourms, J. P. (1981). Viviparity: the maternal-fetal relationship in fishes. *Am. Zool.* **21**, 467–509.

Wright, S. (1968–78). *Evolution and the genetics of populations*, Vols I–IV. (University of Chicago Press, Chicago).

Ydenberg, R. C. (1989). Growth–mortality trade-offs and the evolution of juvenile life histories in the Alcidae. *Ecology* **70**, 1494–506.

Yodzis, P. (1981). Concerning the sense in which maximizing fitness is equivalent to maximizing reproductive value. *Ecology* **62**, 1681–2.

Yoo, B. H. (1980). Long-term selection for a quantitative character in large replicate populations of *Drosophila melanogaster*. I. Response to selection. *Genet. Res.* **35**, 1–17.

Young, T. P. (1981). General model of comparative fecundity for semelparous and iteroparous life histories. *Am. Nat.* **118**, 27–36.

Young, T. P. (1984). Comparative demography of semelparous *Lobelia telekii* and iteroparous *Lobelia keniensis* on Mount Kenya. *J. Ecology* **72**, 637–50.

Young, T. P. (1985). *Lobelie telekii* herbivory, mortality, and size at reproduction: variation with growth rate. *Ecology* **66**, 1879–83.

Young, T. P. (1990). Evolution of semelparity in Mount Kenya lobelias. *Evol. Ecol.* **4**, 157–72.

Zelditch, M. L. (1988). Ontogenetic variation in patterns of phenotypic integration in the laboratory rat. *Evolution* **42**, 28–41.

Ziòlko, M. and Kozlowski, J. (1983). Evolution of body size: an optimization model. *Math. Biosci.* **64**, 127–43.

Zwaan, B. J., Bijlsma, R., and Hoekstra, R. F. (1991). On the developmental theory of aging. I. Starvation resistance and longevity in *Drosophila melanogaster* in relation to preadult breeding conditions. *Heredity* **66**, 29–39.

AUTHOR INDEX

SUBJECT INDEX

Page numbers given in **bold** indicate where a subject is defined.